PLANT AND SOIL: INTERFACES AND INTERACTIONS

Developments in Plant and Soil Sciences

Plant and Soil
Interfaces and Interactions

Proceedings of the International Symposium: Plant and Soil:
Interfaces and Interactions. Wageningen, The Netherlands
August 6−8, 1986

Edited by

A. VAN DIEST, Wageningen

with the collaboration of:

S. DASBERG, Bet Dagan
S.K. DE DATTA, Manilla
R.O.D. DIXON, Edinburgh
M.A. EL-SHARKAWY, Cali
R. RODRIGUEZ- KÁBANA, Auburn, AL

Managing Editor:

A. HOUWERS, Wageningen

First published as *Plant and Soil*, Volume 100 (1987)

1987 **MARTINUS NIJHOFF PUBLISHERS**
a member of the KLUWER ACADEMIC PUBLISHERS GROUP
DORDRECHT / BOSTON / LANCASTER

Distributors

for the United States and Canada: Kluwer Academic Publishers, P.O. Box 358, Accord Station, Hingham, MA 02018-0358, USA
for the UK and Ireland: Kluwer Academic Publishers, MTP Press Limited, Falcon House, Queen Square, Lancaster LA1 1RN, UK
for all other countries: Kluwer Academic Publishers Group, Distribution Center, P.O. Box 322, 3300 AH Dordrecht, The Netherlands

Library of Congress Cataloging in Publication Data

International Symposium: Plant and Soil: Interfaces
    and Interactions (1986 : Wageningen, Netherlands)
    Plant and soil.

    (Developments in plant and soil sciences)
    1. Crops and soils--Congresses.   2. Plant-soil
relationships--Congresses.   3. Crops--Nutrition--
Congresses.   I. Plant and soil.   II. Series.
S596.7.I58 1986   582'.013   87-12222

ISBN-13: 978-94-010-8122-1   e-ISBN-13: 978-94-009-3627-0
DOI: 10.1007/978-94-009-3627-0

Copyright

Contents

The participants of the Symposium

Foreword

Forty years ago, when PLANT AND SOIL first appeared, Europe was still recovering from the devastating effects of World War II. During the war years, work in many centres of agricultural research had come to a virtual standstill. Buildings and equipment were destroyed, scientists were often forced to terminate their research and teaching activities and funds allocated to such work were diverted to other, at that time, more pressing needs.

During the first post-war years reconstruction was undertaken with great zeal and in that light the founding of the new journal PLANT AND SOIL must be viewed. In the pre-war period most agricultural science journals were still primarily national ones and consequently many articles were published in languages mastered by only a limited number of potential readers. In small countries whose languages are not widely understood, the desire arose to publish research findings in one of the major languages. It is therefore understandable that in the early years of the journal's existence, large portions of PLANT AND SOIL were filled with articles from the Scandinavian countries and The Netherlands.

Originally, rather frequent use was made of the opportunity to publish also in German and French, but with the advance of English as a major language of communication, a decline was noticeable in the number of German and French manuscripts submitted. As a consequence the Editorial Board has recently decided to terminate the publishing of articles in these languages.

In the foreword to the first issue of PLANT AND SOIL, the editors justify their intention to start a new journal by stating that 'existing journals are overcrowded and often one has to wait many months or even more than one year before papers, even of universal importance, can be published'. Today we observe that with considerably more journals on the market, this situation has not changed much. PLANT AND SOIL reacted to the ever increasing number of manuscripts submitted by raising the number of volumes issued per year (eight in 1987).

One way to stem the flow of incoming manuscripts is to ask consulting editors to be more critical. About 60% of the submitted manuscripts are presently accepted for publication, but a slight tendency towards more rejections is noticeable.

A glance at the original and present names and residences of consulting editors will make it clear that in its forty years of existence the journal has become a great deal more cosmopolitan. This trend is also noticeable when examining the affiliations of the present authors. They have contributed to making PLANT AND SOIL a truly international journal.

One complication associated with this internationalisation is the growing need for editing of the English used in the original manuscripts. Consulting editors have the right to confine their editorial work to the scientific contents of the manuscripts, and to leave any needed correction of the English to others. Since the employment of a professional corrector is financially unfeasible, the editorial board has decided from now on to enforce the stipulation that contributors bear the responsibility to submit manuscripts written in acceptable English. Any prospective contributor not in full command of the English language is therefore advised to consult a person who has mastered this language.

To celebrate the forthcoming publication of the 100th volume, the Publishers and the Editorial Board decided in 1985 to organize an international symposium in which consulting editors would be given an opportunity to present papers that together would form the content of the 100th volume of the journal. As a result, about 50 scientists attended the INTERNATIONAL SYMPOSIUM PLANT AND SOIL 'INTERFACES AND INTERACTIONS' held August 6–8, 1986 in Wageningen, The Netherlands. Their contributions will fill the double issue of the 100th volume.

During this International Symposium tribute was paid to Dr. E.G. Mulder, one of the founders of the Journal and the only member of the original Editorial Board still serving in this capacity. PLANT AND SOIL is greatly indebted to him for his untiring and expert service to the journal over a period of 40 years. With deference, the Publishers and fellow members of the Editorial Board accepted his wish to terminate his period of active service with PLANT AND SOIL. It is a pleasure to announce that Dr. Mulder was appointed Honorary Member of the Editorial Board.

This 100th volume is dedicated to Dr. J.L. Harley, Dr. S. Larsen, Dr. L. Leyton, Dr. L.K. Wiersum and Dr. W.R.C. Handley, who all were of invaluable assistance to the journal during long periods of service as Members of the Editorial Board and as Consulting Editor, respectively.

Forty years ago scientists in Plant Physiology and Soil Science were in the first stages of discovering the interfaces and interactions of their respective disciplines. The editors of 'PLANT AND SOIL' have constantly tried to contribute to a better understanding of the many interrelationships between these two disciplines and it is their hope that in this respect the journal may also play an important role in the years to come.

It can be announced here that starting with the 101st Volume, the format of the journal will be enlarged to a 21 × 30 cm size, with two columns per page. With this change the legibility, especially of tables, will be improved.

The 100th Volume will be the last one appearing with the so familiar front cover. On the covers of future volumes the name 'PLANT AND SOIL' will appear in a modernized design as an indication that also in this respect the journal will keep abreast of the times.

The Editors

Plant and Soil 100, 1–9 (1987).

Nitrogen fertilization in citrus orchards*

S. DASBERG

Institute of Soils and Water, Agricultural Research Organization, The Volcani Center, Bet Dagan, Israel

Key words   Citrus   N balance   N-15   N nutrition   N requirement   N storage   Oranges

Summary   The purpose of this review was to evaluate critically the results obtained in citrus nitrogen fertilization experiments in Israel and in other parts of the world, in order to increase our understanding of the processes involved and to improve the recommendations to growers. Mature citrus trees contain 1–2 kg N/tree, 30–60% of which is in the annual parts (leaves and fruits). 30 g N is deposited annually in the tree skeleton. Based on these results and on a review of long-term fertilization experiments with citrus from various parts of the world, it was concluded that 200 kg N/ha applied annually is sufficient to sustain good citrus yields and tree development, about half of which is incorporated in the fruits and one-tenth deposited in the tree, the balance being made up by leaching and gaseous losses. Experiments with ^{15}N labeled fertilizer applications showed that the highest N-uptake rate occurred during fruit set and that in winter the uptake was very low. N reserves in the older tissues played an important part in the development of new leaves and flowers in the spring, when the uptake from the soil was still low. It was concluded that the nitrogen contained in the soil organic matter (2 Mg/ha) and in the mature trees (1 Mg/ha) plays an important part in the regulation of N supply to the growing parts of the tree. More N is derived from these parts with low N fertilization than with an abundant supply. The purpose of fertilization is to ensure proper development of the tree, not the current fruit yield.

Introduction

Citrus orchards cover considerable areas of irrigated soils, to which large quantities of N-fertilizer are added annually (100–400 kg N/ha)[13]. Nitrogen is added periodically in order to supply the amounts exported from the orchard with the harvested fruit and in order to replenish the soil N reservoir. However, there is not much experiment evidence of citrus yield response to N fertilization. The annual N demand of fruit trees seems to be much lower than that of field crops. A recent review on fertilizer requirement of deciduous orchards has shown that most of their N requirements may be met from the soil by natural processes without supplementation by fertilizers[16]. Could it be that the situation in citrus orchards is similar to that of deciduous trees? The purpose of this review is to reassess the economic and environmental implications of the N fertilization requirements of citrus orchards, based on the older literature and on some recent experiments with ^{15}N labeled N applications, which provide more insight into the storage and translocation of N in citrus trees.

* Contribution from the Agricultural Research Organization, The Volcani Center, Bet Dagan, Israel. No. 1770-E. 1986 series.

Response of citrus trees to N application

The primary basis for annual N requirement assessment is the N content and its distribution in citrus trees. Few data have been published in the literature on mature trees because of the experimental difficulties involved, especially in the excavation of the root system. Table 1 gives some of the published data on mature citrus trees. In spite of the diversity in tree size and age, some general conclusions can be drawn from these data. The total dry matter of a developed tree ranges between approximately 100 and 300 kg, depending on the age of the tree. 70–90% of the dry matter is of a perennial nature (branches, trunk and root system). The N distribution, however, is different: 30–60% of the total N can be found in the annual parts of the tree, i.e. fruits and leaves (a large part of the citrus leaves are renewed annually). Large fluctuations are apparent in the N content of the annual tree parts, depending on the tree status: 1.02 kg N in the fruits and leaves of a Shamouti orange tree with a high N supply vs 0.65 kg N in a N-starved tree[14] (Table 1, columns 7–8); 0.39 kg N in a Wilking mandarin tree in an "on"-year vs 0.21 kg N in an "off"-year[15] (Table 1, columns 5–6).

Two questions must be answered in order to estimate the N fertilizer requirement from these data: How much N is annually deposited in the skeleton of the tree, and what portion of the N of shed leaves, blossoms, fruits and prunings is recycled into the soil organic N and eventually mineralized. Based on the data in Table 1, an annual deposition of 30–60 g N in the tree skeleton (equivalent to 20 kg/ha) seems a reasonable estimate. This is less than one tenth of the N in the fruits and leaves. The fruits are exported from the orchard and can be regarded as a definite loss of N from the system. Soil organic N tends to reach an equilibrium in the soil under stable climatic and management conditions[26]. It seems safe to assume therefore, that in a 10–20 year old orchard, the amount of N incorporated annually in the soil organic matter will be balanced by an equivalent amount of mineralized N. The question remains what percentage of the shed tree parts decomposes into volatile compounds (NH_3, N_2, N_2O). It was assumed by Embleton and Jones[11] that upon decomposition, 75% of the N of the shed tree parts would be lost to the atmosphere under their conditions. However, this estimate was considered excessively high[22]. In deciduous orchards under temperate conditions, almost complete recycling of the N in the shed tree parts is assumed[16]. Some N losses can be assumed under arid conditions with high soil pH, but in irrigated orchards continuous decomposition of organic matter and downward movement into the soil of NH_4^+ and eventually NO_3^- seems to be the prevalent process.

Citrus yield responses to N application are variable. It is often difficult

Table 1. Dry matter and nitrogen distribution in mature citrus trees

	Grapefruit	Valencia	Valencia	Wilking	Wilking	Shamouti	Shamouti
Variety	Grapefruit	Valencia	Valencia	Wilking	Wilking	Shamouti	Shamouti
Age	19	10	9–10	15	15	20	20
Planting distance						6 × 4	6 × 4
Reference	Barnette et al.	Cameron et al.		Golomb and Goldschmidt		Feigenbaum et al.	
Year	1931	1934	1945	1980	1980	1986	1986
Remarks				On	Off	High N	Low N
Dry matter (kg/tree)	273	80	95	156	144	320	320
Distribution (%)							
Fruits	3	16	–	16	–	13	8
Leaves	6	19	17	6	7	7	6
Trunk and branches	57	49	61	65	67	55	55
Roots	34	16	22	13	25	24	31
Total N (g/tree)	2061	734	732	858	811	2379	2072
Distribution (%)							
Fruits	6	21	–	32	–	20	11
Leaves	18	41	45	14	26	22	20
Trunk and branches	33	28	35	44	52	42	44
Roots	43	10	20	10	22	15	25

to compare results obtained from different tree spacings. We have tried to express the published data on a soil area basis. A long-term experiment with Valencia oranges in Florida showed no effect of N fertilization above the 150 kg/ha level[23]. Subsequent experimentation showed yield response up to 180 kg/ha[25]. More recent data show a definite yield response in three varieties up to a 202 kg/ha annual application[21]. In California, 100–150 kg N/ha applied annually tended to support maximum yields[10]. In Arizona, no yield response was obtained with Valencia oranges above the 100 kg/ha level[24]. In South Africa no yield response beyond the 180 kg/ha were obtained[9]. In Japan, 150–250 kg N/ha is recommended for good Satsuma mandarin quality[28]. A long-term experiment on orchard management in citrus in Australia showed that the trees responded to N up to 200 kg N/ha[5]. All these data obtained from different parts of the world tend to support the notion that in terms of fruit yields, citrus trees do not respond to high N applications, *i.e.* above 200 kg N/ha annually, while the fruit quality deteriorates. However, a positive response to high N doses was obtained in a recent fertigation experiment. The highest yields were obtained at N applications above 300 kg N/ha, probably because of the continuous application with the irrigation water[7]. The timing of N application seems to be less important for the attainment of yield than for its effect on fruit quality[13,18].

N balance studies

Attempts to make an N balance in a citrus orchard were reviewed previously[8] and the findings are presented in Table 2. These data show that even with high yield the N removal by citrus fruits is seldom above 100 kg/ha. A greater N-removal (146 kg N ha^{-1}) was obtained only with very high yields (850 Mg/ha) at excessively high N applications[8]. These data tend to support the belief that it is unnecessary to apply N at rates above 200 kg/ha, taking into account an annual deposition of 20 kg/ha in the tree skeleton and N losses of the same order of magnitude.

The data in Table 2 show that there are cases with a negative balance, which means that more N was leached from the soil than the annual excess of fertilizer input over fruit removed. This means that the soil may supply N by mineralization from the organic N pool. On the other hand, excessive N losses, such as occur with high N applications, may mean either incorporation in the soil organic matter or storage in the tree organs[8].

N storage and translocation

Table 1 indicates the existence of a large N pool in the tree skeleton, ranging from 100–500 kg N/ha. How active is this N in the tree metabol-

Table 2. Nitrogen balance sheet for several citrus orchards ($kg\,ha^{-1}\,yr^{-1}$)

Variety	Reference	Nitrogen				
		Input	Removal	Excess	Leached	Balance*
Valencia oranges	Bingham et al.	156	60	106	67	39
Washington	Embleton and Jones	124	52	72	18	54
Navel oranges		368	61	307	58	249
Lisbon	Embleton et al.	57	73	−16	31	−47
Lemon		165	86	79	109	−30
		486	101	385	239	146
Shamouti oranges	Dasberg	50	51	−1	48	−49
		250	53	197	80	117
Shamouti oranges	Dasberg et al.	108	74	34	21	13
		170	128	42	15	27
		308	146	162	61	101

* Negative numbers mean more N leached than excess of input over removal.

ism and which part is translocated? A partial answer to these questions was provided with the advent of the use of ^{15}N as a tracer for N uptake and translocation. Early work on translocation of N in citrus trees was carried out by Wallace et al.[28] One of the conclusions from their work was that before abscission old orange leaves contained much less N than young leaves and that more than 50% of the leaf N returned to other parts of the tree before abscission. This estimate seems very high according to later studies[13]. Moreover, based on a preliminary experiment in which ^{15}N enriched fertilizer was applied to young citrus trees, it was concluded that only 15% of the N in young leaves was derived from the soil and that a major portion came from tree reserves[28]. This phenomenon is well known in deciduous trees, where the N reserves in the tree bark are of great importance for early spring development[27]. The existence of a similar phenomenon in citrus is less well known.

Subsequent work was carried out by Japanese workers[17]. They found that the highest recovery of fertilizer N occurred from summer fertilization and the lowest from autumn or early spring application. Not only less N is taken up by the tree, but less is translocated from the roots to the branches, leaves and fruits in these seasons. They also showed that fertilizer N was taken up mainly by the new organs, but that there was a large exchange of N between the young parts and the N reserve in the tree[17]. Small citrus trees (Calamondin) in sand culture were used by Legaz et al.[20] as a model for N absorption and translocation studies using ^{15}N. It was concluded that fertilizer N applied at flowering moved preferentially to the new organs, but was also absorbed in the reserve organs. A major part of the N in the developing fruits was withdrawn from reserve organs. During flowering, the trees absorbed 30 mg N daily per kg dry matter. During fruit set the rate of N absorption increased considerably[20]. More elaborate work by the same authors was carried out with four-year old Valencia trees fertilized with labeled N during different periods of the growth cycle[19]. It was found that N uptake was lowest during dormancy, increased during flowering and was highest during fruit set. During dormancy most of the absorbed N accumulated in the roots, while during flowering and fruitset most of the absorbed N was found in the leaves and fruits. In a second group of young trees (their dry weight was less than 1 kg/tree) the labeled N was applied during autumn and the translocation of absorbed N was observed for almost one year afterwards. More than 50% of the N in the old organs (roots, stems and old leaves) was translocated to the new growth during flowering. This proportion increased gradually until it reached almost 90% during the autumn flush. The roots and the old leaves made up the largest part of the N reserves in these small trees. When the trees were supplied with N after the labeling period, the importance of the reserve organs in supplying N to the young growth decreased gradually from 70% during flowering to 30% during autumn flush. N-starved trees, however, relied heavily on their N reserves during all growth periods (more than 80%)[19].

The work with small trees in sand culture[17,19,20] showed clearly the seasonal changes in N uptake and its distribution among the tree organs and the importance of N reserves in the older tissues for the proper development of new leaves and flowers in the spring.

Recently, an experiment was carried out in which labeled N was applied to two large, mature citrus trees during a complete growth season, one N-starved and the other heavily fertilized. The quantities of fertilizer N applied to the two trees were 0.34 and 1.0 kg per tree, respectively. The fate of the fertilizer N was followed quantitatively until fruit ripening[14]. Values of dry matter and total N distribution of these two trees are given in the last two columns of Table 1. Although the trees had similar total dry weights, the N-starved tree had a larger root system and less fruits and leaves. Total N in the N-starved tree was less, especially in the fruits and branches. Table 3, based on the above-mentioned experiment[14], shows the percentages of fertilizer N recovered in the various tree parts and in the soil, and the percentages of N in the tree parts derived from the seasons fertilizer N application. Total N recovery for the two cases was similar, but for the N-starved tree only 5% of the fertilizer N was found in the soil as compared with 16% in the case of high N application. The higher recovery of fertilizer N in the N-starved tree, was especially apparent in the leaves, roots and branches. It seems surprising that even in an overfertilized tree only 23% of the fruit N and 21% of the leaf N were taken up directly from the fertilizer,

Table 3. Percentage recovery of fertilizer N in tree parts and soil and part of total N in tree parts taken up from the seasons fertilizer application

| | Fertilizer N recovery, % | | N taken up from latest N dose (%) | |
	Low N tree	High N tree	Low N tree	High N tree
Fruits	10.7	11.4	16.0	23.4
Leaves	19.2	11.2	15.6	20.9
Branches	15.6	13.7	5.8	13.8
Roots	11.1	4.0	7.4	10.8
Whole tree	56.6	40.3	9.2	16.7
Soil N	5.1	16.1		
Total recovery	61.7	56.4		

while close to 80% of the N in the young tissues was derived from reserves stored in the tree. The permanent tree parts supplied an even higher percentage of the leaf and fruit N in the N-starved tree (84%). This tree also showed a much lower apparent N uptake in roots and branches, which can be explained as a depletion of N from those organs in favor of the developing fruits, twigs and leaves. These data are in agreement with the results obtained with small trees in sand culture as shown previously[17,19,20,28].

A more comprehensive picture of the N pools and their balance in a citrus orchard as based on the data of the last experiment[7,8,14] is shown in Table 4. The largest N pool is contained in the soil organic matter (more than 2 Mg/ha). The mineral N in the form of NO_3, which is directly available to the roots, is only 4–5% of this large N-pool. The N applied as fertilizer to the two experimental trees was 140 and 416 kg/ha, as compared to the recommended rate of 200 kg/ha. The amounts of N taken up by the tree from the fertilizer applied during that year were equivalent to 80 and 166 kg/ha, or 57% and 40% uptake efficiency,

Table 4. Calculated nitrogen pools in a citrus orchard (kg/ha, based on ref. 7, 8, 14)

Experimental treatment	Low N	High N
Soil organic N	2080	2140
Soil NO_3-N	85	113
Total N in trees	861	990
N applied in fertilizer	140	416
N in trees derived from season's fertilizer	80	166
N in fruits derived from season's fertilizer	15	22
N in leaves derived from season's fertilizer	27	47
N leached as NO_3 below root zone	37	64
Uptake efficiency	57%	40%

respectively. These amounts of N comprise only a small part of the total N pool of a mature tree, which is four to five times more than the average annual fertilizer application. The amounts of N derived from the fertilizer in the annual parts of the tree are again small compared with the applied N. All this shows the importance of two major N pools, the soil organic matter and the tree skeleton. Comparing the two experimental trees one can see how N from these pools is used in order to sustain the growth and production in the case of low N fertilization. In spite of the fact that the uptake efficiency of the N starved tree was relatively high and the nitrate in the soil was lower, the N leaching continued, although at a lower rate.

Conclusions

The annual growth and fruit yield of citrus trees contain only a small proportion of the fertilizer N applied during the growth period. More than 80% of the N in the new growth is drawn from the tree reserves. Fertilizer N is therefore not applied directly to the fruits, but to the whole tree and is essential for sustaining the N reserves in the permanent tree parts—trunk, branches and roots. This explains why lack of N application does not always result in a yield decrease and why the trees are able to grow vigorously in the spring when N uptake is still very low. It seems also that some N leaching is unavoidable, since the citrus roots are not able to take up all the available N in the soil. Leaf N is only an indication of the N status of the tree but in itself represents only a minor portion of total N in the tree. In many fertilizer experiments with citrus, differences in leaf N content appeared before effects on fruit yield were noted.

References

1   Barnette R M, DeBusk E F, Hester J B and Jones W W 1931 The mineral analysis of a nineteen year old Marsh seedless grapefruit tree. Citrus Ind. 12, 5–6.
2   Bingham F T, Davis S and Shade E 1971 Water relations, salt balance and nitrate leaching losses of a 960 acre citrus watershed. Soil Sci. 112, 410–418.
3   Cameron S H and Appleman D 1933 The distribution of total nitrogen in the orange tree. Am. Soc. Hort. Sci. Proc. 30, 341–348.
4   Cameron S H and Compton O C 1945 Nitrogen in bearing orange trees. Am. Soc. Hort. Sci. Proc. 46, 260–268.
5   Cary P R and Weerts P C J 1977 Crop management factors affecting growth, yield and fruit composition of citrus. Proc. Int. Soc. Citriculture 1, 39–43.
6   Dasberg S 1978 Nitrogen balance in a citrus grove. Hassadeh 58, 874–877 (*Hebrew, with English summary*).
7   Dasberg S, Bielorai H and Erner Y 1983 Nitrogen fertigation of Shamouti oranges. Plant and Soil 75, 41–51.
8   Dasberg S, Erner Y and Bielorai H. 1984 Nitrogen balance in a citrus orchard. J. Environ. Qual. 13, 353–356.

9   duPlessis S F 1977 Soil analysis as a necessary complement to leaf analysis for fertilizer advisory purposes. Proc. Int. Soc. Citriculture 1, 15–20.

10 Embleton T W and Jones W W 1977 Impact of research on California citrus fertilization. Proc. Int. Soc. Citriculture 1, 1–5.

11 Embleton T W and Jones W W 1978 Nitrogen fertilizer management programs, nitrate pollution potentials and orange productivity. In Nitrogen in the Environment. Eds. D R Nielsen and T G MacDonald. 1, 275–297, Academic Press, New York.

12 Embleton T W, Pallares C O, Jones W W, Summers L L and Matsumura M 1981 Nitrogen fertilizer management of vigorous lemons and nitrate pollution potential of ground water. Calif. Water Res. Center, Univ. of California Contr. 182, 1–30.

13 Embleton T W, Reitz H J and Jones W W 1973 Citrus fertilization. In The Citrus Industry. Ed. W Reuther 3, 122–181.

14 Feigenbaum Sala, Bielorai H, Erner Y and Dasberg S 1986 The fate of ^{15}N labeled nitrogen applied to mature citrus trees. Plant and Soil 97, 179–187.

15 Golomb A and Goldschmidt E 1980 The mineral balance in the biannual bearing mandarin Wilking. Alon Hanotea 35, 639–648 (Hebrew, with English summary).

16 Greenham D W P 1980 Nutrient cycling: The estimation of orchard nutrient uptake. In Mineral Nutrition of Trees. Eds. D Atkinson, K E Jackson, R O Sharpless and W M Waller. pp 345–353 Butterworths, London.

17 Iwakiri T and Nakahara M 1981 Nitrogen fertilization programs in Satsuma mandarin groves in Japan. Proc. Int. Soc. Citriculture 2, 571–574.

18 Koo R C J 1980 Results of citrus fertigation studies. Proc. Fla. St. Hort. Soc. 93, 33–36.

19 Legaz F, Primo-Millo E, Primo-Yufera E X and Gil C 1981 Dynamics of ^{15}N labeled nitrogen nutrients in Valencia orange trees. Proc. Int. Soc. Citriculture 2, 575–582.

20 Legas F, Primo-Milloo E, Primo-Yafera E, Gil E and Rubio J L 1982 Nitrogen fertilization in citrus I. Absorption and distribution of nitrogen in Calamondin trees (Citrus mitis) during flowering, fruit set and initial fruit development periods. Plant and Soil 66, 334–351.

21 Reese R L and Koo R C J 1975 N and K fertilization effects on leaf analysis, tree size and yield of tree major Florida orange cultivars. J. Amer. Soc. Hort. Sci 100, 195–198.

22 Reitz H J 1978 Criutique of nitrogen fertilizer management programs, nitrate pollution potential and orange productivity. In Nitrogen in the Environment. Eds. D R Nielsen and T G MacDonald. 1, 297–302, Academic Press, New York.

23 Reuther W, Smith P F, Scudder G K and Hrnciar G 1957 Response of Valencia orange trees to timing, rates and ratios of nitrogen fertilization. Am. Soc. Hort. Sci. Proc. 70, 223–230.

24 Sharpless G C and Hilgeman R H 1969 Influence of differential nitrogen fertilization on production, trunk growth, fruit size, quality and foliage composition of Valencia orange trees in central Arizona. Proc. First Int. Citrus Symp. 3, 1564–1578.

25 Smith P F 1971 Effects of time and application of N and K and of N rate on performance of nucellar Valencia orange trees on two stocks. J. Am. Soc. Hort. Sci. 96, 568–571.

26 Stevenson F J 1982 Origin and distribution of nitrogen in soils. In Nitrogen in Agricultural Soils. Ed. F J Stevenson. Agron. 22, 1–43. ASA, Madison, WI.

27 Titus J S and Kang Seong-Mo 1982 Nitrogen metabolism, translocation and recycling in apple trees. Hort. Rev. 4, 204–246.

28 Wallace A, Zidan Z I, Mueller R T and North C P 1954 Translocation of nitrogen in citrus trees. Am. Soc. Hort. Sci. Proc. 64, 87–105.

29 Yuda E 1977 Nutritional problems in citrus culture in Japan. Proc. Int. Soc. Citriculture 1, 5–9.

Plant and Soil 100, 11–19 (1987).
© 1987 *Martinus Nijhoff Publishers, Dordrecht.*

Ms. 100-03

Soil-oxygen and plant-root interaction: An electrical analog study

M.B. KIRKHAM
Evapotranspiration Laboratory, Kansas State University, Manhattan. KS 66506, USA

Key words Depth of plow layer Depth of root zone Electrical analog Nutrient uptake Oxygen conductance Row crop

Summary The objective of this experiment was to determine, by electrical analog, how oxygen movement to plant roots is affected by two variables: the depth of the plow layer (depth of tillage) and the depth of the roots in the plow layer. The analog was developed for a row crop that had a long, narrow, vertical sheet of roots. Oxygen moved from the cultivated soil surface to the sheet of roots. Plow-layer depth (depth of soil cultivated) was 20, 10, 5 or 2.5 cm. Depth of the root sheet was 20, 10, 5, 2.5, 1.25, 0.625, or 0 cm. Dimensionless values for flow of oxygen were determined and compared to theoretical values.

Experimentally determined dimensionless flow values agreed with those determined theoretically, which showed that the model simulated expected values for oxygen transport. As the depth of the root sheet increased, movement of oxygen to the roots increased curvilinearly. Roots that penetrated only a small distance into the plow layer took up large amounts of oxygen. As the depth of the plow layer increased, oxygen movement to the root sheet increased. The results suggested that, for maximum flow of oxygen to the root sheet, a deep plow layer is more important than a deep root sheet.

Introduction

Aeration of soil is necessary for the maximum absorption of nutrients by plant roots[16]. Insufficient supply of oxygen to roots influences the uptake of N, P, K, Ca, Mg, Cl, B, Zn, Cu, Mn, and Fe[8], all of which are essential inorganic elements. Many elements are taken up actively[3]. Energy is required for active uptake of nutrients and aerobic respiration is the major supplier of this energy. During respiration, oxygen is taken up and carbon dioxide is given off. For adequate aeration, roots of plants usually need at least 10% by volume air in the soil to survive[22]. Therefore, gas exchange between air in the pore space of the soil and air in the atmosphere is needed for assimilation of inorganic elements[1].

Analog experiments have long been used to study the movement of water in soil and other linear-flow phenomena in porous media[13]. The analogs are usually based on Ohm's law, which describes the linear flow of electricity in a conducting medium. The movement of one gas into another by diffusion, such as oxygen into air by a concentration gradient, is, like Ohm's law, a linear-flow process. Fick discovered the linear-flow law of diffusion, which is analogous to Ohm's law. Because of the analogy between the two laws, electrical analogs can be constructed to describe the process of oxygen flow in porous media, such as soil[8,12].

Contribution No. 87-5-J from the Kansas Agricultural Experiment Station.

Oxygen diffusion rate is considered to be the best index of soil oxygen availability to plant roots[8].

Most plant and soil scientists who have worked with electrical analogs have been concerned with the novement of water in soil. Few have considered analogs to study the movement of oxygen in soil. The objective of this experiment was to determine, by electrical analog, how oxygen transfer to roots is influenced by two variables: the depth of the plow layer and the depth of the roots in the plow layer. Another objective was to see if the results from the experiment checked with theoretical values.

The model

Figure 1 shows the schematic model. Roots were assumed to be hanging vertically in sheets (curtains) below equally spaced furrows (plant rows). This would be similar to a situation in which roots are pruned on each side of a row to control parasitic weeds of roots or to reduce rates of transpiration[5,6,10,18]. When roots are pruned in this way, the remaining roots form a long, narrow curtain under the row of plants. The soil surface on each side of the plants in a row was assumed to be sealed (closed to the movement of oxygen). The rest of the soil surface was assumed to be cultivated (open to the movement of oxygen through the surface of the soil). A barrier (boundary), to be described in the paragraph after next, lay at the bottom of the model.

Figure 1 shows only one unit of the model. The rows of plants were assumed to be 50 cm (2L) apart. The symbols used in the model were the same as those used by Warrick and Kirkham[20] in a drainage problem, which in a special case (their distance 's' in their Figure 1 taken to be zero) yields the theory for the diffusion problem of this study. By so adapting Warrick and Kirkham[20], the data obtained in the present experiment could be compared with theoretical curves derived using their method. The advantage of doing this is that the electrical analog could be verified mathematically, and, conversely, the results from the analog could be used to confirm the mathematics.

The depth of the plow layer (h) was variable. It was given values, which, in a field situation, would be 20, 10, 5, and 2.5 cm. Five cm of soil surface on each side of the plants ($2\varepsilon = 10$ cm in Fig. 1) in a row were assumed to be sealed. The cultivated area between two rows was assumed to be 40 cm (50 cm row spacing minus 10 cm sealed soil surface = 40 cm). The barrier at the bottom of the plow layer (Fig. 1) could be clay, bedrock, a water table, compacted soil, or any barrier

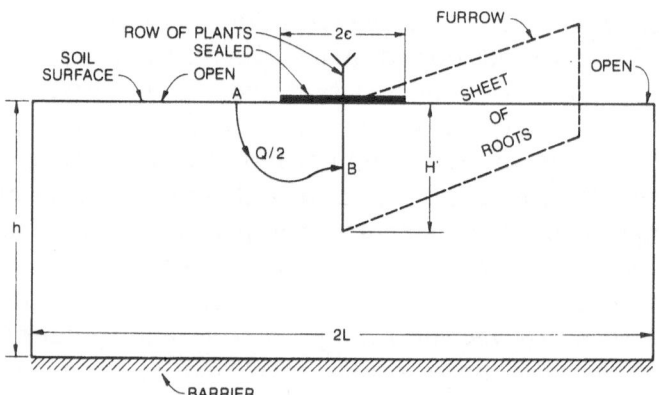

Fig. 1. Vertical cross section of the model for transport Q/2 of oxygen from the soil surface to one side of a row of a vertical sheet of roots. Width of root sheet is small compared with other model dimensions. Q/2 enters each side of the sheet. Symbols 2ε, 2L, H' are as in reference no. 20.

impervious to diffusion of oxygen. The 20 cm maximum depth for the plow layer is approximately the average plow-layer depth in the USA (17 cm)[9]. Row width of maize (*Zea mays* L.), soybean [*Glycine max* (L.) Merr.], and grain sorghum [*Sorghum bicolor* (L.) Moench] varies from 50 to 100, 15 to 100, and 30 to 100 cm, respectively[14,15,17]. Thus, the 50-cm row spacing in the model fell within ranges of row widths used for maize, soybean, and sorghum. The analog does not specify a crop.

In the model, the row spacing (50 cm) and the noncultivated space (10 cm) (or cultivated space = 40 cm) were assumed to be constant. The two variables were the depth of the plow layer (h = 20, 10, 5, 2.5 cm) and the depth of the root sheet (H' = 20, 10, 5, 2.5, 1.25, 0.625 cm). The third dimension of the model (dimension 1 = plow-layer depth; dimension 2 = row width) was taken as a length perpendicular to the plane of the cross section of the model shown in Figure 1. The third dimension was the length of the furrow (root sheet). In theoretical expression, as was done by Warrick and Kirkham[20], the length of the furrow was considered to be a unit length. One line of flow of oxygen from the cultivated soil to the active sheet of roots is shown in Figure 1. Additional lines could be sketched by symmetry. Lines of symmetry were below the center of the cultivated surface. Other lines of symmetry were below the plant rows. In the model, the physical situation from row to row repeated itself in units that could be added together to determine a cumulative effect. Assumptions inherent in the model included the following: a homogenous soil, oxygen diffusion in the gas phase, and a uniform and limited root distribution.

Materials and methods

A rectangular ceramic container, 9 cm wide, 25 cm long, and 9 cm tall (inside dimensions), was used to hold the electrolyte, which was tap water. In the analog, the tap water represented the flow medium, the porous soil, in which oxygen diffuses. The model was upside-down compared to field conditions. The bottom of the container represented the surface of the soil. The analog in its vertical cross section was 2.22 times smaller than the field conditions that it represented. Tap water was added to the container until it stood to one of four different heights (nominal depths): 9, 4.5, 2.25, and 1.125 cm. These heights of water simulated the following depths of plow layer in the field (multiply by 2.22): 20, 10, 5, and 2.5 cm. The depth of the cultivated soil (plow layer) was the depth of the tap water. The tap water was allowed to come to constant temperature (24°C) before it was added to the container.

Copper electrodes simulated cultivated soil which had the concentration of oxygen in the air (20.95%)[21]. (The air is the source of oxygen, and it can move into the soil through the cultivated soil.) The copper electrodes were cut by hand with scissors from a copper sheet (thickness = 0.8 mm). They were burnished with sandpaper to eliminate any resistance caused by corrosion, layers of oxide, grease, or dirt. Two electrodes simulating the interface of atmospheric oxygen and cultivated soil were used. Each measured 9 cm × 9 cm. The two electrodes were placed (because the laboratory model was upside-down) on the bottom of the ceramic container. A third electrode simulated a sheet of roots which actively takes up oxygen. The root sheet varied in depth. Six different depths (six electrodes) were used (all electrodes were 9 cm wide): 9, 4.5, 2.25, 1.125, 0.5625, and 0.28125 cm (nominal depths). Since the model was 2.22 times smaller than field conditions, these depths of electrodes represented root sheets in the field that were 20, 10, 5, 2.5, 1.25, and 0.625 cm deep. The electrode that represented the sheet of roots was placed vertically into the container. It was perpendicular to the two electrodes that represented the unsealed cultivated soil-air interface.

The electrodes representing the sheets of roots were placed, one at a time, into the water, which was at one of four depths (see above). The two electrodes representing the cultivated soil were connected together and attached to one end of the 'unknown' terminal of a Wheatstone bridge (Leeds and Northrup Model No. 4760 connected to Beede Model No. ES5OUADC galvanometer and a 6-volt dry-cell battery). The 'root-sheet' electrode was attached to the other end of the unknown terminal of the Wheatstone bridge.

Figure 2 shows a side view of the experimental setup. To visualize the model simulating field conditions, one must think of a row of plants located at the center of the container. (As stated before, the model was upside-down). The bottom of the model, where the two electrodes lay (the unsealed, cultivated soil-oxygen interface), represented the surface of the soil in contact with atmospheric oxygen.

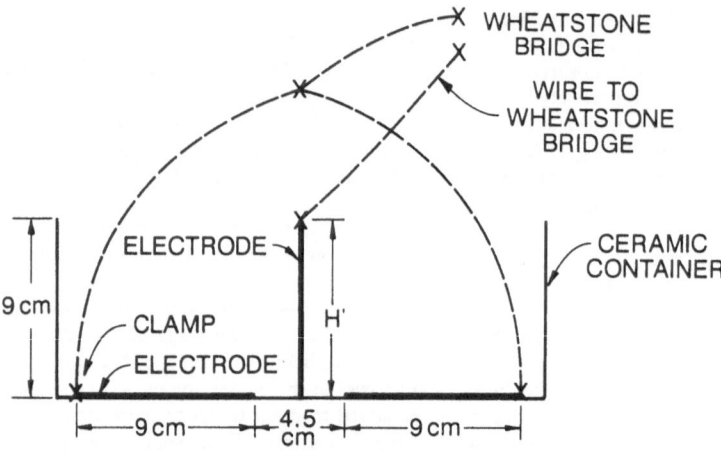

Fig. 2. Side view of the experimental setup. For details, see text.

Three runs were done for each geometry (different depths of tap water representing the plow layer and different depths of electrodes representing the sheet of roots). Only one sheet of roots or no sheet was in the container for a measurement. The depth of the plow layer (h) varied from 2.5 to 20 cm, while the depth of the root sheet (H') varied from 0 to 20 cm (field conditions). Resistance was measured in ohms (Ω) and the reciprocal of ohms was calculated to obtain conductance in Siemens (S) ($1\,S = 1/\Omega = 1$ mho). [The Système International (SI) unit for mho is Siemen.] Conductance (not to be confused with conductivity) is symbolized by the letter G. Means and standard deviations of the three replications were calculated.

The specific electrical conductivity (σ) (defined as the electrical conductance measured in a cubic cell containing electrodes 1 cm × 1 cm and separated by a distance of 1 cm)[7], of the tap water was determined by using the law of resistance:

$$R = \frac{\varrho L}{A}$$

where R = resistance (ohms), L = length (cm), A = area (cm^2), and ϱ = resistivity (ohm-cm). Conductivity, σ, is the reciprocal of resistivity: $\sigma = 1/\varrho$ (mho/cm or S/cm). The specific electrical conductivity was determined by plotting 1/R (y-axis) versus A/L (x-axis). The slope of this line is σ. Sigma, so calculated, was 0.000350 S/cm, which was a value similar to one determined independently by using a salinity meter (Model No. 5000-A Soil Salinity Sensor and Model No. 5500 Salinity Bridge, Soilmoisture Equipment Corporation, Santa Barbara, California, USA), which was 250 mg/L (250 ppm) $\cong 0.0004$ S/cm[19].

Dimensionless flow values were calculated to determine transport of oxygen to the root sheet. The dimensionless flow values calculated by Warrick and Kirkham[20] were $Q/2k\delta$, where Q is the quantity of oxygen transported per unit time per unit length of furrow (root sheet in this study) (cm^3/sec/cm); k = transport coefficient (cm/sec); δ = driving-potential difference (cm) that is the potential difference across points of flow as A and B of the system (Fig. 1). $Q/(2k\delta)$ is equal to $G/(2\sigma)$. [$Q/(2k\delta) = I/(2\sigma V) = (1/2\sigma)(1/R) = (1/2\sigma)G$, where I = current in amperes, V = voltage in volts, and R = resistance in ohms]. The G values determined in this experiment were multiplied by $1/(2\sigma)$ to obtain the dimensionless flow values shown in the next section.

Results and discussion

Figure 3 and Table 1 show the experimental results. The curves obtained experimentally were similar to those determined theoretically

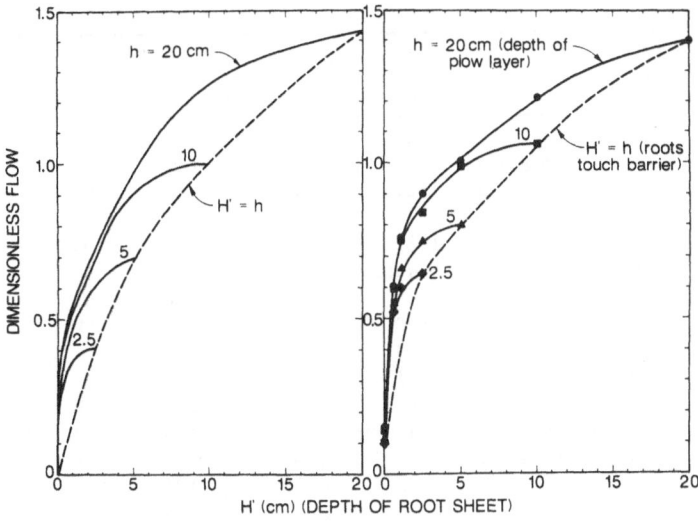

Fig. 3. Dimensionless flow of oxygen *versus* length H′ of vertical sheet of roots. *Left*: theoretical values. *Right*: experimental values.

(compare Fig. 3, right and left). As the depth of the root sheet (H′) increased, transport of oxygen to the roots increased curvilinearly. Roots that penetrated only a small distance into the plow layer took up significant amounts of oxygen. For example, roots that extended only 5 cm into the soil, when the plow-layer depth was 20 cm, had a dimensionless flow value of 1.01 (experimentally determined; see Table 1), which was not much smaller than the dimensionless flow value of 1.40, which occurred when the roots penetrated the entire depth of the plow layer.

As the depth of the plow layer (h) decreased, oxygen movement to the root sheet decreased. Just scratching the soil surface [h = 2.5 cm or (2.5 cm/20 cm) × 100 = 12.5% of the plow layer tilled] increased the dimensionless flow value by a large amount (0.11 to 0.64, experimentally determined). These results are similar to those found in a previous experiment[13], in which an electrical analog was used to determine the effect of the amount of soil surface tilled (depth to barrier) on oxygen movement to plant roots. In that experiment, dimensionless ratios of G/G_o were calculated and compared to theoretical values. Similar dimensionless ratios are shown for this experiment in Table 1. When 20% of the soil surface was tilled (constant value in this experiment) and 50% of the plow layer was tilled, $G/G_o = 0.91$ for the earlier experiment[13] and $G/G_o = 0.85$ for this experiment (theoretical values). The difference is due to the different geometries in the two experiments. In the earlier experiment, the roots were located in a horizontal sheet at the bottom of

Table 1. Electrical analog data

Case	h[†] (cm)	Plow layer tilled (%)	H′[†] (cm)	G ± s.d.[†] (mS)	Dimensionless flow[‡]		G/G$_o$*	
					Theory	Experiment	Theory	Experiment
1	20	100	20	4.1 ± 0.6	1.427	1.40	1.00	1.00
2	20	100	10	3.8 ± 0.5	1.257	1.21	1.00	1.00
3	20	100	5	3.2 ± 0.3	0.969	1.01	1.00	1.00
4	20	100	2.5	2.8 ± 0.3	0.721	0.90	1.00	1.00
5	20	100	1.25	2.4 ± 0.2	0.583	0.76	1.00	1.00
6	20	100	0.625	1.9 ± 0.2	Not done	0.60	1.00	1.00
7	20	100	0	0.5 ± 0.04	0	0.15	1.00	1.00
8	10	50	10	3.3 ± 0.2	1.070	1.06	0.85	0.88
9	10	50	5	3.1 ± 0.3	0.908	0.97	0.94	0.96
10	10	50	2.5	2.7 ± 0.1	0.701	0.84	0.96	0.93
11	10	50	1.25	2.4 ± 0.3	0.544	0.75	0.93	0.99
12	10	50	0.625	1.9 ± 0.2	Not done	0.60	0.98	1.00
13	10	50	0	0.4 ± 0.1	0	0.14	0	0.93
14	5	25	5	2.5 ± 0.03	0.692	0.80	0.71	0.79
15	5	25	2.5	2.2 ± 0.1	0.606	0.75	0.85	0.83
16	5	25	1.25	2.1 ± 0.1	0.492	0.66	0.84	0.87
17	5	25	0.625	1.7 ± 0.2	0.410	0.55	0.89	0.92
18	5	25	0	0.4 ± 0.1	0	0.12	0	0.80
19	2.5	12.5	2.5	2.0 ± 0.04	0.410	0.64	0.57	0.71
20	2.5	12.5	1.25	1.9 ± 0.1	0.376	0.60	0.66	0.79
21	2.5	12.5	0.625	1.7 ± 0.2	0.328	0.52	0.70	0.87
22	2.5	12.5	0	0.3 ± 0.1	0	0.11	0	0.73

[†] h = depth of plow layer; H′ = depth of root sheet. Dimensions are for field conditions. The electrical analog, in its cross section, was 2.22 times smaller than field conditions. G ± s.d. = conductance ± standard deviation.

[‡] Dimensionless flow = $Q/(2k\delta) = G/(2\sigma)$ (see text). Theoretical values calculted by D Kirkham[11]. Experimental values determined as follows:

1. The resistance value in ohms measured on the Wheatstone bridge was divided by two because the analog measured Q (oxygen flow to both sides of root sheet) and not Q/2 (oxygen flow to one side of root sheet), for which the theory was developed. The circuit was in parallel and to compare measured values with theoretical values, the resistance had to be halved to find the resistance in just one branch of the circuit.

2. Conductance, G, was calculated from the reciprocal of resistance.

3. G was multiplied by $1/(2\sigma)$.

4. The value obtained in Part 3 was divided by the furrow length (length of plant row).

* Ratios of dimensionless flow were obtained as follows: $[G/(2\sigma)]/[G_o/(2\sigma)] = G/G_o$. G_o = maximum conductance possible when 100% of the plow layer (20 cm) was tilled or open to the movement of oxygen.

the plow layer. In this experiment, the roots were oriented in a vertical sheet below the plant row. The results show that there was slightly more movement of oxygen to the roots oriented horizontallycompared to the roots oriented vertically, when the plow-layer depth and amount of soil surface tilled were constant.

One can compare the effect of plow-layer depth and root-sheet depth on the dimensionless flow values by using the theoretically determined

Table 2. Comparison of dimensionless flow values between one depth of the plow layer or root sheet and another depth, one-half as deep

Decrease		Dimensionless flow	
		Plow layer	Root sheet
From (cm)	20	1.43	1.43
To (cm)	10	1.07	1.26
		Difference 0.36	0.17
From (cm)	10	1.07	1.26
To (cm)	5	0.69	0.97
		Difference 0.38	0.29
From (cm)	5	0.69	0.97
To (cm)	2.5	0.41	0.72
		Difference 0.28	0.25

ones shown in Table 1. A comparison between one depth and another depth, one-half as deep, is shown in Table 2. In each case, the dimensionless flow value was decreased more when the plow-layer depth was halved than when the root-sheet depth was halved. The difference was more than double (0.36 versus 0.17) at the deepest depths (20 and 10 cm). These results suggested that, for maximum movement of oxygen to the root sheet, it is more important to plow deeply than to have deep root sheets.

The values shown in Fig. 3 are for flow from the soil surface to the root surface, where the root surface may include a film of water. That is, the values are valid only for movement of oxygen through the porous medium, the soil. Because it is generally believed that there is water around roots, unless the soil is near the permanent wilting point (-1.5 to -2.0 MPa)[2,4], the oxygen must flow from the soil through a film of water to enter the root. This could be a large resistance to the transport of oxygen, since diffusion of oxygen is about 10^4 times faster through air than through water[1,8]. However, in most cases, this film is probably small compared to the distance that oxygen molecules must move in going from the soil surface to the root. The analog applies to situations in which the major resistance ($1/G$) is between the soil surface and the film of water at the root surface.

The results of this experiment may have application to fertilizer-placement studies. We need to know where to apply fertilizer for optimum uptake of the nutrients. As stated in the Introduction, active nutrient uptake requires oxygen. Presumably more nutrients would be taken up where the dimensionless flow values are high than where the values are low (Fig. 3).

As stated in the Introduction, electrical analog studies apply to any linear-flow process. The results of this experiment, therefore, could be applied to the flow of water, heat, or gas (*e.g.*, oxygen, carbon dioxide). The analog considered in this paper is intended to apply mainly to oxygen movement when the root sheets are in water-unsaturated soil, which contains air-filled pores. But it can also apply to water-saturated soil if an oxygen concentration difference exists between the soil surface and the plant root, although the oxygen movement may be slow. The analog can apply to heat movement when a temperature difference occurs between the surface of the soil and the sheet of roots under the plant row. The 'sealed' part of the surface in Fig. 1 would then become insulated. The analog can apply to water movement (rather than oxygen diffusion) when the root sheets are in water-saturated soil. Oxygen in dissolved surface water could be carried with the water to the root sheet. But as far as nutrient uptake is concerned, the analog probably best applies to oxygen and heat movement. Both are important for absorption of minerals. Warm and aerated conditions provide a better environment for uptake of elements than cool and anoxic conditions.

Acknowledgement   I thank Don Kirkham for providing theoretical values for Figure 3 and Table 1.

References

1   Barber S A 1984 Soil Nutrient Bioavailability. A Mechanistic Approach. John Wiley and Sons, New York, 398 p. (See p. 153)
2   Bernstein L, Gardner W R and Richards L A 1959 Is there a vapor gap around plant roots? Science 129, 1750 and 1753 (two pages).
3   Clarkson D T 1974 Transport and Cell Structure in Plants. John Wiley and Sons, New York, 350 p. (See p. 410 and 282)
4   Gardner W R and Ehlig C F 1962 Impedance to water movement in soil and plant. Science 138, 522–523.
5   Geisler G and Ferree D C 1984 Response of plants to root pruning. Hortic. Rev. 6, 155–188.
6   Geisler G and Maarufi D 1976 Investigations on the importance of the root systems of cultivated plants. II. The influence of root trimming on plant growth, root morphology, transpiration and nitrogen uptake in relation to the soil water content and nitrogen fertilization. J. Agron. Crop Sci.143, 1–17. (*In Ger., Eng. sum*)
7   Gilmont R and Wechsler L I 1986 Electrical resistivity of absolute water calculated from the ions. Am. Lab. 18, 70–82.
8   Gliński J and Stepniewski W 1985 Soil Aeration and Its Role for Plants. CRC Press, Inc., Boca Raton, Florida, 229 p. (See p. 46, 48, 152–161, 189).
9   Higgs R, Heidenreich C, Loberger R, Cropp R and Mitchell M 1981 Agricultural Mathematics. Second Ed. The Interstate Printers and Publishers, Inc., Danville, Illinois, 297 p. (See p. 69)
10 Insley H and Buckley G P 1985 The influence of desiccation and root pruning on the survival and growth of broadleaved seedlings. J. Hortic. Sci. 60, 377–387.
11 Kirkham D 1986 Unpublished.
12 Kirkham D and Powers W L 1984 Advanced Soil Physics. Reprint Ed. (Revised). Robert E. Krieger Publishing Co., Malabar, Florida, 534 p. (See p. 46, 74–75)
13 Kirkham M B 1986 Simulation of oxygen movement to plant roots as affected by tillage width and depth. Soil Tillage Res. 7, 221–231.

14   Mitchell R L 1970 Crop Growth and Culture. The Iowa State University Press, Ames, Iowa, 349 p. (See p. 109, 118–120)

15   Peterson, V 1981 Optimum planting practices. *In* Grain Sorghum Handbook. Coop. Ext. Service Pub. No. C-494 (Revised). Kansas State Univ., Manhattan, Kansas, pp 5–7.

16   Taylor S A 1949 Oxygen diffusion in porous media as a measure of soil aeration. Soil Sci. Soc. Am. Proc. 14, 55–61.

17   Thompson C A 1983 Super thick sorghum management. Rep. of Progress No. 437. Agr. Exp. Sta., Manhattan, Kansas, 24 pp.

18   United States Department of Agriculture 1985 Agronomic practices for protection from *Striga asiatica*. *In* 1984 Annual Report, Whiteville Methods Development Center. U.S. Dep. Agr., Anim. Plant Health Inspection Service, Plant Protection and Quarantine, Whiteville, North Carolina, pp 98–99.

19   United States Salinity Laboratory Staff 1954 Diagnosis and Improvement of Saline and Alkali Soils. Agr. Handbook No. 60. U.S. Dep. Agr., Washington, D.C., 160 p. (See p. 11)

20   Warrick A W and Kirkham D 1969 Two-dimensional seepage of ponded water to full ditch drains. Water Resources Res. 5, 685–693.

21   Weast R C 1964 Handbook of Chemistry and Physics. 45th Ed. The Chemical Rubber Co., Cleveland, Ohio, (pages not numbered sequentially). (See p. F-84)

22   Wesseling J and van Wijk W R 1957 Soil physical conditions in relation to drain depth. *In* Drainage of Agricultural Lands. Ed. J N Luthin. Am. Soc. Agronomy, Madison, Wisconsin, pp 461–504.

Plant and Soil 100, 21–34 (1987).

Ms. 100-04

Effect of calcium on the absorption and translocation of heavy metals in excised barley roots: Multi-compartment transport box experiment

T. KAWASAKI and M. MORITSUGU

Institute for Agricultural and Biological Sciences, Okayama University, Kurashiki 710, Japan

Key words Absorption Barley Cd Excised roots Mn Multi-compartment transport box Radioisotope Translocation Zn

Summary The effect of Ca on the absorption and translocation of Mn, Zn and Cd in excised barley roots was studied using a multi-compartment transport box technique. A radioisotope (^{54}Mn, ^{65}Zn or ^{115m}Cd)-labelled test solution was supplied to the apexes of excised roots and the distribution pattern in the roots was examined in the absence or presence of Ca. Results obtained were as follows.

Addition of Ca to the test solution reduced the absorption of Mn and inhibited drastically its translocation in excised roots. With increasing concentrations of Ca in test solutions, its inhibitory effects on the absorption and translocation of Mn became severe.

Similar results were observed for the absorption and translocation of Zn. Ca in the test solution decreased the absorption and inhibited drastically the translocation of Zn; as in the case of Mn, higher concentrations of Ca had severe effects on these functions.

It was also evident that the addition of Ca to the test solution reduced the absorption of Cd at all levels of Cd concentration (1, 10, and 100 μM). Cd absorption decreased with increasing concentrations of Ca in the test solution. However, Ca accelerated the translocation of Cd in excised roots supplied with test solutions containing up to 10 μM Cd. At 100 μM Cd, addition of Ca caused a negligibly small acceleration of Cd translocation.

The accelerating effect of Ca on Cd translocation, especially "xylem exudation", decreased markedly with the addition of 2,4-dinitrophenol, but not with the addition of chloramphenicol or p-chloromercuribenzene sulphonic acid. When barley plants were supplied with only CaSO₄ during the entire growing period, that is, plants were not supplied with nutrient solution on the last day of this period, Ca had no accelerating effect on Cd translocation in excised roots.

Introduction

Mn and Zn have been established to be essential microelements for higher plants, while their excesses give significant injurious effects on plant growth[20]. Numerous investigations have been carried out on the absorption of Mn[19,21,23,24] and Zn[1,2,4,5,6,26,27] by plant roots. Several researchers have also reported on microelement transport from roots to shoots[4,10,25]. However, there have been no reports about the translocation of Mn and Zn in plant roots.

On the other hand, it is well known that Cd is not only harmful to humans[18], but also toxic to plants[20]. Though many researchers have been carried out on the absorption of Cd by plants[8,9,11,12,13,22], there has been also no report thus far dealing with the transport pattern of Cd in plant roots.

To investigate in more detail the absorption and translocation of heavy metals, the distributions of Mn, Zn and Cd in plant roots were examined in relation to Ca supply in the present experiment, using a multi-compartment transport box[16,17].

Materials and methods

Excised roots of 4-day-old barley plants (*Hordeum vulgare* L., cv. Akashinriki) were used as experimental materials in this investigation. Seeds of barley were allowed to germinate for 24 h in aerated water at 25°C. The germinating seeds were spread over a layer of plastic screen, and grown with 0.25 mM CaSO$_4$ for 48 h. Thereafter, a nutrient solution was supplied for 24 h in the dark at 25°C under continuous aeration. Composition of the nutrient solution used was as follows: KNO$_3$ 4.0 mM, NH$_4$H$_2$PO$_4$ 1.0 mM, CaCl$_2$ 1.0 mM, MgSO$_4$ 1.0 mM, Fe 1.0 ppm (as Fe-citrate), B 0.5 ppm (as H$_3$BO$_3$), Mn 0.5 ppm (as MnCl$_2$), Zn 0.05 ppm (as ZnSO$_4$), Cu 0.02 ppm (as CuSO$_4$) and Mo 0.01 ppm (as (NH$_4$)$_6$Mo$_7$O$_{24}$). The roots of the seedlings were excised, washed thoroughly with deionized water and used for transport experiments.

In the present investigation, a multi-compartment transport box, the description of which was given in detail in the previous papers[16,17], was used. As can be seen in Fig. 1, the apparatus consists of 4 compartments, each of which is about 10 mm long, 50 mm wide and 15 mm deep, with plexiglass barriers between the compartments. Roman numerals in the figure indicate the position of each compartment.

Excised barley roots were set horizontally so that the apical part of the root was put in compartment I and the basal cut end in compartment IV (see Fig. 1). Then the upper half of each barrier was put on the roots without crushing them. The barriers were sealed with vaseline to prevent leakage of the test solution. In all of the experiments, 8 roots were used for each treatment.

^{54}Mn, ^{65}Zn and ^{115m}Cd were used to label manganese, zinc and cadmium, respectively, in the test solutions. Compartment I was supplied with a radioisotope-labelled test solution, and compartments II, III and IV with a non-labelled test solution having the same chemical composition as that of the radioisotope-labelled one. In all treatments, sodium tartrate was added at 1.0 mM concentration to avoid pH changes. The pH of all the test solutions was checked and adjusted to 5.0. The absorption treatment was continued for about 20 h at 25°C in the dark.

After the absorption period, solutions in each compartment were placed in sample tubes. The roots were cut at the barriers between each compartment and each segment was put into a separate sample tube. Only the apexes of roots in compartment I, which were supplied with a radioisotope-labelled test solution, were sampled after about 5 min of desorption treatment with non-labelled

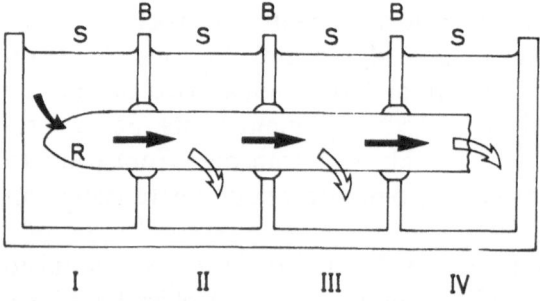

Fig. 1. Multi-compartment transport box. *Black arrows* show diagrammatically the absorption and translocation of ions in a plant root, and *white arrows* show the efflux of ions from a plant root. **I, II, III** and **IV**: the position of each compartment; **R**: root; **S**: test solution; **B**: plexiglass barrier. For a further description, see Materials and methods.

solutions having the same chemical composition as that of the radioisotope-labelled one. Radioactivity of the samples was measured with a well-type scintillation counter for ^{54}Mn and ^{65}Zn, and by means of Cerenkov radiation[3] with a liquid scintillation counter for ^{115m}Cd. The amounts of Mn, Zn and Cd in each part of the excised roots and in the solution of each compartment were calculated, based on the radioactivities of ^{54}Mn, ^{65}Zn and ^{115m}Cd, respectively. Results of each treatment are shown as the mean values of three to eight replicates, and the results are expressed in μmol/g fresh weight of roots/24 h. These results show the distribution of Mn, Zn and Cd through the roots, and the sum of these results gives total amounts of Mn, Zn and Cd absorbed in the apex of the roots.

Results

Mn absorption and translocation in excised roots

The absorption and distribution of Mn in excised barley roots were examined using various concentrations of Mn in the absence or presence of Ca (500 μM). The results for 1, 10 and 100 μM treatments of Mn in test solutions are presented in Figs. 2, 3 and 4, respectively. In the figures, the Roman numeral under the abscissa shows the position of each compartment in the transport box, and an asterisked numeral indicates the compartment supplied with a radioisotope-labelled test solution. Experimental treatments are presented under the numerals. A shadowed bar indicates the amount of Mn in the roots and an empty bar shows that in the solution of each compartment. The total amount of Mn absorbed is also presented in the figures.

From the data presented in Figs. 2 to 4, it is evident that the addition of 500 μM Ca to the test solution decreased Mn absorption and translocation in excised roots, as compared with the results in the absence of Ca.

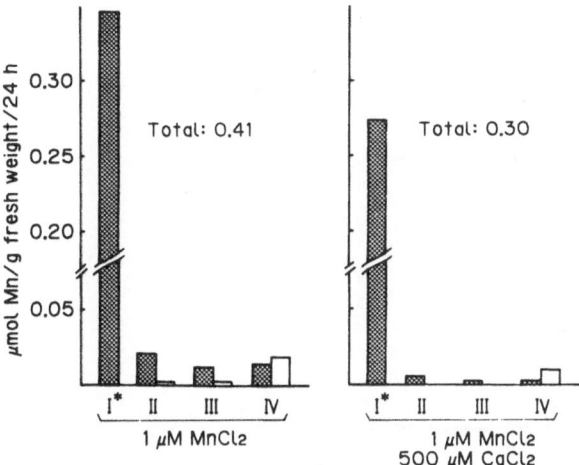

Fig. 2. Effect of Ca on the absorption and translocation of Mn in excised barley roots (Mn 1 μM). An *asterisked numeral* indicates the compartment supplied with a radioisotope-labelled test solution. *Shadowed bars* show the amount of Mn in the roots and *empty bars* that in the solution.

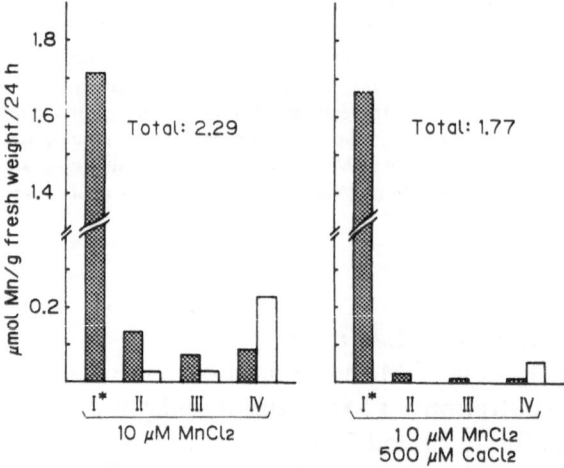

Fig. 3. Effect of Ca on the absorption and translocation of Mn in excised barley roots (Mn 10 μM).
The *symbols* are the same as in Fig. 2.

The distribution pattern of ions in an excised plant root can be classified qualitatively as follows: the portion of ions retained in the apex, the portion translocated and redistributed in the root, the portion moved outward across the cortex and the portion exuded into the external solution through the cut end of the root. These four portions are named tentatively "accumulation", "redistribution", "cortical efflux" and "xylem exudation", respectively, and represented schematically in Fig. 5.

With this distribution pattern in mind, the results presented in Figs. 2, 3 and 4 are summarized in Table 1, with an emphasis on comparison between the absence and presence of Ca in test solutions. In Table 1, the

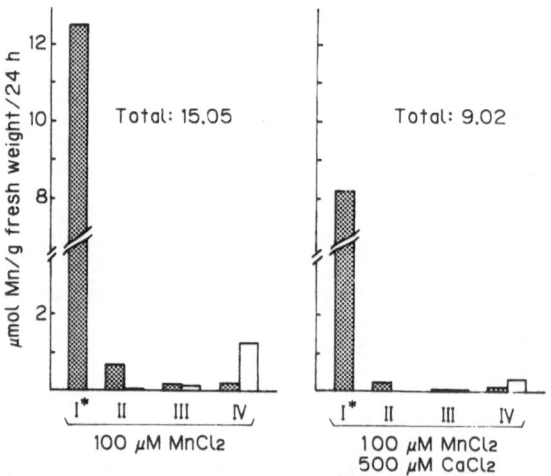

Fig. 4. Effect of Ca on the absorption and translocation of Mn in excised barley roots (Mn 100 μM).
The *symbols* are the same as in Fig. 2.

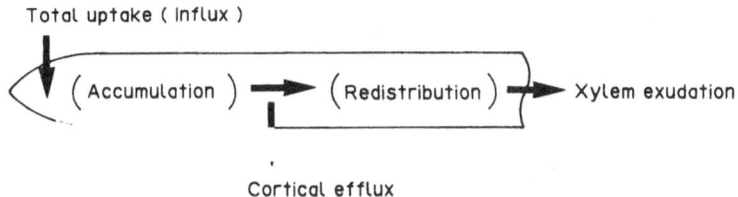

Fig. 5. Schematic representation of the absorption and distribution of ions in excised roots.

results are shown as the percent of each portion to total Mn absorbed. As Table 1 illustrates Ca (500 μM) in test solutions inhibited the absorption and decreased severely the translocation of Mn in excised roots.

The effects of various concentrations of Ca on the absorption and translocation of Mn were examined at 10 μM treatment of Mn in test solutions (Table 2). Results clearly show that Ca at 5 or 50 μM in test solutions had no effect on the absorption of Mn, while 500 μM of Ca inhibited Mn absorption. A 50 μM Ca concentration reduced slightly the translocation of Mn and 500 μM Ca decreased it drastically.

Zn absorption and translocation in excised roots

Using various concentrations of Zn, its absorption and translocation

Table 1. Percent expression of the absorption and distribution of Mn in excised barley roots in the absence and presence of Ca

Treatment (μM)		Total uptake (%)	Accumulation (%)	Redistribution (%)	Cortical efflux (%)	Xylem exudation (%)
Mn	Ca					
1	0	100 (0.41)*	84	11	0.9	4
	500	100 (0.30)	93	4	0.1	4
10	0	100 (2.29)	75	13	2	10
	500	100 (1.77)	94	2	0.1	3
100	0	100 (15.1)	83	7	1	9
	500	100 (9.0)	91	4	0.7	4

* Numbers in parentheses show total amounts of Mn absorbed as $\mu mol/g$ fresh weight/24 h.

Table 2. Percent expression of the absorption and distribution of Mn in excised barley roots at various Ca concentrations

Treatment (μM)		Total uptake (%)	Accumulation (%)	Redistribution (%)	Cortical efflux (%)	Xylem exudation (%)
Mn	Ca					
10	0	100 (2.29)*	75	13	2	10
10	5	100 (2.58)	75	14	2	9
10	50	100 (2.33)	81	8	0.7	11
10	500	100 (1.77)	94	2	0.1	3

* Numbers in parentheses show total amounts of Mn absorbed as $\mu mol/g$ fresh weight/24 h.

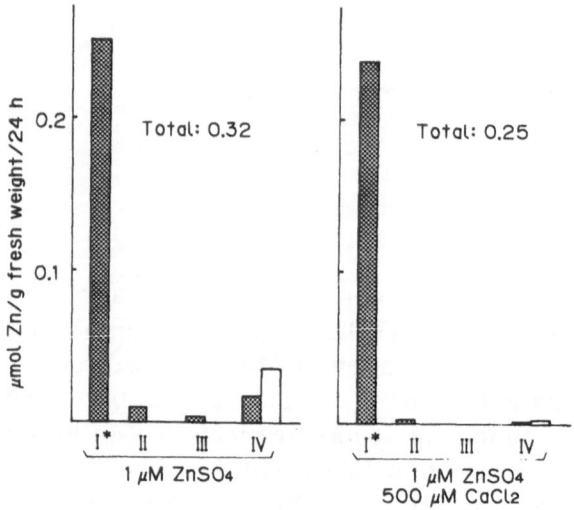

Fig. 6. Effect of Ca on the absorption and translocation of Zn in excised barley roots (Zn $1 \mu M$). An *asterisked numeral* indicates the compartment supplied with a radioisotope-labelled test solution. *Shadowed bars* show the amount of Zn in the roots and *empty bars* that in the solution.

in excised barley roots were examined, as in the case of Mn, in the absence and presence of Ca ($500 \mu M$). Figs. 6, 7 and 8 show the results of 1, 10 and $100 \mu M$ Zn treatments in test solutions. From Figs. 6 to 8, it is clear that the addition of $500 \mu M$ Ca to the test solution decreased Zn absorption and translocation in excised roots.

Figs. 6 to 8 are summarized in Table 3 with an emphasis on comparison between the absence and presence of Ca in test solution, and the

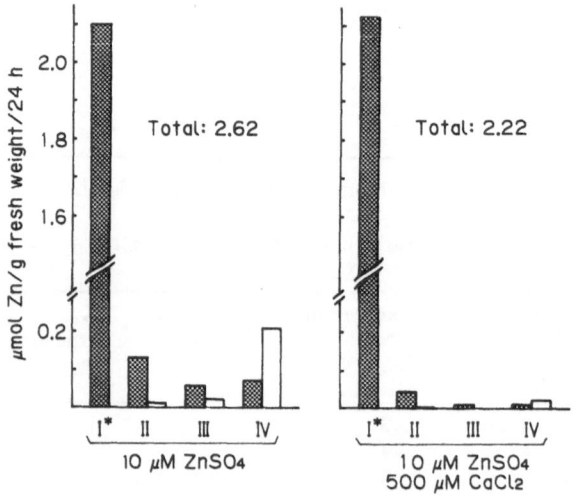

Fig. 7. Effect of Ca on the absorption and translocation of Zn in excised barley roots (Zn $10 \mu M$). The *symbols* are the same as in Fig. 6.

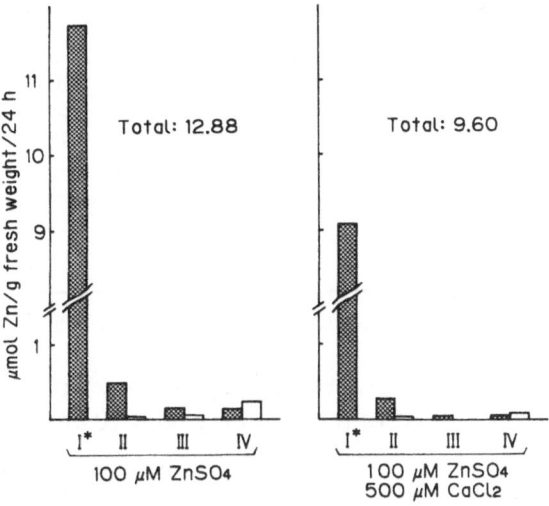

Fig. 8. Effect of Ca on the absorption and translocation of Zn in excised barley roots (Zn $100\,\mu M$). The *symbols* are the same as in Fig. 6.

results are shown as the percent of each portion to total Zn absorbed. Table 3 illustrated clearly that a $500\,\mu M$ concentration of Ca in test solution inhibited Zn absorption and translocation. These functions decreased with increasing concentrations of Ca in test solutions (Table 4).

Cd absorption and translocation in excised roots

Effect of Ca on the absorption and translocation of Cd in excised roots. The absorption and translocation of Cd in excised barley roots were examined using various concentrations of Cd in the absence and presence of Ca. The results for 1, 10 and $100\,\mu M$ treatments of Cd in test

Table 3. Percent expression of the absorption and distribution of Zn in excised barley roots in the absence and presence of Ca

Treatment (μM)		Total uptake (%)	Accumulation (%)	Redistribution (%)	Cortical efflux (%)	Xylem exudation (%)
Zn	Ca					
1	0	100 (0.32)*	78	10	0.5	11
	500	100 (0.25)	97	2	0	1
10	0	100 (2.62)	80	10	1	8
	500	100 (2.22)	96	3	0.1	1
100	0	100 (12.9)	91	6	0.7	2
	500	100 (9.6)	95	4	0.4	1

* Numbers in parentheses show total amounts of Zn absorbed as μmol/g fresh weight/24 h.

Table 4. Percent expression of the absorption and distribution of Zn in excised barley roots at various Ca concentrations

Treatment (μM)		Total uptake (%)	Accumulation (%)	Redistribution (%)	Cortical efflux (%)	Xylem exudation (%)
Zn	Ca					
10	0	100 (2.62)*	80	10	1	8
10	5	100 (2.48)	84	8	0.8	7
10	50	100 (2.29)	92	4	0.4	3
10	500	100 (2.22)	96	3	0.1	1

* Numbers in parentheses show total amounts of Zn absorbed as μmol/g fresh weight/24 h.

solutions are shown in Figs. 9, 10 and 11, respectively. Ca, at various concentrations up to 500 μM, was added to test solutions at a 10 μM Cd concentration (Fig. 10).

The addition of Ca to test solutions drastically decreased the absorption of Cd (Figs. 9–11), and this inhibition became severer with increasing Ca concentrations (Fig. 10). On the other hand, it appears that Ca increased the translocation of Cd in excised roots, except at a 100 μM Cd concentration (Figs. 9–11).

As we mentioned previously, the distribution pattern of ions in excised roots is classified qualitatively into four portions (Fig. 5). Keeping in mind this distribution pattern, the data in Fig. 10 are summarized in Table 5. In Table 5, the results are shown as the percent of each portion of ions to total Cd absorbed.

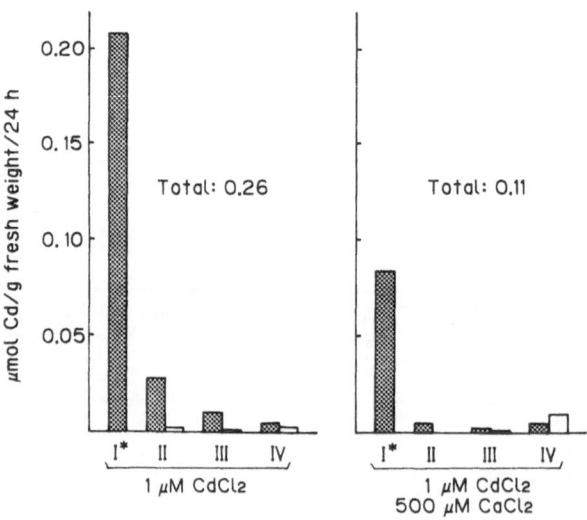

Fig. 9. Effect of Ca on the absorption and translocation of Cd in excised barley roots (Cd 1 μM). An *asterisked numeral* indicates the compartment supplied with a radioisotope-labelled test solution. *Shadowed bars* show the amount of Cd in the roots and *empty bars* that in the solution.

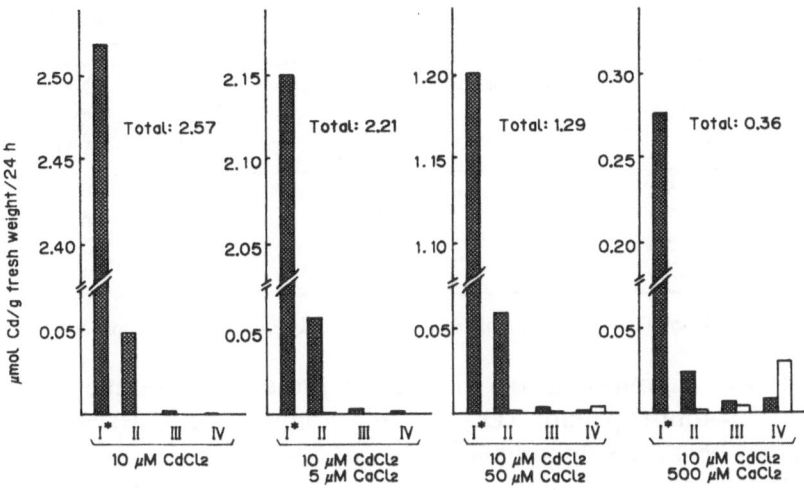

Fig. 10. Effect of Ca on the absorption and translocation of Cd in excised barley roots (Cd 10 μM). The *symbols* are the same as in Fig. 9.

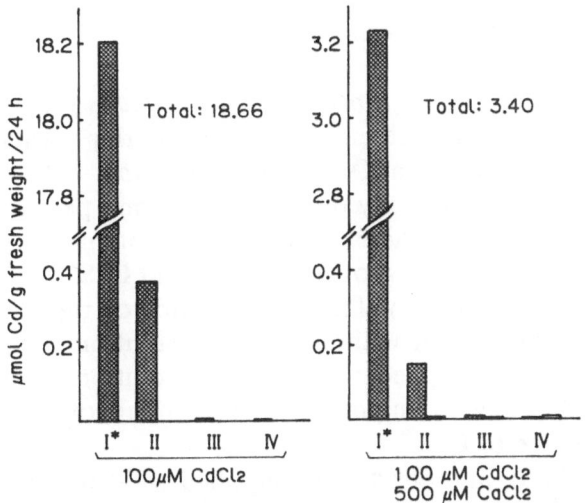

Fig. 11. Effect of Ca on the absorption and translocation of Cd in excised barley roots (Cd 100 μM). The *symbols* are the same as in Fig. 9.

Ca in test solutions inhibited the absorption, while it accelerated the translocation of Cd, especially "xylem exudation" across the cut end of the roots into the external solution (Table 5). These effects of Ca on Cd absorption and translocation became bigger with increasing Ca concentrations in test solutions.

Effects of metabolic inhibitors on the Ca-induced acceleration of Cd translocation. The accelerating effect of Ca on Cd translocation in ex-

Table 5. Percent expression of the absorption and distribution of Cd in excised barley roots at various Ca concentrations

Treatment (μM)		Total uptake (%)	Accumulation (%)	Redistribution (%)	Cortical efflux (%)	Xylem exudation (%)
Cd	Ca					
10	0	100 (2.57)*	98	2	0	0
10	5	100 (2.21)	97	3	0	0
10	50	100 (1.29)	94	5	0.2	0.3
10	500	100 (0.36)	78	12	2	9

* Numbers in parentheses show total amounts of Cd absorbed as $\mu mol/g$ fresh weight/24 h.

cised roots was examined in the absence or presence of the metabolic inhibitors, 2,4-dinitrophenol (DNP), chloramphenicol (CP) and p-chloromercuribenzene sulphonic acid (PCMBS). The results are shown in Table 6.

It was found that the accelerating effect of Ca on Cd translocation, especially xylem exudation, decreased with the addition of $10 \mu M$ DNP, though $1 \mu M$ DNP had no effect. CP and PCMBS had no effect or slightly stimulating effects on Cd translocation (Table 6).

Distribution patterns of Cd in excised roots from barley plants grown under varied culture conditions. In the present experiment, barley plants were usually supplied with $0.25 \, mM$ $CaSO_4$ for two days and, subsequently, with nutrient solution for one day during the growing period. To investigate the effect of culture conditions on the distribution patterns of Cd in excised roots, barley plants were grown with $0.25 \, mM$ $CaSO_4$ for the entire growing period and no addition of nutrient solution on the last day. Excised roots from plants grown without nutrient solution were compared to those from plants supplied with nutrient solution. The results are presented in Table 7, with an emphasis on comparison between the absence and presence of Ca.

As was seen earlier (Table 5), in excised roots supplied with nutrient solution on the last day of the growth period, the addition of Ca to test solutions inhibited Cd absorption, but accelerated its translocation. By contrast, in excised roots supplied with $CaSO_4$ solution (without nutrient solution), the addition of Ca did not accelerate Cd translocation, though inhibited its absorption.

Discusion

It is well known that Ca plays an essential role in the ion absorption of plants, especially in the selectivity of cation absorption[14,15], and that it inhibits the absorption of heavy metals by plant roots[5,7,10,25]. In addi-

Table 6. Percent expression of the absorption and distribution of Cd in excised barley roots in the absence and presence of Ca and metabolic inhibitors

Treatment (μM)				Total uptake (%)	Accumulation (%)	Redistribution (%)	Cortical efflux (%)	Xylem exudation (%)
Cd	Ca	Inhibitor						
10	0	DNP	0	100 (2.57)*	98	2	0	0
10	500	DNP	0	100 (0.34)	74	12	0.8	14
10	500	DNP	1	100 (0.30)	72	13	0.9	14
10	500	DNP	10	100 (0.25)	86	11	0.4	3
10	0	CP	0	100 (2.56)	98	2	0	0
10	500	CP	0	100 (0.34)	78	11	2	10
10	500	CP	5	100 (0.38)	73	10	1	16
10	500	CP	100	100 (0.40)	68	12	1	18
10	0	PCMBS	0	100 (2.88)	98	2	0	0
10	500	PCMBS	0	100 (0.33)	74	12	1	13
10	500	PCMBS	10	100 (0.35)	64	13	2	21
10	500	PCMBS	100	100 (0.31)	71	12	1	16

* Numbers in parentheses show total amounts of Cd absorbed as $\mu mol/g$ fresh weight/24 h.

Table 7. Percent expression of the absorption and distribution of Cd in excised barley roots grown under varied culture conditions

Culture condition*	Treatment (μM)		Total uptake (%)	Accumulation (%)	Redistribution (%)	Cortical efflux (%)	Xylem exudation (%)
	Cd	Ca					
A	10	0	100 (2.54)**	98	2	0	0
	10	500	100 (0.35)	78	11	1	10
B	10	0	100 (2.00)	97	3	0	0
	10	500	100 (0.56)	96	3	0.2	0.1

* A: $CaSO_4$ → Nutrient solution; B: $CaSO_4$ → $CaSO_4$
** Numbers in parentheses show total amounts of Cd absorbed as $\mu mol/g$ fresh weight/24 h.

tion, it was reported that Ca depressed the translocation of Zn from roots to shoots[10,25].

In the present investigation, it is clear that Mn absorption in excised barley roots was inhibited (Figs. 2–4, and Tables 1 and 2), as was that of Zn (Figs. 6–8, and Tables 3 and 4), by the presence of Ca in the test solutions. This presence, moreover, inhibited the translocation of Mn and Zn, the effect being slightly severer in the case of Zn.

The amount of ions translocated in excised roots is regarded as a sum of ions participating in the "redistribution", "cortical efflux" and "xylem exudation" in Fig. 5. When we compare the effect of Ca on the total absorption to those on the translocation of Mn or Zn, we see that its inhibitory effects were more drastic on the latter function (Tables 1–4). These results provide a possible explanation for the effect of Ca on the transport pattern of heavy metals in intact plants.

On the other hand, Cd is very similar in its chemical properties to Zn. However, Cd is not an essential element and unlike Zn, is toxic to both plants and animals[20]. There have been many reports on the absorption of Cd by plant roots[8,9,12] and intact plants[9,11,13,22]. During the course of studies on Cd absorption, it has been found that Ca in nutrient solution depressed the Cd absorption in intact plants[13] and excised plant roots[12].

Our investigation revealed that while Ca in test solutions inhibited Cd absorption (Figs. 9–11 and Table 5), it accelerated Cd translocation in excised roots at 1 and 10 μM Cd concentrations (Figs. 9 and 10, and Table 5). The results contrast with those of studies on Mn and Zn mentioned above, in which Ca inhibited both absorption and translocation of these elements.

At a 100 μM Cd concentration, the accelerating effect of Ca on the Cd translocation was negligibly small (Fig. 11). This may be due to the toxicity of Cd itself at high concentrations.

There are some conflicting results on Cd absorption in plants, in which one report concluded that the Cd absorption process was non-metabolic[8], while another suggested active absorption at low external concentrations of Cd^9. In the present experiment, the addition of DNP reduced the accelerating effect of Ca on Cd translocation in excised roots, though CP and PCMBS had no effect (Table 6). These results suggest that this accelerating effect might be involved in the metabolic pathway, perhaps in energy production. Further investigations should be carried out.

In excised roots grown under culture conditions without a nutrient solution (with only $CaSO_4$ solution) Ca did not accelerate Cd translocation (Table 7). In the future, a detailed examination of the effects of culture conditions, especially nutrient supply, on Cd translocation would be of value and interest.

Acknowledgements The multi-compartment transport box used in the present investigation was designed fundamentally by J M Stassart, Instituut voor Moleculaire Biologie, Vrije Universiteit Brussel. The authors thank Stassart for his valuable suggestions. The authors also wish to thank Miss E Kimoto for her excellent technical assistance.

The present work was partly supported by a Grant-in-Aid for Scientific Research (No. 60560073) from the Ministry of Education, Science and Culture, Japan.

References

1 Bajaj Y P S, Rathore V S, Wittwer S H and Adams M W 1970 Effect of dimethyl sulfoxide on zinc^{65} uptake, respiration, and RNA and protein metabolism in bean (*Phaseolus vulgaris*) tissues. Am. J. Bot. 57, 794–799.

2 Bowen J E 1973 Kinetics of zinc absorption by excised roots of two sugarcane clones. Plant and Soil 39, 125–129.

3 Braunsberg H and Guyver A 1965 Automatic liquid scintillation counting of high-energy β emitters in tissue slices and aqueous solutions in the absence of organic scintillator. Anal. Biochem. 10, 86–95.

4 Chandel A S and Saxena M C 1980 Mechanism of uptake and translocation of zinc by pea plants (*Pisum sativum* L.). Plant and Soil 56, 343–353.

5 Chaudhry F M and Loneragan J F 1972 Zinc absorption by wheat seedlings: I. Inhibition by macronutrient ions in short-term experiments and its relevance to long-term zinc nutrition. Soil Sci. Soc. Amer. Proc. 36, 323–327.

6 Chaudhry F M and Loneragan J F 1972 Zinc absorption by wheat seedlings: II. Inhibition by hydrogen ions and by micronutrient cations. Soil Sci. Soc. Am. Proc. 36, 327–331.

7 Chaudhry F M and Loneragan J F 1972 Zinc absorption by wheat seedlings and the nature of its inhibition by alkaline earth cations. J. Exp. Bot. 23, 552–560.

8 Cutler J M and Rains D W 1974 Characterization of cadmium uptake by plant tissue. Plant Physiol. 54, 67–71.

9 Fujimoto T and Uchida Y 1979 Cadmium absorption by rice plants. I. Mode of the absorption. Soil Sci. Plant Nutr. 25, 407–415.

10 Giordano P M, Noggle J C and Mortvedt J J 1974 Zinc uptake by rice, as affected by metabolic inhibitors and competing cations. Plant and Soil 41, 637–646.

11 Greger M and Lindberg S 1986 Effects of Cd^{2+} and EDTA on young sugar beets (*Beta vulgaris*). I. Cd^{2+} uptake and sugar accumulation. Physiol Plant. 66, 69–74.

12 Hardiman R T and Jacoby B 1984 Absorption and translocation of Cd in bush beans (*Phaseolus vulgaris*). Physiol. Plant. 61, 670–674.

13 Iwai I, Hara T and Sonoda Y 1975 Factors affecting cadmium uptake by the corn plant. Soil Sci. Plant Nutr. 21, 37–46.

14 Jacobson L, Hannapel R J, Moore D P and Schaedle M 1961 Influence of calcium on selectivity of ion absorption process. Plant Physiol. 36, 58–61.

15   Kawasaki T and Hori S 1973 The role of calcium in selective cation uptake by plant roots. II. Effects of temperature, desorption treatment and sodium salt on rubidium uptake. Ber. Ohara Inst. Landw. Biol. Okayama Univ. 16, 19–28.

16   Kawasaki T, Shimizu G and Moritsugu M 1983 Effects of high concentrations of sodium chloride and polyethylene glycol on the growth and ion absorption in plants. II. Multi-compartment transport box experiment with excised roots of barley. Plant and Soil 75, 87–93.

17   Kawasaki T, Moritsugu M and Shimizu G 1984 The absorption and translocation of ions in excised barley roots: A multi-compartment transport box experiment. Soil Sci. Plant Nutr. 30, 417–425.

18   Kobayashi J 1969 An investigation on the cause of "itai-itai disease". I–III. Kagaku (Science) 39, 286–293, 369–375, 424–429 (*In Japanese*)

19   Landi S and Fagioli F 1983 Efficiency of manganese and copper uptake by excised roots of maize genotypes. J. Plant Nutr. 6, 957–970.

20   Mengel K and Kirkby E A 1978 Principles of Plant Nutrition, Intern. Potash Inst., Berne, p. 441–458 and 519–520.

21   Page E R and Dainty J 1964 Manganese uptake by excised oat roots. J. Exp. Bot. 15, 428–443.

22   Petit C M, Ringoet A and Myttenaere C 1978 Stimulation of cadmium uptake in relation to the cadmium content of plants. Plant Physiol. 62, 554–557.

23   Ramani S and Kannan S 1975 Manganese absorption and transport in rice. Physiol. Plant. 33, 133–137.

24   Riedell W E and Schmid W E 1986 Physiological and cytological aspects of manganese toxicity in barley seedlings. J. Plant Nutr. 9, 57–66.

25   Sadana U S and Takkar P N 1983 Effect of calcium and magnesium on ^{65}Zn absorption and translocation in rice seedlings. J. Plant Nutr. 6, 705–715.

26   Schmid W, Haag H P and Epstein E 1965 Absorption of zinc by excised barley roots. Physiol. Plant. 18, 860–869.

27   Veltrup W 1978 Characteristics of zinc uptake by barley roots. Physiol. Plant. 42, 190–194.

Plant and Soil 100, 35–45 (1987).

Influence of differences in nutrition on important quality characteristics of some agricultural crops

K. MÜLLER and J. HIPPE

Institut für Agrikulturchemie der Universität Göttingen, v.-Siebold-Str. 6,D-3400 Göttingen, BRD

Key words Cauliflower Lettuce Mineral nutrition Organic compounds Potato Quality Spinach Tomato

Summary Over a period of several years, quality characteristics of plant products have been studied intensively. Next to variety, climate and location, the types and amounts of fertilizers used appeared to exert strong influences on quality parameters. In recent research it could be shown that especially N-, but also P-, K- and Mg-supply affected dry matter content and furthermore starch, crude protein, amino acids, nitrate, sugars and citric acid in cauliflower, lettuce, potato, spinach and tomato. Altogether the results showed that the influence of variations in nutrient supply on the quality of important compounds in usable plant parts is relatively high, but that different plants react not similarly to increasing or decreasing amounts of nutrients. These relations are of special importance when the qualitative and nutritive values of plant products are evaluated for consumption and processing purposes.

Introduction

For already a number of years quality characteristics of plant products, used for either direct consumption or for processing, have been studied intensively. Special attention is given to compounds with health-promoting and nutritive value, influencing aroma and taste, and to those lending themselves to industrial processing (*e.g.* starch, sugars, organic acids). Vegetative quality characteristics such as contents of injurious and toxic substances in plant parts (nitrate, heavy metals, toxins *etc.*) have also been taken into consideration.

After development of more sensitive equipment (*e.g.* high perfor-mance liquid chromatography, gas chromatography) and more rapid analytical techniques, we can now determine important plant com-pounds present in low concentrations like amides, biogenic amines, volatile substances a.o. Through this analytical progress the spectrum of detectable substances has become more complete and changes in the concentrations can be determined more exactly.

Next to varietal and climatic factors, type and quality of fertilizers applied exert strong influences on the concentrations of many plant compounds. This is documented in a number of publications[6,7,8,9,10] and journals.

Over a period of in all seven years pot experiments were conducted in which the influences of variations in N, P, K and Mg on the chemical composition of plants were investigated. Nitrogen was applied in organic and inorganic form. The selection of plant species was based on morphological differences of the plant parts used for consumption: on the one side vegetative parts harvested in a state of active metabolism like leafs and tubers, on the other side generative and mature parts (fruits).

Materials and methods

Plant species
> Cauliflower (*Brassica oleracea* var. *botrytis*)
> Lettuce (*Lactuca sativa* var. *capitata*)
> Potato (*Solanum tuberosum*)
> Spinach (*Spinacia oleracea*)
> Tomato (*Lycopersicum esculentum*)

Pot trials
The plants were grown in Mitscherlich pots filled with a mixture of 5 kg sand and 2 kg loamy clay soil (pH 7.39, available P_2O_5:5.3 mg/100 g, K_2O:7.5 mg/100 g). N was applied as NH_4NO_3 or in organic (horn shavings) form, P as $Ca(H_2PO_4)_2 \cdot 2H_2O$, K as KCl and K_2SO_4 (50/50) and Mg as $MgSO_4 \cdot 7H_2O$. The various treatments and combinations are presented in Table 1.

The following quantities of minor elements were added always to each pot: 25 mg $FeCl_3 \cdot 6H_2O$, 10 mg $MnSO_4 \cdot H_2O$, 10 mg boric acid, 1 mg $CuSO_4 \cdot 5H_2O$, 1 mg $ZnSO_4 \cdot 7H_2O$, 0.25 mg (NH_4) $Mo_7O_{24} \cdot 4H_2O$. All fertilizer compounds were mixed in the dry soil as solutions except for P, which was added as pure salt. Moisture content was adjusted at about 60% of capacity and held at this level by irrigating during the experiment. The plants were grown under natural light and temperature conditions, except for periods of rain (rolled into greenhouse).

Immediately after harvesting, all plant materials were cut, frozen, freeze-dried and ground. Only vitamin C was determined in fresh material. The following compounds were determined: dry matter calculated from the difference before and after freeze-drying, considering the remaining water determined by oven-drying at 105°C, starch polarimetrically[11], crude protein (Kjeldahl)[1], total amino acids after acid hydrolysis (amino acid analyzer, Biotronic), asparagine and glutamine (HPLC)[3], nitrate (ion-specific electrode)[4,5], sugars (glucose, fructose, sucrose) and citric acid enzymatically[2], vitamin C (2.6 dichlorphenol-indophenol)[12].

Depending on plant species, experiments were repeated two to five times, each time with 4–10 replicates per treatment (Table 1). From year to year, the results did not show much variability. Therefore only the figures of the year 1980 are presented in this paper, if not mentioned otherwise in the captions.

Table 1. Combinations and amounts of nutrients used as fertilizers in the pot experiments (values are grams per pot)

Nutrient	Treatment no.									
	1	2	3	4	5	6	7	8	9	10
N	0.75	1.5	3.0	6.0	1.5	1.5	1.5	6.0	4.5	9.0
P_2O_5	1.5	1.5	1.5	1.5	1.5	1.5	6.0	6.0	1.5	1.5
K_2O	3.0	3.0	3.0	3.0	6.0	3.0	3.0	6.0	3.0	3.0
MgO	0.6	0.6	0.6	0.6	0.6	2.0	0.6	2.0	0.6	0.6

To the treatment nos. 1–8, N was added in inorganic form, to 9 and 10 in organic (horn shavings) form.

Results and discussion

The results presented in Figure 1 show that the dry matter and starch contents of potato tuber increase with increasing nitrogen supply, except when the highest quantity of N was applied in horn shavings form. The percentages dry matter and starch reach a maximum in treatment 3 and are reduced by high amounts of K fertilizer (Treatments 5 and 8). Trebling magnesium (Treatment 6) appears to have a small positive effect on both contents. Except for tomato fruit, the dry matter contents of all other plant species react similarly to increasing nitrogen supply (Fig. 2), reaching their maximum at treatment 10. In case of cauliflower and spinach, high amounts of P, in case of lettuce, those of K, Mg and P applied separately lower the dry matter content.

It can be seen from Figure 3 that the crude protein contents are closely correlated with the quantities of N applied. The differences in crude protein contents of the treatments 1 and 4 are smallest in case of the generative tomato fruit. Unbalanced fertilization with high amounts of K, Mg and P clearly influences the crude protein contents. Such interactions are important, as they affect the quality of the products. When the crude protein content increases, the concentrations of soluble amino substances, such as amino acids, amides, biogenic amines and other

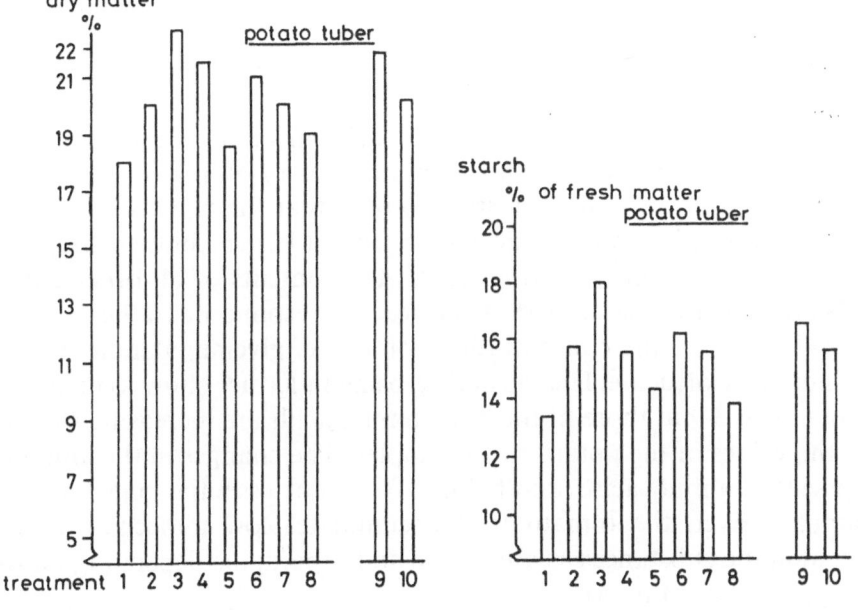

Fig. 1. Dry matter and starch content of potato tuber as functions of variations in the nutrient supply.

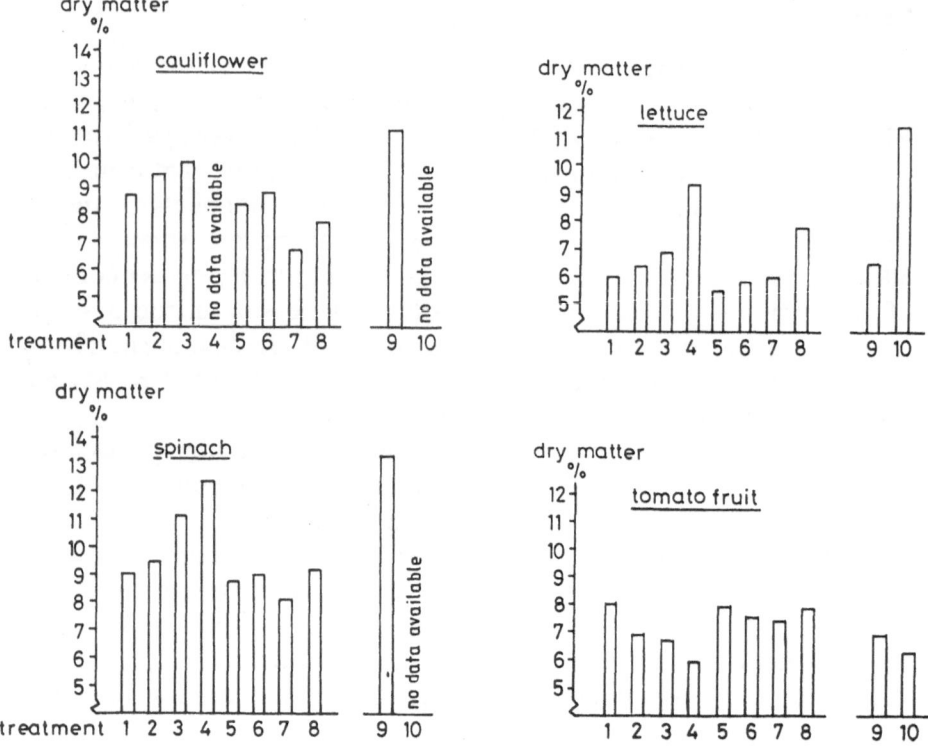

Fig. 2. Dry matter contents in cauliflower, lettuce, spinach, and tomato fruit after different nutrient supplies to the plants.

increase, which in turn can influence certain criteria, such as taste and aroma.

As an example, the contents of amino acids in lettuce and potato tuber samples of the treatments with raised amounts of N in inorganic forms are given (Table 2). It is interesting to note that the highest concentrations are reached with glutamic acid in case of lettuce, but with aspartic acid in case of potato tuber, although the contents of all other amino acids are higher in the samples of lettuce. Corresponding to the amounts of crude protein, the concentrations of serine, glycine, alanine, valine, methionine, isoleucine, leucine and tyrosine in lettuce show a maximum in treatment 3, but those of the other amino acids are increasing steadily like the values of the potato tuber samples. The changes in the amounts of essential amino acids don't differ from not essential ones in these examples. Even the tomato fruit accumulates low molecular weight N-containing substances as glutamine and asparagine depending on the nitrogen supply (Fig. 4).

In some countries there are limits set to the concentrations of nitrate in vegetable, because of its poisonous effect on infants and the possible

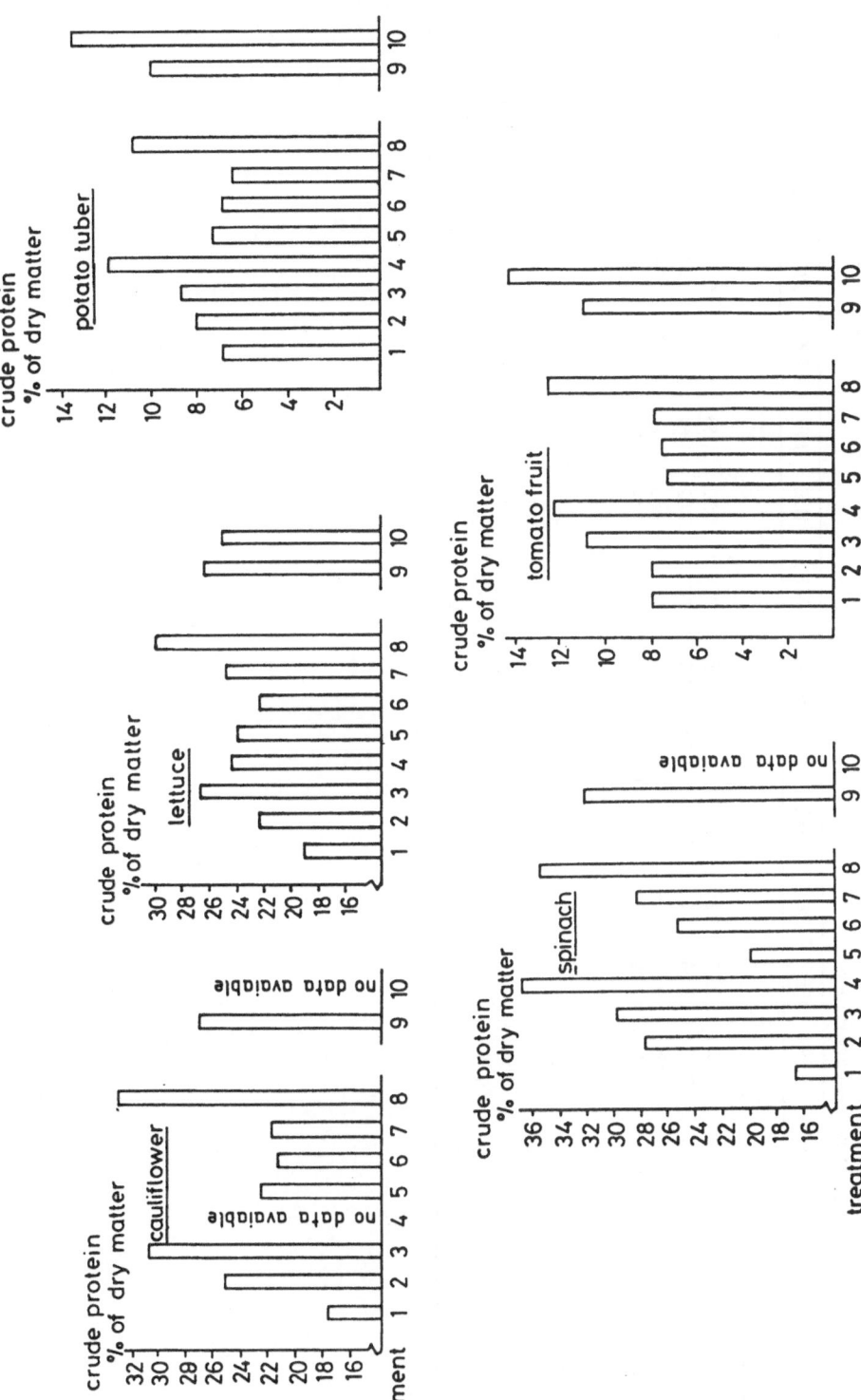

Fig. 3. Contents of crude protein in cauliflower, lettuce, potato tuber, spinach and tomato fruit as influenced by variations in nutrient supply.

Table 2. Effect of N rate on amounts of total amino acids in lettuce and potato (average values of two years)[3]

| | Amino acids, % of dry matter | | | | | | | | | | | | | | | |
	ASP	THR	SER	GLU	PRO	GLY	ALA	VAL	MET	ILE	LEU	TYR	PHE	LYS	HIS	ARG
Lettuce																
Treatment																
1	1.76	0.88	0.75	2.63	0.88	1.00	1.01	1.17	0.16	1.22	1.50	0.53	0.89	1.08	0.33	0.88
2	2.22	1.00	0.96	3.65	1.02	1.13	1.18	1.33	0.17	1.02	1.64	0.59	1.02	1.26	0.39	1.03
3	2.77	1.17	1.08	6.31	1.34	1.36	1.48	1.92	0.16	1.30	2.15	0.48	1.24	1.25	0.48	1.56
4	3.55	1.29	1.07	12.43	1.94	1.30	1.37	1.68	0.13	1.28	1.92	0.45	1.25	1.66	0.53	2.93
Potato tuber																
Treatment																
1	1.93	0.38	0.36	1.18	0.37	0.35	0.37	0.55	0.13	0.45	0.57	0.39	0.47	0.50	0.22	0.34
2	2.26	0.37	0.33	1.27	0.37	0.36	0.37	0.57	0.15	0.46	0.56	0.37	0.46	0.49	0.19	0.35
3	3.62	0.47	0.39	2.32	0.45	0.40	0.37	0.68	0.14	0.49	0.64	0.39	0.55	0.55	0.20	0.44
4	4.17	0.55	0.46	2.91	0.53	0.44	0.46	0.77	0.16	0.55	0.79	0.46	0.61	0.73	0.22	0.54

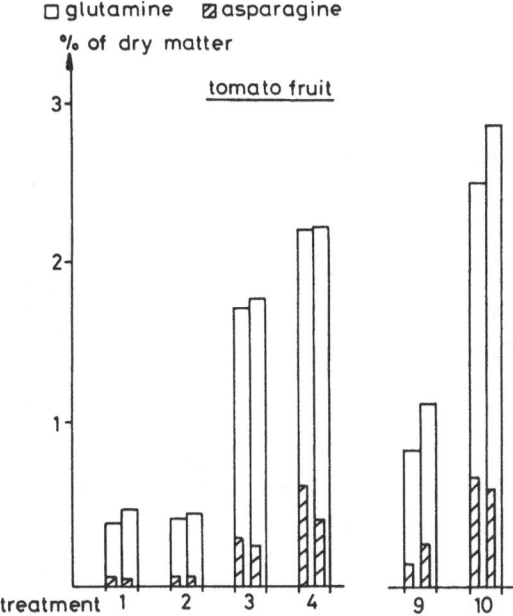

Fig. 4. Influences of variation in nitrogen supply on the contents of glutamine and asparagine in tomato fruit (1st year: left blocks, 2nd year: right blocks)[3].

increased risk of cancer by nitrosamines forming in the human metabolism. Nitrate contents are strongly affected by nitrogen fertilization in case of leaf material and tubers (Fig. 5 lettuce, potato tuber and spinach). Cauliflower contains nitrate in a medium range, while potato tuber and tomato fruits accumulate only small amounts of nitrate. It is of special interest to note that in lettuce and potato tuber nitrogen supplied in organic form caused the highest concentrations of nitrate and that except for the treatments 5–7 of potato tuber and the treatment 5 of spinach there is only a small influence exerted by other nutrients.

Many compounds are very important for processing, e.g. glucose and fructose in potato tubers to be used for chips production, because of their darkening effect during frying (Maillard reaction). It can be observed (Fig. 6), that an inverse relationship exists between the concentrations of the sugars and the quantities of N applied in the treatments 1–3. However, with a further increase in N supply (Treatments 4 and 8), the sugar concentrations rise again, but less so when the nitrogen is applied in organic form. Glucose and fructose production appears to be stimulated by magnesium (Treatment 6).

Organic acids in potato tubers are also important as a quality characteristic both for consumption and processing purposes. High concentrations of citric and ascorbic acid decrease the darkening of potato meat

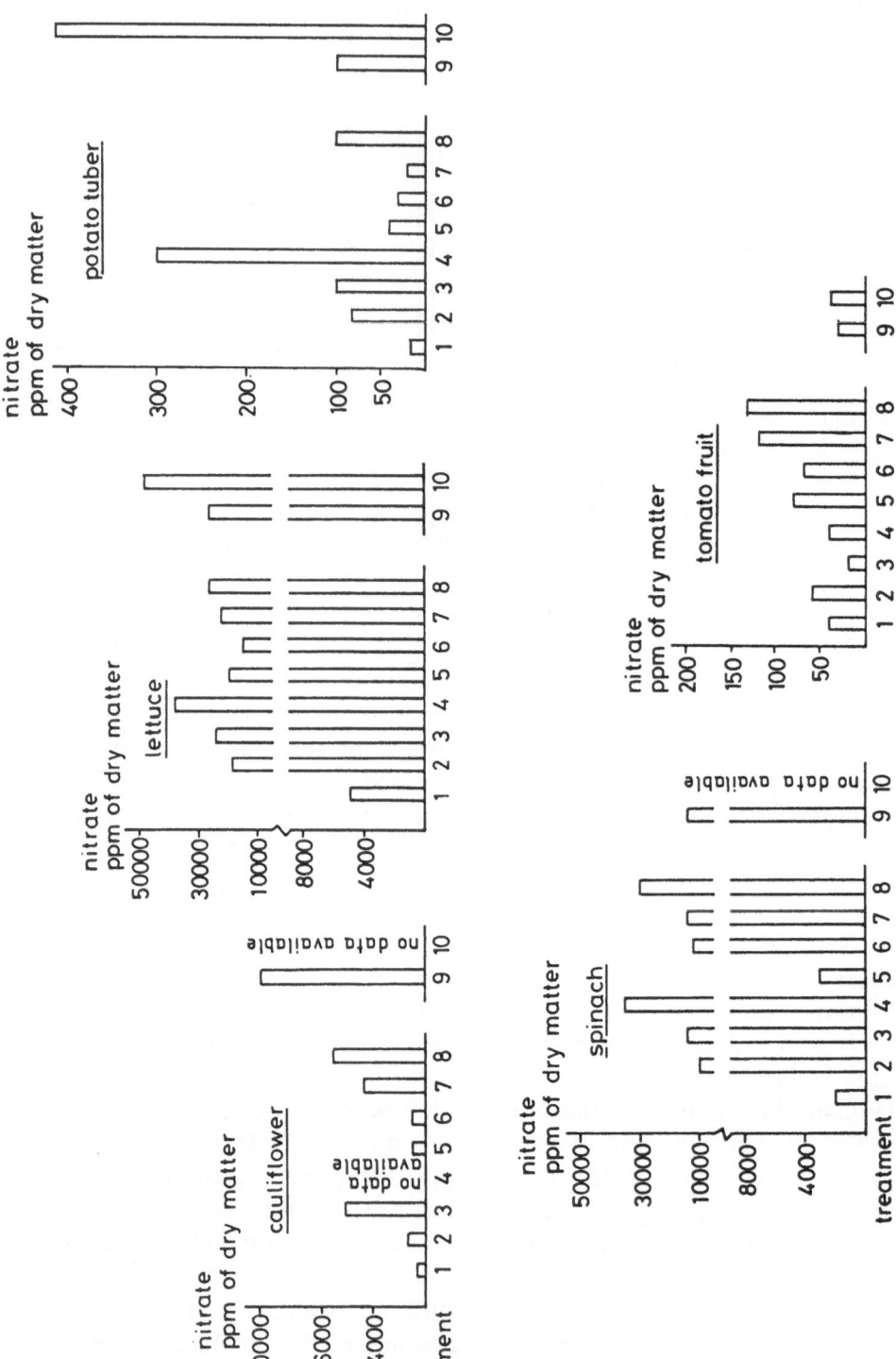

Fig. 5. Nitrate contents in cauliflower, lettuce, potato tuber, spinach and tomato fruit as a result of variation in nutrient supply.

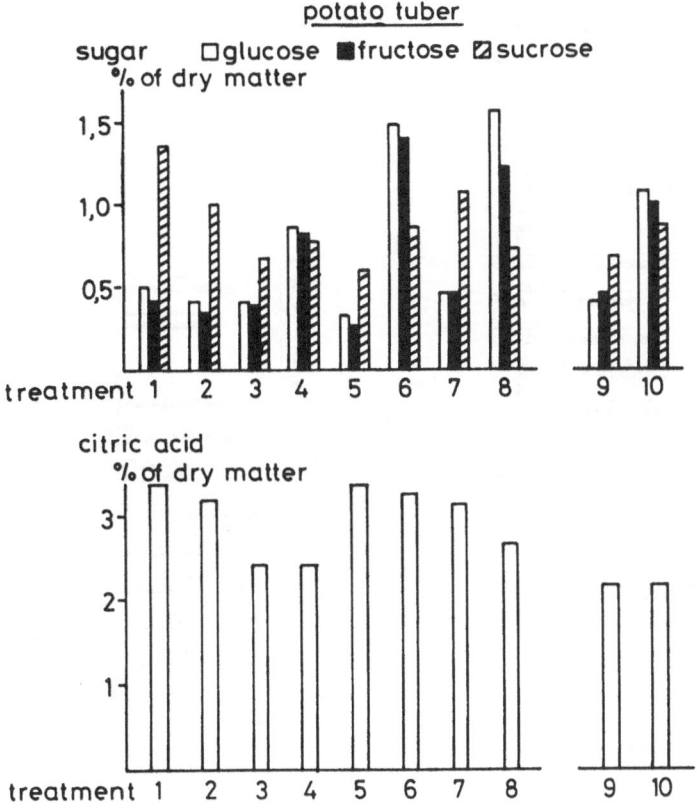

Fig. 6. Sugar and citric acid contents in potato tuber as a result of variation in nutrient supply.

and pulp. The information given in Fig. 6 shows that an inverse relationship exists between citric acid production and nitrogen supply. A high and separately applied amount of K-fertilizer promotes the content of citric acid.

Figure 7 shows the concentration of vitamin C and its components ascorbic and dehydroascorbic acid. In cauliflower an inverse relationship between nitrogen fertilization and vitamin C content seems to be present, whereas in lettuce the vitamin C concentration appears to increase with increasing N supply up to 3 mg N per pot. Beyond that N level the content declines again. As can be seen in Figure 7, the amounts of vitamin C in potato tubers are little affected by variation in concentration of nutrients except for potassium, but the relations between ascorbic and dehydroascorbic acid change in the samples of the treatments 1–4, which affects the darkening processes mentioned above. In cauliflower and spinach potassium and in potato tubers magnesium stimulates the vitamin C production. In tomato fruits the vitamin C content is also

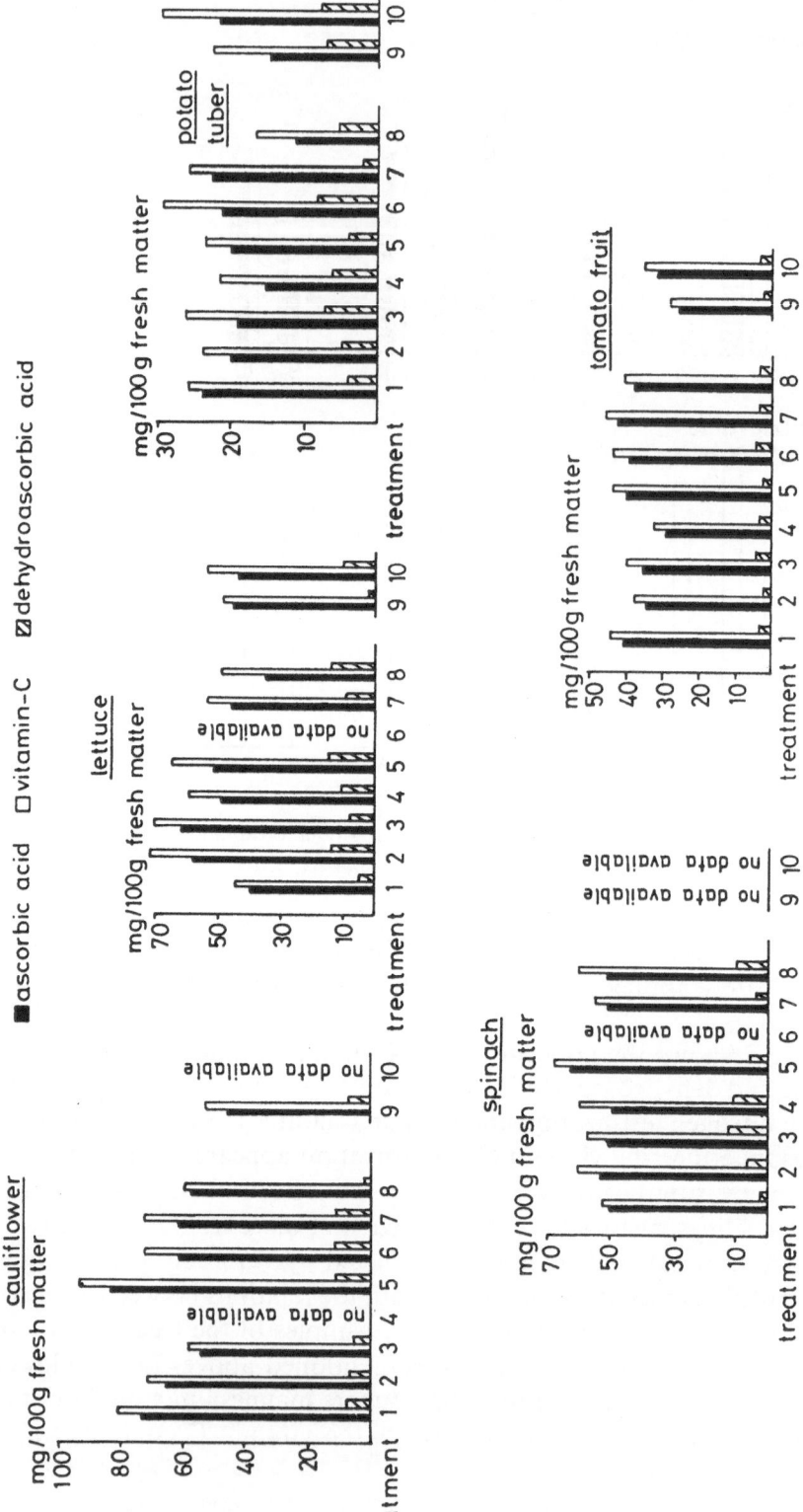

Fig. 7. Vitamin C contents (ascorbic and dehydroascorbic acid) in cauliflower, lettuce, potato tuber, spinach and tomato fruit as a result of variation in nutrient supply to the plants.

inversely correlated with the N supply and there are only very small amounts of dehydroascorbic acid present.

In conclusion it can be remarked that in general a considerable effect of variation in nutrient supply on quality characteristics of plant products can be discerned, but that crop species can differ widely in their response to these variations. It is therefore clear that each crop has to be investigated individually to establish relationship between nutrient supply and crop parameters determining the storage-, consumption-, and processing characteristics.

References

1   Beythien A and Diemair W 1963 Laboratoriumsbuch für den Lebensmittelchemiker. Verlag Th. Steinkopff, Dresden/Leipzig.
2   Boehringer A 1984 Methoden der enzymatischen Lebensmittelanalytik, Fa. Boehringer-Mannheim GmbH.
3   Hippe J 1984 Einfluss stark differenzierter Nährstoffgaben auf die Bildung von N-Nitrosaminen, die Konzentration ihrer Vorstufen sowie auf die Gehalte einiger anderer wertbestimmenden Inhaltsstoffe in Kartoffeln, Kohlrabi, Kopfsalat und Tomaten unter besonderer Berücksichtigung der quantitativen analytischen Erfassung von in diesem Zusammenhang bedeutenden niedermolekularen Stickstoffverbindungen. Dissertation Göttingen.
4   Kolbe H und Müller K 1984 Über die quantitative Bestimmung von Nitrat mit Hilfe einer nitratsensitiven Elektrode (für Serienanalysen, aufgezeigt am Beispiel von Kartoffelknollen). Landw. Forsch. 1984, 434–444.
5   Künsch U, Schärer H and Temperli A 1981 Eine Schnellmethode zur Bestimmung von Nitrat in Frischgemüse mit Hilfe der ionensensitiven Elektrode. Mitteilungen der Eidgenössischen Forschungsanstalt für Obst-, Wein- und Gartenbau, Wädenswil/Schweiz, Flugschrift Nr. 106
6   Mulder E G 1949 Mineral nutrition in relation to the biochemistry and physiology of potatoes. Plant and Soil 2, 59–121.
7   Mulder E G 1956 Nitrogen magnesium relationship in crop plants. Plant and Soil 7, 341–376.
8   Mulder E G 1958 Nitrogen magnesium relationships in crop plants. Stikstof 2, 39–48.
9   Mulder E G and Bakema K 1956 Effect of the nitrogen, phosphorus, potassium and magnesium nutrition of potato plants on the content of free amino acid composition of the protein of tubers. Plant and Soil 7, 135–166.
10 Müller K 1975 Veränderungen wertgebender Inhaltsstoffe in der Kartoffelpflanze und-knolle im Verlauf der Vegetation und Lagerung, ihre Bedeutung für die Qualität der Knolle. Hefte für den Kartoffelbau 17.
11 Müller K und Cervenkova I 1978 Die Ermittlung des Stärke- und Trockensubstanzgehaltes in Kartoffelknollen nach Bestimmung des Unterwassergewichtes an Hand modifizierter Tabellenwerte. Die Stärke 30, 12.
12 Paech K und Tracey M V 1955 Moderne Methoden der Pflanzenanalyse. Band II. Springer-Verlag Berlin.

Plant and Soil 100, 47–69 (1987).

Ms. 100-06

Nitrogen transformation processes in relation to improved cultural practices for lowland rice

S.K. DE DATTA

Agronomy Department, International Rice Research Institute, Los Baños, P.O. Box 933, Manila, Philippines

Key words Ammonia volatilization Denitrification Nitrogen Rice

Summary Inappropriate method and timing of N fertilizer application was found to result in 50–60% N losses. Recent nitrogen transformation studies indicate that NH_3 volatilization in lowland rice soils is an important loss mechanism, causing a 5–47% loss of applied fertilizer under field conditions. Estimated denitrification losses were between 28 and 33%. Ammonia volatilization losses from lowland rice can be controlled by i) placement of fertilizer in the reduced layer and proper timing of application, ii) using phenylphosphorodiamidate (PPD) to delay urease activity in flooded soils, and iii) using algicides to help stabilize changes in floodwater pH.

Appropriate fertilizer placement and timing is probably the most effective technique in controlling denitrification at the farm level. The effectivity of nitrification inhibitors as another method is still being evaluated.

With 60–80% of N absorbed by the crop derived from the native N pool, substantial yield gains in lowland rice are highly possible with resources already in the land. Extensive studies on soil N and its management, and an understanding of soil N dynamics will greatly facilitate the decrease in immobilization and ammonium fixation in the soil and the increase in N availability to the rice crop. Critical research needs include greater emphasis on N transformation processes in rainfed lowland rice which is grown under more harsh and variable environmental regimes than irrigated lowland rice.

Introduction

The number of people to be fed in 1990 is likely to be about 25% higher than in 1980, even with the expected slow down of population growth in most of Asia. By the year 2000, it will be about 50% higher[1]. Modern rice varieties, increased area under irrigation and greater fertilizer use have dramatically increased rice production in the region. Traditional rice importing countries are faced with surplus and storage problems while rice exporting countries are looking for markets for their surpluses.

Past experience suggests that the use of nitrogenous fertilizers can substantially increase rice yields. Returns to producers, however, depend very much on the mode of fertilizer application. Unless high losses and low N fertilizer use efficiency are corrected, most of the potential benefits of increased N fertilizer use may not be realized.

Research results showed that N losses ranged from 50 to 60% when applied with inappropriate method and timing[17]. These include N lost to the atmosphere via ammonia volatilization, through nitrification and denitrification, retained in the soil as a result of immobilization and ammonium fixation, and leached.

Nitrogen transformation processes in lowland rice soils were reviewed by Savant and De Datta[46] and Keeney and Sahrawat[32]. An understanding of these processes will greatly help minimize N losses from lowland rice soils to increase N availability to rice.

This paper deals with current information from basic and applied studies on increased N fertilizer efficiency in lowland rice soils.

Nitrogen transformation processes

Even a slight improvement in N fertilizer efficiency will save energy costs and foreign exchange in N fertilizer-importing countries. Hence, the increased interest in N transformation processes in lowland rice in recent years.

The behavior of nitrogen in lowland soil markedly differs from that in upland soil. Flooding the soil results in ammonium accumulation, ni-

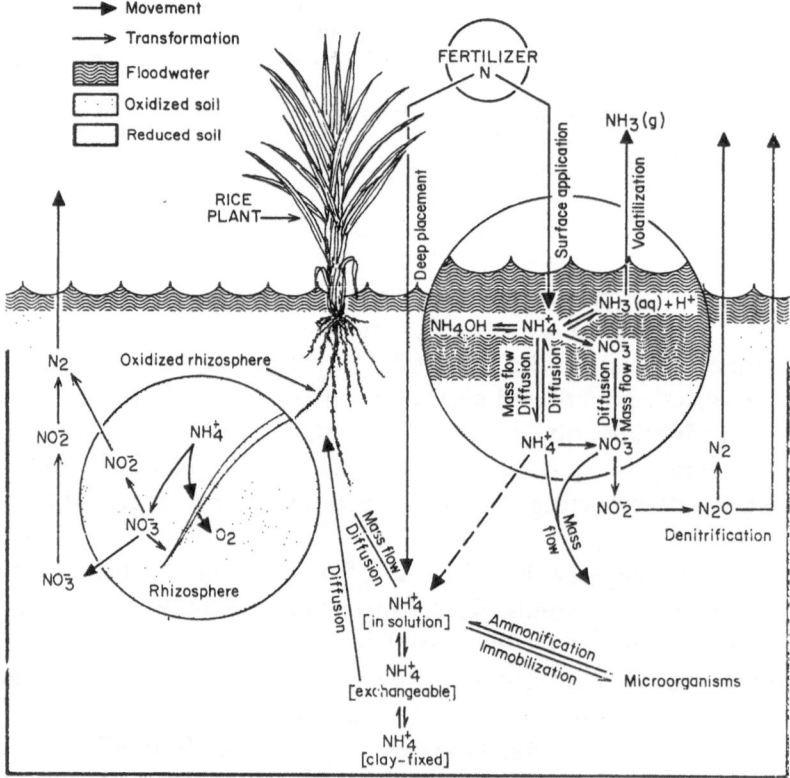

Fig. 1. Nitrogen transformations in submerged rice soils.[13,46]

trate instability, and low N requirement for organic matter decomposition. Ammoniacal nitrogen is subject to fixation by clay and loss by volatilization, nitrification and denitrification, leaching, runoff, and seepage. Significant inorganic N is assimilated into the organic fraction of lowland rice soil although its magnitude is lower than in an upland soil. Various fractions in the inorganic N pool of a lowland rice soil constitute an extremely dynamic N system in the soil, which is affected by physical, chemical, and microbiological reactions. These reactions are summarized by Savant and De Datta[46] in a schematic form (Fig. 1).

Ammonium dynamics

In lowland rice, 60–80% of N absorbed by the crop is derived from the native nitrogen pool[5]. Recent results suggest that approximately 60% of the rice yields, ranging between 2 and 4 t/ha, are produced without N fertilizer. With increased study on soil N and its management, substantial yield gains are possible with resources already on land[2].

Ammonium nitrogen, the dominant form of mineral N in lowland soils, exists in three major fractions:
· Ammonium in soil solution
· Ammonium in exchange sites
· Ammonium in nonexchangeable forms

In a recent investigation, a highly significant correlation was recorded between the exchangeable soil NH_4^+ measured before the experiment was conducted and N uptake of rice[49]. Soil solution NH_4^+ tends to equilibrate with the exchangeable soil NH_4^+, considered as the most important fraction for rice nutrition. Results of Schön et al.[49] fully supported the view that the initial exchangeable soil NH_4^+ behaved liked fertilizer N as shown in the linear relationship between N uptake by the crop and initial exchangeable NH_4^+ + fertilizer N. The coefficient of determination $R^2 = 0.91$ suggests that other processes, such as ammonification, N_2 fixation, and denitrification during the growing season have only minor influence on N uptake in soils with different pH, organic N, and organic C.

In lowland rice, the nonexchangeable NH_4^+ fraction plays an important role in N nutrition[34]. Subsequent studies by Keerthisinghe et al.[34] showed that exchangeable NH_4^+ decreases to a very low level during the crop growing season. In most cases, this coincides with the highest rate of N uptake by the crop (Fig. 2). Some of the exchangeable NH_4^+, however, may be immobilized by soil microorganisms or lost through volatilization[46].

Keerthisinghe et al.[34] reported that in some soils net release of exchangeable and nonexchangeable NH_4^+ was higher than 50% of the total

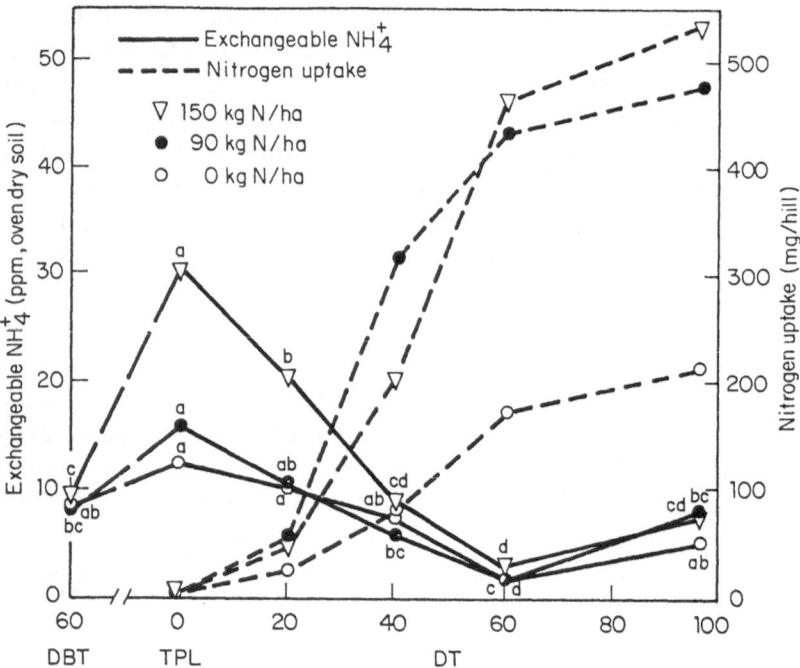

Fig. 2. Effect of nitrogen fertilizer application on exchangeable NH_4^+ in soil and nitrogen uptake by IR36 rice in Maligaya silty clay loam, 1981 dry season[33]. At the same N rate, means with common letters are not significantly different at the 5% level. DBT = days before transplanting, TPL = at transplanting, DT = days after transplanting. Reproduced from G. Keerthisinghe, S.K. de Datta and K. Mengel, Importance of exchangeable soil ammonium in nitrogen nutrition of lowland rice, *Soil Science*, Vol. 140, No. 3, pp. 194–201, by Williams and Williams Company, 1985.

N uptake of crop. It was concluded that the exchangeable NH_4^+ and, in some soils, the nonexchangeable NH_4^+ are the most important soil N fractions easily available to lowland rice at early to mid-crop stages. Moreover, organic soil N may contribute substantially to N supply at the spikelet-filling stage. In soils rich in vermiculite the nonexchangeable NH_4^+ should also be considered for fertilizer N recommendation in lowland rice[37].

Biological nitrogen fixation

From extensive data, Watanabe *et al.*[57] found that a rice crop used an average of 40–50 kg N/ha from the soil pool, when no N fertilizer was applied. In most instances, soil N did not decrease, indicating that used N was compensated for by mechanisms among which biological nitrogen fixation (BNF) is most important. In a review by Lowendorf[36], the ranges of fixed nitrogen per crop are: legumes, 25–61; blue–green algae (BGA), 0.2–39; azolla green manure, 25–121; azolla grown with rice, 2–7.5; and soil and rhizosphere fixers, 1.2–18.3 kg N/ha. Roger and

Watanabe[42] reported that only N_2 fixing legume and azolla green manures are currently used. Straw incorporation, which enhances heterotropic N_2 fixers, is not widely practiced in farmer's fields. BGA inoculation techniques have been developed and their use advocated, but adoption by farmers is still limited.

Green manures have been used in lowland rice in India and China for a long time. Some of the *Sesbania* and *Crotolaria* spp. are known to be high N producers. A considerable interest in *Sesbania rostrata*, a green manure crop from Senegal, has been aroused. It forms N_2 fixing nodules on the roots and the stems, and has 5–10 times more nodules than most legumes. Because of its stem nodules, *S. rostrata* can fix N even under waterlogged conditions[22].

Mineralization–immobilization

Mineralization of organic N to NH_4^+ is the key process in the N nutrition of lowland rice[46]. The important environmental factors affecting mineralization–immobilization are temperature, soil moisture regime, and soil drying; important soil characteristics include pH, organic matter, C/N ratio, and amount and quality of organic residues[32].

Broadbent and Tusneem[6] reported that soil N mineralization was higher in the soil planted to rice than that unplanted, because the presence of active rice roots decreased N loss due to nitrification–denitrification.

Immobilization is important since it affects N availability and supply to rice. The ammonification–immobilization process assumes practical significance in the N nutrition of rice because rice plants heavily depend on soil N for their N requirement. However, information is lacking on ammonification–immobilization processes in actual lowland rice fields.

Nitrification

Nitrification or the biological oxidation of NH_4^+ to NO_3^- in lowland soil is common but undesirable because it leads to N loss. Nitrification, strictly an aerobic process, occurs in the floodwater and in the thin, oxidized, surface soil layer. Studying nitrification *in situ* in a flooded soil system is difficult because as soon as NO_3^- is formed, it diffuses down to the reduced layer and is lost from the system by dentrification, or reduced to NH_4^+ by dissimilatory NO_3^- reduction[46]. Nevertheless, nitrification is recognized as a mechanism of N loss via nitrification–denitrification in lowland soils. Nitrate has been established as an inefficient N source in lowland rice culture.

Occurrence of nitrification in the rhizosphere of a rice plant, which Savant and De Datta[46] referred to as Site II, is a subject of speculation since no data are available on *in situ* nitrification.

Urea hydrolysis

Very few studies have emphasized urea hydrolysis in lowland rice soils. Sahrawat[43] found that the urease activity in 10 Philippine lowland rice soils was highly correlated with total N and organic C, but was not significantly correlated with other soil properties. Vlek *et al.*[54] studied urea hydrolysis in three flooded soils and reported that urea hydrolysis occurred at the floodwater-soil interface. Urease activity in the flooded soils was dynamic and was affected by the length of presubmergence period. The use of fertilizer additives to control urea hydrolysis in lowland rice soils is recently receiving increased attention. Delaying urea hydrolysis has been sought to minimize ammonia volatilization loss in lowland rice soils.

Leaching

Nitrate produced in the oxidized surface layer of a flooded soil moves easily by diffusion and percolation into the underlying reduced layer, where it is rapidly denitrified.

Ammonium nitrogen is less subject to leaching than nitrate because of its adsorption on the cation exchange complex. Losses by leaching occur mainly in coarse-textured soils with low cation exchange capacity. Krishnappa and Shinde[35] found that when urea was incorporated into a flooded lowland soil under tropical field conditions, about 8% of N applied was lost through leaching.

Surface runoff

Information is limited on the quantity of nutrient losses by runoff from lowland rice fields. Under certain situations, N losses through runoff were from 13 to 16%[13].

Denitrification

Special soil and environmental conditions in a flooded rice field support the redox N processes, *i.e.*, nitrification (oxidation of NH_4^+ to NO_3^-) and denitrification (reduction of NO_3^- to N_2). Potentially, these reactions occur in continuously flooded lowland and upland rice fields, the latter being subjected to alternate flooding and draining cycles.

Nitrification (k $= 3.18 \, \mu g$ N/cm^3 per day where K $=$ kinetic constant) takes place in the surface aerobic layer. This results in a concentration gradient of NH_4^+-N across the aerobic layer and anaerobic layer and causes the NH_4^+ present in the anaerobic layer to diffuse ($\bar{D} = 0.216 \, cm^2$/day) into the aerobic layer where it is nitrified. The NO_3^--N formed diffuses back into the anaerobic layer ($\bar{D} = 1.33 \, cm^2$/day) where it is easily denitrified (k $= 15.0 \, mg$ N/cm^3 per day). Many factors governing denitrification were described earlier[13].

In a lowland rice soil Savant and De Datta[44,45] and Savant et al.[47] studied the movement of N fertilizers, (prilled urea [PU], urea super-granules [USG], sulfur-coated urea [SCU], and urea placed in mudballs) placed in the root-zone (10 cm deep). The studies indicated that NH_4^+ movement was downward > lateral > upward. However, when urea was surface applied, Savant and De Datta[45] measured a significant amount of NH_4^+ at the 12–14 cm depth, 4 weeks after application. Thus, downward movement of plant available N can result in a significant loss from coarse-textured soils. Vlek et al.[54] measured significant loss in N fertilizer after USG placement in lowland soils with a percolation rate of 5 cm/day.

In soils with high organic matter content, the oxidized layer is thin. Under these conditions NH_4^+ tends to diffuse into the floodwater. Nitrification is indicated by the significant concentration of NO_3^- formed in the floodwater of an organic soil[41].

Under O_2-free conditions, many microorganisms utilize NO_3^- as a terminal electron acceptor. This process is called NO_3^- respiration or dissimilatory NO_3^- reduction. The pathway involving the NO_3^- reduction to gaseous end products ($NO_3^- \rightarrow NO_2 \rightarrow HNO \rightarrow NO \rightarrow N_2O \rightarrow N_2$) is known as denitrification. Denitrification is a dominant process in soils with Eh values of 200–300 mV, while NO_3^- reduction to NH_4^+ occurs in soils with Eh values of < -100 mV[7].

Nitrogen loss due to denitrification was also found significant in soils planted to rice. However, the magnitude of N loss was lower compared to that for systems without plants[40]. Smith and De Laune[52] measured significantly greater $N_2O + N_2$ production in the first 2 days after fertilizer application to lowland soil planted to rice as compared to non-planted lowland soil. The controlling factor of N loss from the rhizosphere is the competition for NO_3 uptake between denitrifying bacteria and rice roots.

Measurement of denitrification Several methods have been developed to measure denitrification losses from flooded soils. The most frequently used technique is to develop an ^{15}N balance then attribute the lost ^{15}N fraction to denitrification[29]. The gaseous products of denitrification, which include N_2, N_2O, and possibly NO, must directly be measured to distinguish denitrification from other losses. The elemental nitrogen gas produced from denitrification is difficult to measure against a background of 78% N in the atmosphere. However, nitrous oxide, the other product of denitrification, is normally only 300 parts/billion in the atmosphere and, therefore, losses as nitrous oxide can be measured easily. Figure 3 shows some measurements of nitrous oxide flux at the

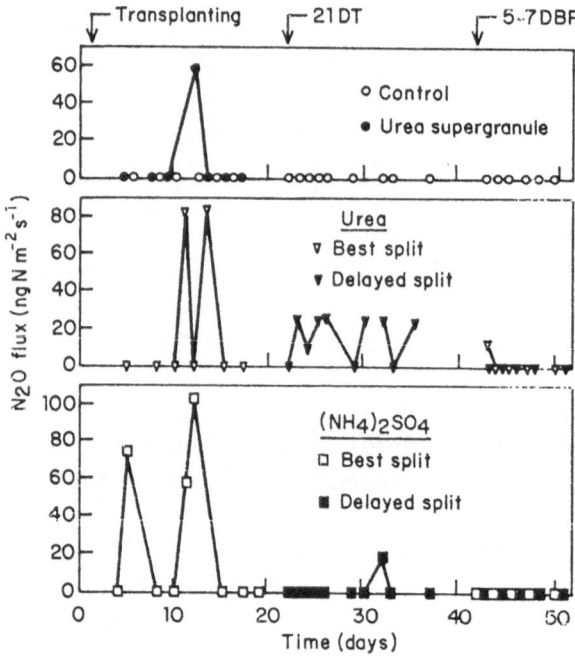

Fig. 3. Effect of time and mode of N fertilizer application on N_2O flux. IRRI, 1979 dry season (adapted from Craswell and De Datta[11]) DT = days after transplanting, DBPI = days before panicle initiation.

IRRI farm. These measurements, the first in the tropics, prove that denitrification occurs in lowland rice fields[11]. The nitrous oxide fluxes are short-lived and occur soon after fertilizer application. Nitrous oxide seems to be only a minor product of nitrification–denitrification.

Freney et al.[27] and Smith et al.[51] attempted to measure N_2O emission under field conditions to provide direct evidence for nitrification and denitrification. Nitrous oxide emissions amounted to less than 0.1% of the applied ammonium sulfate or urea.

Because of great demands for electron acceptors in the reduced zones of lowland soils, N_2O produced during denitrification is rapidly converted to N_2 during microbial respiration. Since N_2O is only the minor and product of this process, the low evolution rate does not represent the total loss due to denitrification[40].

Recently, Craswell et al.[10] developed a new method for measuring the ^{15}N content of mixture of N_2 and O_2 gases with a nonrandom N_2 isotopic distribution. The method involves a high voltage arc passed between a pair of tungsten electrodes sealed in a glass vessel to fix some of the N_2 as NO_x species, thereby redistributing the N isotopes of the N_2 into an NO_x pool. The NO_x species are absorbed and oxidized by acidic permanganate, reduced to NH_4^+, and distilled by routine procedures[10].

This methodology was evaluated in a field study on flooded rice soil[23]. Only about 1% of the applied urea N was determined to be lost through denitrification.

Ammonia volatilization

Nitrogen loss through ammonia volatilization from flooded soils was reported by various researchers and reviewed by Mikkelsen and De Datta[38] and Vlek and Craswell[55].

Conditions in lowland rice field are conducive to ammonia volatilization loss. During the volatilization process, gaseous ammonia is formed: $NH_4^+ \leftrightarrows NH_3 + H^+$. That is, H^+ is released for each conversion of $NH_4^+ \rightarrow NH_3$ (g). The pH and buffering capacity of oxidized, as well as reduced, soil layers, thus influence the process.

The overall inorganic ammonium nitrogen (equilibrium) system (N_t) in floodwater can be written as:

$$N_t = NH_3(aq) + NH_4^+$$

The concentration of $NH_3(aq)$ changes in direct proportion with NH_4^+. It increases about tenfold per unit increase in pH of an aqueous system up to pH 9. Floodwater pH in lowland rice soils is markedly influenced by PCO_2 and alkalinity, which in turn are determined by algal photosynthetic activity and respiration[39]. Wetselaar *et al.*[58] observed significant correlation in the relationship:

$$(pH) = \frac{(NH_3) \text{ volatilized}}{[NH_3(aq) + NH_4^+] \text{ floodwater}}$$

Other soil factors (pH, CEC, PCO_2, buffering capacity, and alkalinity) and environmental factors (temperature, wind velocity, *etc.*) as well as the value and amount of N fertilizer applied and plant canopy size affect NH_3 volatilization loss in flooded soil[46].

Summarized data of recent field studies on ammonia volatilization in lowland rice soils show a large variation (5–47%) in the amounts of ammonia volatilized (Table 1). The magnitude of losses are due to the different measurement techniques used by the researchers, the chemical nature of fertilizer, and the varying fertilizer application methods. Nevertheless, data from these recent field studies indicate that ammonia volatilization losses from lowland rice fields could be substantial. The advantages and disadvantages of these methods have been discussed by various workers but most recently by Vlek and Craswell[55], Savant and De Datta[46], and Fillery *et al.*[25].

The aerodynamic and micrometeorological models suggest that the environmental parameters such as air temperature, wind speed (directly),

Table 1. Field experiments measuring loss of N fertilizer via NH_3 volatilization from lowland rice fields

Reference	Estimated ammonia lost % of N applied	Fertilizer material	Fertilizer rate (kg N/ha)	Method of measurement	Site
Mikkelsen et al.[39]	20	$(NH_4)_2SO_4$ broadcast 10 DT	90	Chamber with enclosed acid trap and natural ventilation	IRRI farm, Phils. dry season
	6	Urea broadcast	90	Chamber with enclosed acid trap and natural ventilation	IRRI farm, Phils. wet season
	7	$(NH_4)_2SO_4$ broadcast	90	Chamber with enclosed acid trap and natural ventiliation	IRRI farm, Phils. wet season
Freney et al.[27]	5	$(NH_4)_2SO_4$ broadcast and incorporated	80	Micrometeorological	IRRI farm, Phils. wet season
	11	$(NH_4)_2SO_4$ topdressed at panicle initiation	40	Micrometeorological	IRRI farm, Phils. wet season
Fillery et al.[25]	47	Urea, topdressed into the floodwater, 14 DT	80	Micrometeorological	Maligaya (Muñoz), Phils., dry season
	27	Urea, topdressed into the floodwater, 21 DT	60	Micrometeorological	IRRI farm, Phils. dry season
Simpson et al.[50]	11	Urea, broadcast evenly into the floodwater	80	Micrometeorological	Griffith, N.S.W. Australia
Fillery and De Datta[24]	41–43	Urea or $(NH_4)_2SO_4$, topdressed into the floodwater, 10 DT	58	Micrometeorological	Maligaya (Muñoz), Phils., dry season
De Datta et al.[18]	27	Urea, broadcast into floodwater, 12 DT	53	Simple micrometeorological	Calauan, Phils., dry season
	31	Urea, broadcast into floodwater, 12 DT	80	Simple micrometeorological	Calauan, Phils., dry season

and solar radiation (indirectly) influence ammonia loss. Ammonia volatilization rate seems to follow the diurnal pattern of water temperature[27]. High wind speed or rapid air exchange at the water-air interface results in a marked decrease in ϱNH_3 in air, which is in immediate contact with the body of floodwater, and thus enhances ammonia volatilization[3]. Based on results of Bouwmeester and Vlek[4], rice plant canopy, which modifies microclimate, will have a marked effect on ammonia volatilization from lowland rice fields. The two main effects of the rice plant canopy are (1) decreased transfer coefficient of NH_3 within the canopy due to restricted air movement and (2) reduced floodwater pH and temperature presumably due to shading[46].

Ammonia volatilization measurement techniques Most of the studies which reported ammonia volatilization losses of 10% or lower were conducted in laboratory or greenhouse using distilled or deionized water and measurement techniques considered to retard NH_3 volatilization[26]. Mikkelsen et al.[39], however, reported ammonia losses of up to 20% in field studies using closed systems with air exchange and acid trap.

Micrometeorological measurements are much preferred since they do not disturb the environmental or surface processes which influence gas exchange. They permit continuous monitoring, thus facilitating the measurement of environmental effects. Further, they provide a measure of the average flux density over a large area.

Measurements of ammonia volatilization loss Using micrometeorological technique and ammonium sulfate as N source (broadcast and incorporated), Freney et al.[27] reported a 5% ammonia volatilization loss. Losses could have been greater if the ammonium sulfate was surface applied into floodwater 2–3 weeks after transplanting using similar techniques.

Fillery et al.[25] reported a 47% loss in Muñoz, Central Luzon, Philippines when urea was surface broadcast into the floodwater. A lower rate of NH_3 loss, 27% of N applied, was found in a similar experiment conducted as Los Baños (IRRI farm) in the same season, primarily because wind speed was lower at Los Baños than at Muñoz.

From their studies in Australia, Simpson et al.[50] reported that some NH_3 loss (11% of N applied) occurred after applying urea to an 8-week-old direct-seeded rice crop, despite high pH values and elevated urea concentrations in floodwater, and high wind speeds. Low urease activity at the soil–floodwater interface and floodwater depth ($\cong 15\,cm$) may have retarded the ammoniacal N accumulation in floodwater.

Ammonia loss measurements using simple techniques Wilson et al.[59],
Denmead[19], and Freney et al.[28] recently discussed a micrometeorological
technique more appropriate for smaller experimental areas, which have
stringent measurement requirements.

With this area size the vertical ammonia flux density F (in μg N/m^2 per
second) was calculated from

$$F = 0.91 \bar{u}_{0.8} \bar{c}_{0.8}$$

where $\bar{u}_{0.8}$ is the mean wind speed at a height of 0.8 m and $\bar{c}_{0.8}$ is the excess
concentration of NH$_3$ (in μg N/m^3) over the background NH$_3$ concentration measured at a height of 0.8 m at the center of a fertilized circle
of 25 m radius. Then fluxes were converted to kilogram per hectare per
hour.

Using the above technique, high NH$_3$ volatilization rates were recorded between 1400 and 1600 h in the 3 days following urea surface
application into floodwater (Fig. 4).

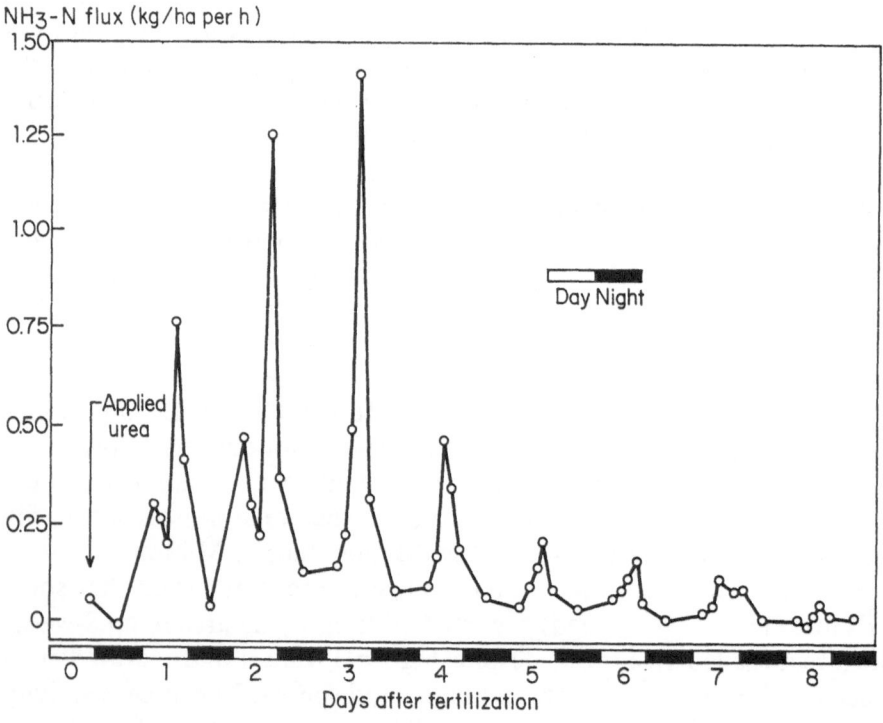

Fig. 4. Ammonia fluxes from the circle as measured by simplified aerodynamic technique. Mabitac,
Laguna, Philippines, 1985 dry season[18].

Floodwater dynamics and potential ammonia volatilization loss Vlek and Craswell[55] found that aqueous NH_3 content in floodwater increases about tenfold per unit increase in pH within the pH range of 7.5–9.0, and that it increases almost linearly with increasing temperatures at a given total concentration of ammoniacal N. In field systems, floodwater pH displays a diurnal pattern (Fig. 5) seemingly synchronized with the cycles of photosynthesis and net respiration or the depletion and addition of CO_2 to floodwater[39].

Cao *et al.*[8] reported that high pH and high total N (urea $+$ NH_4^+-N) concentration in floodwater occurred after urea split application, indicating that greater NH_3 volatilization and denitrification losses may result from this method. In contrast, urea supergranules (USG) point placement resulted in low floodwater pH and low total N concentration (5 mg/kg), indicating that this method can reduce N losses in lowland rice.

In a recent study[14], the equilibrium vapor pressure of NH_3 (ϱNH_3) in floodwater was used as an indicator of volatilized N. At each sampling time, water depth and floodwater pH and temperature were recorded. Relative potential N loss through ammonia volatilization was determined using the following equations:

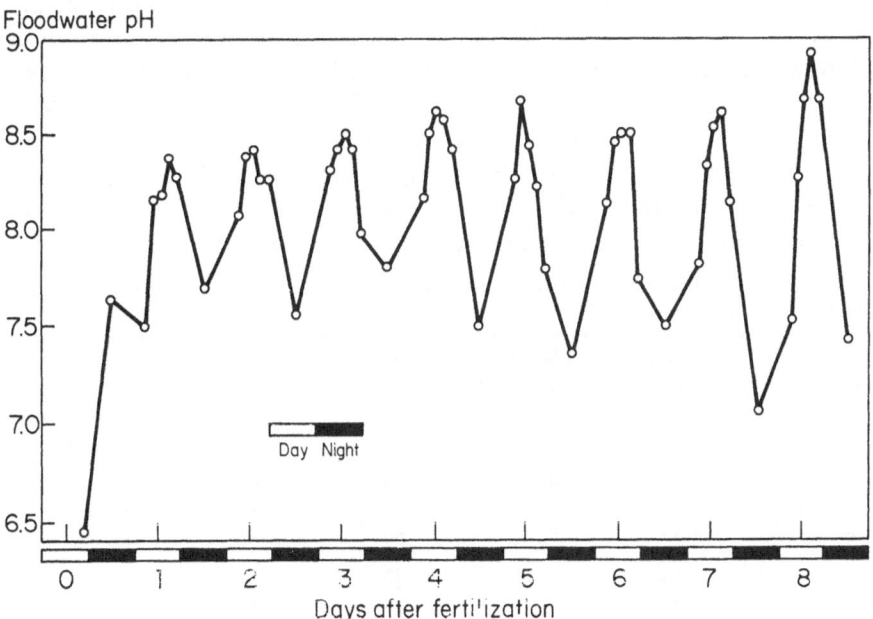

Fig. 5. Diurnal fluctuations in floodwater pH in circle. Mabitac, Laguan, Philippines, 1985 dry season[18].

$$A = \frac{C}{1 + 10^{(0.09018 + 2729.92/T\text{-pH})}} \text{ (Eq. 1, Denmead } et\ al.^{20})$$

$$\varrho NH_3 = \frac{0.00594\,AT}{10^{(1477.8/T-1.6937)}} \times 100 \text{ (Eq. 2, Denmead } et\ al.^{20,21})$$

where A = aqueous NH_3 $(g\,N\,m^{-3})$, C = ammoniacal N $(g\,N\,m^{-3})$, T = temperature (°K), and ϱNH_3 = partial pressure of ammonia (in pascal).

Results obtained by De Datta et al.[14] suggest that ϱNH_3 in the floodwater was lower when fertilizer was incorporated without standing water than when incorporated with 5 cm water. The equilibrium vapor pressure was high 1–4 days after urea was broadcast and incorporated into 5 cm water, suggesting that high volatilization loss occurred during this period.

Contribution of NH_3 loss to total N loss Quantifying the ammonia volatilization loss in relation to total N loss in a lowland rice soil is critical when evaluating N fertilizer source and application techniques. From Australia, Simpson et al.[50] reported that NH_3 fluxes accounted for 24% of the ^{15}N lost. They attributed the balance of ^{15}N loss to nitrification–denitrification.

Fillery et al.[25] and Fillery and De Datta[24] reported that NH_3 volatilization accounted for about 92–100% N loss when urea or ammonium sulfate was topdressed into floodwater 15–20 days after transplanting in Muñoz, Philippines. In Los Baños, NH_3 fluxes from topdressed N accounted for 45% of the ^{15}N lost. Recent studies[18] in Mabitac, Philippines showed that the greatest total losses for 53 and 80 kg N/ha application rates were about 60% when urea was surface broadcast. Estimated denitrification losses were between 28–33% (Table 2).

Table 2. Relationship between total N loss and NH_3 volatilization at the Mabitac site in Laguan, Philippines, 1985, late dry season[18]

Application method	Rate (kg/ha)	Water depth (cm)	Total N loss (%)	NH_3 loss (%)	Estimated nitrification–denitrification loss (%)[a]
Researchers' split[b]	53	0	33	6	27
Researchers' split[b]	53	5	54	22	32
Farmers' split[b]	53	5	60	27	33
Researchers' split[b]	80	0	32	7	25
Researchers' split[b]	80	5	58	27	31
Farmers' split[c]	80	5	59	31	28
Farmers' split[d]	80	5	55	31	24

[a] By difference. [b] 2/3 basal + 1/3 at 5–7 DBPI. [c] 2/3 at 10 DT + 1/3 at booting. [d] Circular plot.

^{15}N balance

Until recently, very few field studies reported on ^{15}N balance sheets for lowland rice. Such information is urgently needed to evaluate the performance of alternative fertilizer N materials and their application techniques most suitable to lowland rice culture. Further, earlier studies on N fertilizer rates in lowland rice were limited to ammonium sulfate. However, urea is currently more widely used than ammonium sulfate in lowland rice.

In field experiments using various application techniques, ^{15}N labelled urea was incorporated into puddled soil[9]. ^{15}N balance data showed that with deep point placement of urea supergranules, only 4–5% of applied N was unaccounted for after harvest of the dry and wet season crops. In contrast, when prilled urea was basally broadcast and incorporated and topdressed at panicle initiation, 11–35% of applied N was not recovered at harvest. This indicates substantial N fertilizer loss, especially when poor incorporation resulted in high floodwater mineral-N levels. The data further showed that 22–56% of the applied urea-N still remained in the soil after the final harvest.

In a recent study with broadcast-seeded flooded rice in Muñoz, Philippines (S.K. De Datta, R.J. Buresh, and M.I. Samson, IRRI, *unpublished data*), N losses were less with $(NH_4)_2SO_4$ than with urea (Fig. 6). Basal deep placement of urea completely eliminated N loss.

Total ^{15}N balance Past results[9,12] suggest that significant amounts of applied N not taken up by the crop or lost by ammonia volatilization and denitrification remain in the soil and organic matter fraction. Schnier *et al.*[48] reported some of the recent work on the subject. Their results showed that the total amount of ^{15}N fertilizer recovered from the soil after 10 weeks of incubation varied between 61 and 66% in four soils. ^{15}N recovery from Maligaya silty loam (Vertic Tropaquept) from Central Luzon, Philippines was 52% of fertilizer applied. From total N uptake at maturity, only 20–30% were derived from N fertilizer.

Cultural practices to minimize nitrogen losses

In recent years, a deeper understanding of the mechanisms causing poor N utilization helped develop cultural practices that improve N fertilizer use efficiency in lowland rice. Basic research results suggest that it is wasteful to apply N in the floodwater between transplanting and early tillering, as commonly practiced by farmers in Southeast Asia. Early N uptake is, however, essential for high tiller production.

Alternative technologies are therefore needed to improve N application efficiency. Improved nitrogen management should maximize N

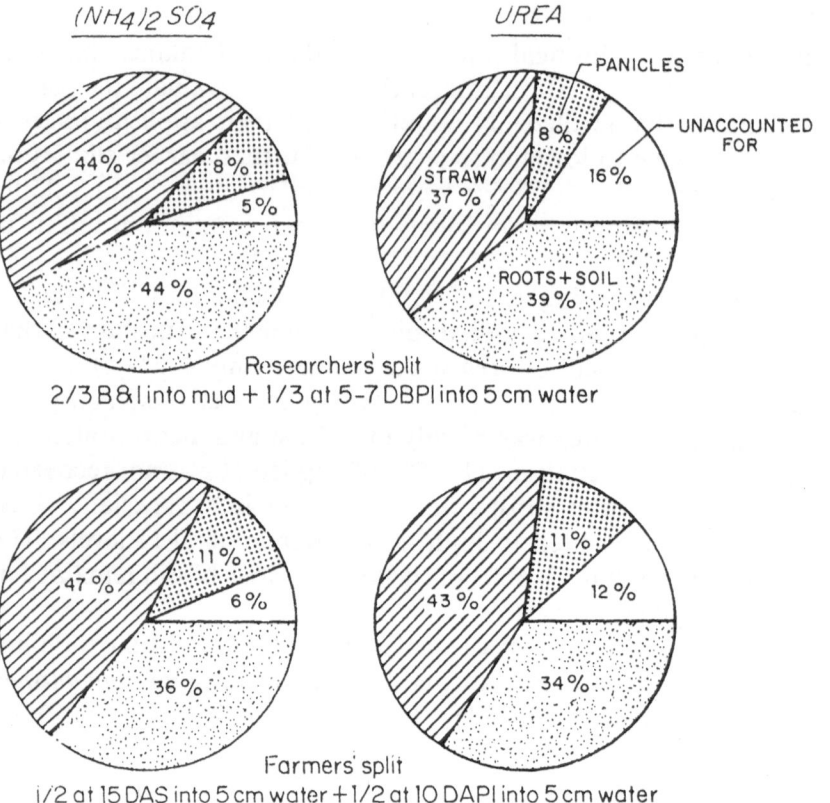

Fig. 6. [15]N recovery in broadcast-seeded IR60 at flowering as affected by N source at 87 kg N/ha. Maligaya Rice Research and Training Center, Nueva Ecija, Philippines, 1985 dry season (S.K. De Datta, R.J. Buresh, and M.I. Samson, IRRI, *Unpublished data*). B & I = broadcast and incorporated; DBPI = days before panicle initiation.

uptake at critical growth stages and minimize transformation processes that lead to N losses or temporary losses from soil-water systems[15]. Ensuring that N absorbed by the rice plant is used for grain production is equally important.

Timing of nitrogen application

Applying N when it is less vulnerable to high losses minimizes N loss and maximizes N use efficiency. Basal dressing without proper incorporation in the soil or early tillering stage leads to high N losses and should be avoided. In the 1986 dry season trials in three farmers' fields in the Philippines, researchers' timing at two N levels gave between 0.4 and 0.6 t/ha more grain yield than did farmers' timing (Table 3). These and other data[14,15] suggest that proper timing of urea application increases the potential for high N use efficiency.

Table 3. Effects of source and method of nitrogen application on grain yield of transplanted rice in farmers' fields, Nueva Ecija, Philippines, 1986 dry season

Treatment	N fertilizer applied (kg N/ha)	Grain yield[a] (t/ha)
No fertilizer N	0	4.3 e
Farmers'split, PU[b]	58	5.6 c
Researchers' split, PU[c]	58	6.0 b
Point-placement, USG	58	6.1 b
Press wedge, USG	58	5.3 d
Plunger auger, PU	58	5.7 c
Farmers' split, PU	87	5.7 c
Researchers' split, PU	87	6.3 a

[a] Av of 3 farms.
[b] 1/2 topdressed at 15 days after transplanting, and 1/2 topdressed after panicle initiation.
[c] 2/3 basal broadcast and incorporated without standing water, and 1/3 topdressed 5–7 days before panicle initiation.

Effect of water depth and management on ammonia loss

Effects of water depth and fertilizer application method on floodwater urea-N + NH_4^+ concentrations and pH were evaluated in a farmer's field in Nueva Ecija, Philippines[30].

The yield response of IR58 to fertilizer was 2.6 t/ha with researchers' split (Table 4). Grain yield was significantly lower with 5 cm water than with no standing water or 2.5 cm water. In fact, the grain yield with point placement was similar to that with broadcast and incorporated (B & I) treatment without standing water and with 2.5 cm water.

Topdressing N at 10 days after transplanting (DT) gave similar grain yield at different water depths (as practiced by most rice farmers in Southeast Asia) but significantly lower than with B & I fertilizer (Table 4).

Table 4. Effects of water depth and fertilizer application method at 87 kg N/ha on IR58 grain yield. San Jose City, Nueva Ecija, Philippines, 1984 dry season

Application method	Water depth (cm)	Yield[a] (t/ha)
No fertilizer N (control)	–	3.4 d
Researchers' split[b]	0	6.0 a
Researchers' split[b]	2.5	5.9 a
Researchers' split[b]	5.0	5.3 b
Farmers' split[c]	0	4.6 c
Farmers' split[c]	5.0	4.6 c
Farmers' split[c]	10.0	4.3 c
Point placement (10 cm soil depth)	0	6.3 e

[a] Means followed by a common letter are not significantly different at the 5% level by Duncan Multiple Range Test. [b] 2/3 basal plus 1/3 at 5–7 days before panicle initiation. [c] 2/3 at 10 days after transplanting plus 1/3 at booting.

Urea, modified urea materials, and their application methods

Urea is likely to remain the leading N source in the world until the end of the century. The large potential for new N capacity in Asia where rice is mostly grown primarily attests to this. However, some changes in the form of urea can be expected. Most urea is produced and sold as dry prills, yet recently production shifted towards that of granular urea. New granular urea production facilities are being built in Malaysia, Thailand, and Mexico[53].

Because substantial N losses occur within 1 to 7 days after N fertilizer application, using slow release and slow dissolving N fertilizers, and enzymic or microbial inhibitors can possibly control these early losses.

In Japan, ureaform, mainly consisting of methyleneurea polymers such as methylenediurea with some free urea, has been developed. Also developed were isobutylidene diurea (IBDU) and 2-oxo-4-methyl-6 ureidohexahydropyrimidine (CDU). Of the coated materials, sulfur-coated urea (SCU) is the most widely tested and reported on[14,16].

Chemists at the International Fertilizer Development Center (IFDC) developed a simple method of manufacturing ureaforms. It has the additional advantage of producing shorter molecular chains of urea as compared with those commercially processed. Results in a farmer's field in the Philippines showed that yield obtained with two ureaforms was comparable to that with researchers' split of urea.

Recently, several newly developed slow release N fertilizers were compared with SCU. Prilled urea mixed with 10% dicyandiamide (DCD) was also evaluated. A single or split dose of these modified urea products did not show a yield advantage over improved method and timing of N application[31].

Urease inhibitor

The potential of urease inhibitor in reducing ammonia volatilization losses from broadcast urea in lowland rice was evaluated jointly by IRRI and IFDC. In earlier IRRI field studies, PPD (Phenylphosphorodiamidate) delayed the hydrolysis of urea broadcast into floodwater at 20 days after transplanting (DT). Although PPD retarded urea hydrolysis, it did not increase grain yield and total N uptake when compared with control without urease inhibitor[15]. Recently, Fillery and De Datta[24] reported that PPD was effective in reducing N loss, as determined by direct measurement of NH_3 loss using micrometerological technique (Fig. 7).

In an IRRI−IFDC joint project at IRRI, urea with and without 1% PPD was broadcast into the floodwater 18, 28, or 38 DT. PPD reduced N loss as determined by the ^{15}N balance. However, yields with or without

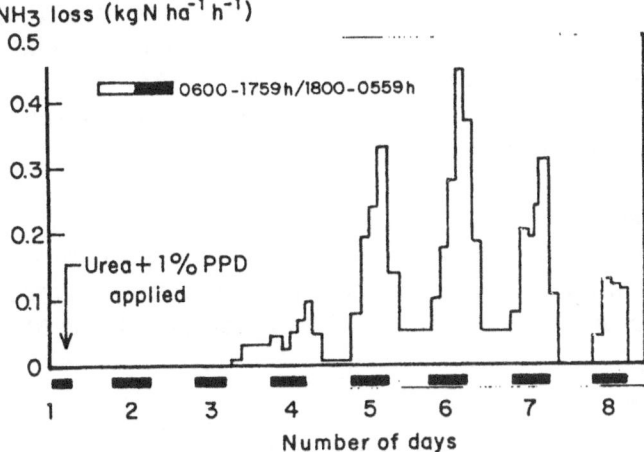

Fig. 7. Ammonia fluxes following the application of urea amended with PPD (1% wt/wt) to floodwater, Maligaya Rice Research and Training Center, Nueva Ecija, Philippines[24]. Reproduced from Soil Science of America Journal, Vol. 50, No. 1, Jan.–Feb. 1986 by permission of Soil Science of America, Inc.

PPD used at different times of N application, did not significantly increase beyond $5.2 t/ha^{31}$.

Deep placement

Cao et al.[9] critically evaluated the deep placement technology using ^{15}N-labelled urea. Rice crop efficiently used the applied N with deep placement resulting in low N fertilizer losses.

Table 5. Effect of urea source and method of application on the grain yield of a traditional variety Binato and a modern breeding line IR29723-143-3-2-1. IRRI, 1986 dry season.

Urea source[a]	N rate (kg/ha)		Application method[b]	Grain yield[c] (t/ha)	
	Binato	IR29723-143-3-2-1		Binato	IR29723-143-3-2-1
–	0	0	No fertilizer N	3.7 a	4.8 d
PU	29	58	Farmers' split	4.1 a	5.5 c
PU	29	58	Researchers' split	3.8 a	6.4 ab
SCU	29	58	Basal, broadcast & incorporated	3.9 a	6.1 bc
USG	29	58	Point placement by hand	3.9 a	6.9 a
USG	30	58	Point placement by press wedge	4.0 a	5.6 c
PU	29	58	Band placement by plunger-auger	4.4 a	6.7 ab
PU	58	87	Researchers split	3.9 a	6.9 a

[a] PU = prilled urea, SCU = sulfur-coated urea, USG = urea supergranules.

[b] Farmers' split = equal split-doses at 10 days after transplanting and 10 days after panicle initiation. Researchers' split = 2/3 basal, broadcast and incorporated into mud plus 1/3 topdressed at 5–7 days before panicle initiation.

[c] In a column, means followed by a common letter are not significantly different at the 5% level by DMRT.

In a recent study with two rices, researchers' split of prilled urea (PU) with good water management gave grain yields similar with that of USG point placement with the modern breeding line IR29723-143. The traditional variety Binato lodged heavily and did not show any significant N response over nonfertilized control (Table 5).

Machine deep placement Several applicators developed for N fertilizer deep placement have been tested by IRRI and national programs in Asia. Although some progress has been made in machine deep placement of PU and USG[14], results have not always been consistent (Table 5).

Critical research needs

Considerable progress has been made in quantifying the magnitude of N losses as affected by various N sources and their management. It was clearly established that ammonia volatilization is an important loss mechanism in lowland rice system. If soil, water, and atmosphere conditions are favorable, ammonia volatilization loss could be serious (up to 50% loss). While ammonia volatilization losses are minimized by appropriate cultural practices, other losses, probably by nitrification and denitrification, are often substantial in irrigated rice. Quantifying ammonia volatilization loss in relation to total N losses from irrigated lowland rice fields is therefore critical in evaluating various N fertilizer sources and their application techniques.

While N transformation processes in irrigated lowland rice have been fairly well understood, hardly any research has been done on rainfed lowland rice. Rainfed conditions are more harsh and variable, so that denitrification losses may be high. Research should measure denitrification losses in relation to ammonia volatilization losses in rainfed lowland rice using simultaneous measurement techniques.

In recent years greater emphasis has been placed on N transformation processes and N fertilizer management in lowland rice. Likewise, more effort is warranted in studying soil N dynamics and their management which could result in substantial yield-gains with resources already on land. Biological nitrogen fixation (BNF) research should focus not only on sustaining N fertility but also on enhancing BNF in lowland rice fields by proper manipulation of ecology and cultural practices.

References

1   ADB (Asian Development Bank) 1985 Agriculture in Asia. Its performance and prospects. The role of ADB in its development. A bank staff working paper, March 1985.
2   Bouldin D R 1986 The chemistry and biology of flooded soils in relation to the nitrogen economy in the fields. A double special issue on Nitrogen Economy of Flooded Rice Soils, Eds. S K De Datta and W H Patrick, Jr. Fert. Res. 9, 1–14.

3   Bouldin D R and Alimagno B V 1976 NH_3 volatilization losses from IRRI paddies following broadcast applications of fertilizer nitrogen. International Rice Research Institute, Los Baños, Philippines (*Unpubl. mimeo.*).

4   Bouwmeester R G B and Vlek P L G 1981 Rate control of ammonia volatilization from rice paddies. Atmos. Environ. 15, 131–140.

5   Broadbent F E 1979 Mineralization of organic nitrogen in paddy soils. *In* Nitrogen and Rice. Ed. I Watanabe, International Rice Research Institute, Los Baños, Philippines, pp 105–118.

6   Broadbent F E and Tusneem M E 1971 Losses of nitrogen from some flooded soils in tracer experiments. Soil Sci. Soc. Am. Proc. 35, 922–926.

7   Buresh R J and Patrick Jr W H 1981 Nitrate reduction to ammonium and organic nitrogen in an estuarine sediment. Soil Biol. Biochem. 13, 279–283.

8   Cao Z H, De Datta S K and Fillery I R P 1984 Effect of placement methods on floodwater properties and recovery of applied nitrogen (^{15}N-labeled urea) in wetland rice. Soil Sci. Soc. Am. J. 48, 196–203.

9   Cao Z H, De Datta S K and Fillery I R P 1984 Nitrogen-15 balance and residual effects of urea-N in wetland rice fields as affected by deep placement techniques. Soil Sci. Soc. Am. J. 48, 203–208.

10   Craswell E T, Byrnes B H, Holt L S, Austin E R, Fillery I R P and Strong W M 1985 Nitrogen-15 determination of nonrandomly distributed dinitrogen in air. Soil Sci. Soc. Am. J. 49, 664–668.

11   Craswell E T and De Datta S K 1980 Recent developments in research on nitrogen fertilizers for rice. IRRI, Res. Paper Ser. No. 49.

12   Craswell E T and Vlek P L G 1979 Fate of fertilizer nitrogen applied to wetland rice. *In* Nitrogen and Rice. Ed. I Watanabe, International Rice Research Institute, Los Baños, Philippines, pp 175–192.

13   De Datta S K 1981 Principles and Practices of Rice Production. John Wiley and Sons, New York. 618 p.

14   De Datta S K 1986 Improving nitrogen fertilizer efficiency in lowland rice in tropical Asia. A double special issue on Nitrogen Economy of Flooded Rice Soils. Eds. S K De Datta and W H Patrick, Jr. Fert. Res. 9, 171–186.

15   De Datta S K, Fillery I R P and Craswell E T 1983 Results from recent studies on nitrogen fertilizer efficiency in wetland rice. Outl. Agric. 12, 125–134.

16   De Datta S K and Gomez K A 1981 Interpretive analysis of the international trials on nitrogen fertilizer efficiency in wetland rice. *In* Fertilizer International. pp 1–5, The British Sulphur Corp. Ltd., Parnell House, England.

17   De Datta S K, Magnaye C P and Moomaw J C 1968 Efficiency of fertilizer nitrogen (N^{15}-labelled) for flooded rice. Proc 9th Int. Soil Sci. Congr. Vol. 4, 67–76.

18   De Datta S K, Trevitt A C F, Obcemea W N, Freney J R and Simpson J R 1986 Comparison of total N loss and ammonia volatilization in lowland rice using simple techniques. *In* Agron. Abstr., p 197, Am. Soc. Agron., Madison, WI.

19   Denmead O T 1983 Micrometeorological methods for measuring gaseous losses of nitrogen in the field. *In* Gaseous Loss of Nitrogen from Plant–Soil Systems. Eds. J R Freney and J R Simpson, Martinus Nijhoff/Dr. W Junk Publishers, Dordrecht pp 133–157.

20   Denmead O T, Freney J R and Simpson J R 1982 Dynamics of ammonia volatilization during furrow irrigation of maize. Soil Sci. Soc. Am. J. 46, 149–155.

21   Denmead O T, Freney J R and Simpson J R 1983 Dynamics of ammonia volatilization during furrow irrigation of maize. Soil Sci. Soc. Am. J. 47, 618.

22   Dreyfus B L and Dommergues Y R 1980 Non-inhibition de la fixation d'azote atmospherique che une legumineuse a nodules caulinaires, *Sesbania rostrata*. CR Acad. Sci. Paris, D. 291, 767–770.

23   Fillery I R P and Byrnes B H 1984 Concurrent measurement of ammonia loss and denitrification in flooded rice fields. *In* Agronomy Abstracts, p. 204, American Society of Agronomy, Madison, WI.

24   Fillery I R P and De Datta S K 1986 Ammonia volatilization from nitrogen sources applied to rice fields. I. Methodology, ammonia fluxes, and nitrogen-15 loss. Soil Sci. Soc. Am. J. 50, 80–86.

25   Fillery I R P, Simpson J R and De Datta S K 1984 Influence of field environment and fertilizer management on ammonia loss from flooded rice. Soil Sci. Soc. Am. J. 48, 914–920.

26   Fillery I R P and Vlek P L G 1986 Reappraisal of the significance of ammonia volatilization as an N loss mechanism in flooded rice fields. A double special issue on Nitrogen Economy of Flooded Rice Soils, Eds. S K De Datta and W H Patrick, Jr. Fert. Res. 19, 15–38.

27   Freney J R, Denmead O T, Watanabe I and Craswell E T 1981 Ammonia and nitrous oxide losses following applications of ammonium sulfate to flooded rice. Aust. J. Agric. Res. 32, 37–45.

28   Freney J R, Leuning R, Simpson J R, Denmead O T and Muirhead W A 1985 Estimating ammonia volatilization from flooded rice fields by simplified techniques. Soil Sci. Soc. Am. J. 49, 1049–1054.

29   Hauck R D 1979 Methods for studying N transformations in paddy soils: review and comments. *In* Nitrogen and Rice. Ed. I Watanabe, International Rice Research Institute, Los Baños, Philippines, pp. 73–91.

30   IRRI (International Rice Research Institute) 1985 Annual Report for 1984. Los Baños, Philippines. 504 p.

31   IRRI (International Rice Research Institute) 1986 Annual Report for 1985. Los Baños, Philippines, 555 p.

32   Keeney D R and Sahrawat K L 1986 Nitrogen transformations in flooded soils. A double special issue on Nitrogen Economy of Flooded Rice Soils. Eds. S K De Datta and W H Patrick, Jr. Fert. Res. 9, 15–38.

33   Keerthisinghe G, De Datta S K and Mengel K 1985 Importance of exchangeable and nonexchangeable soil NH_4^+ in nitrogen nutrition of lowland rice. Soil Sci. 140, 194–201.

34   Keerthisinghe G, Mengel K and De Datta S K 1984 The release of nonexchangeable ammonium (^{15}N labelled) in wetland rice soils. Soil Sci. Soc. Am. J. 48, 291–294.

35   Krishnappa A M and Shinde J E 1978 The fate of an initial ^{15}N pulse under field conditions of the tropical flooded rice culture. All-India Coordinated Rice Improvement Project working paper presented at the 4th Research Coordination Meeting of FAO/IAEA/GSF Coordinated Programme on N Residue, Piracicaba, Brazil (*Unpubl. mimeo.*).

36   Lowendorf H S 1982 Biological nitrogen fixation in flooded rice fields. Cornell Int. Agri. Mimeo. 96.

37   Mengel K, Schön H G, Keerthisinghe G and De Datta S K 1986 Ammonium dynamics of puddled soils in relation to growth and yield of lowland rice. A double special issue on Nitrogen Economy of Flooded Rice Soils. Eds. S K De Datta and W H Patrick, Jr. Fert. Res. 9, 117–130.

38   Mikkelsen D S and De Datta S K 1979 Ammonia volatilization from wetland rice soils. *In* Nitrogen and Rice. Ed. I Watanabe, International Rice Research Institute, Los Baños, Philippines, pp 135–156.

39   Mikkelsen D S, De Datta S K and Obcemea W N 1978. Ammonia volatilization losses from flooded rice soils. Soil Sci. Soc. Am. J. 42, 725–730.

40   Reddy K R and Patrick Jr W H 1986 Denitrification losses in flooded rice fields. A double special issue on Nitrogen Economy of Flooded Rice Soils. Eds. S K De Datta and W H Patrick, Jr. Fert. Res. 9, 99–116.

41   Reddy K R and Rao P S C 1983 Nitrogen and phosphorus fluxes from flooded organic soil. Soil Sci. 136, 300–307.

42   Roger P A and Watanabe I 1986 Technologies for utilizing biological nitrogen fixation in wetland rice: potentialities, current usage, and limiting factors. A double special issue on Nitrogen Economy of Flooded Rice Soils. Eds. S K De Datta and W H Patrick, Jr. Fert. Res. 9, 39–78.

43   Sahrawat K L 1983 Relationships between soil urease activity and other properties of some tropical wetland rice soils. Fert. Res. 4, 145–150.

44   Savant N K and De Datta S K 1979 Nitrogen release patterns from deep placement sites of urea in a wetland rice soil. Soil Sci. Soc. Am. J. 43, 131–134.

45   Savant N K and De Datta S K 1980 Movement and distribution of ammoniacal-N following deep placement of urea in wetland rice soil. Soil Sci. Soc. Am. J. 44, 559–565.

46 Savant N K and De Datta 1982 Nitrogen transformations in wetland rice soils. Adv. Agron. 35, 241–302.

47 Savant N K, De Datta S K and Craswell E T 1982 Distribution patterns of ammonium nitrogen and N uptake by rice after deep placement of urea supergranules in wetland soil. Soil Sci. Soc. Am. J. 46, 567–573.

48 Schnier H F, De Datta S K and Mengel K 1987 Dynamics of ^{15}N labeled ammonium sulfate in various inorganic and organic soil fractions of wetland rice soils Biol. Fert. Soils (In press).

49 Schön H G, Mengel K and De Datta S K 1985 The importance of initial exchangeable ammonium in the nitrogen nutrition of lowland rice soils. Plant and Soil 86, 403–413.

50 Simpson J R, Freney J R, Wetselaar R, Muirhead W A, Leuning R and Denmead O T 1984 Transformation and losses of urea nitrogen after application to flooded rice. Aust. J. Agric. Res. 35, 189–200.

51 Smith C J, Brandon M and Patrick Jr W H 1982 Nitrous oxide emission following urea-N fertilization of wetland rice. Soil Sci. Plant Nutr. 28, 161–171.

52 Smith C J and De Laune R D 1984 Effect of rice plants on nitrification–denitrification loss of nitrogen. Plant and Soil 78, 287–290.

53 Stangel P J 1985 Changes in the form and price of fertilizer—A global perspective. ASPAC Food & Fertilizer Technology Center Extension Bulletin No. 217, Taiwan. 20 p.

54 Vlek P L G, Byrnes G H and Craswell E T 1980 Effect of urea placement on leaching losses of nitrogen from flooded rice soils. Plant and Soil 54, 441–449.

55 Vlek P L G and Craswell E T 1981 Ammonia volatilization from flooded soils. Fert. Res. 2, 227–245.

56 Vlek P L G, Stumpe J M and Byrnes B H 1980 Urease activity and inhibition in flooded soil systems. Fert. Res. 1, 191–202.

57 Watanabe I, Craswell E T and App A A 1981 Nitrogen cycling in wetland rice fields in south-east and east Asia. In Nitrogen Cycling in South East Asian Wet Monsoonal Ecosystem. Ed. R. Wetselaar, Aust. Acad. Sci. Canberra, pp 4–17.

58 Wetselaar R T, Shaw T, Firth P, Oupathum J and Thitiopoca H 1977 Ammonia volatilization from variously placed ammonium sulphate under lowland rice field conditions in central Thailand. Proc. Int. Seminar SEFMIA, Tokyo, Japan, Society of Science of Soil and Manure, Tokyo, Japan.

59 Wilson J R, Thurtell G W, Kidd G E and Beauchamp E G 1982 Estimation of the rate of gaseous mass transfer from a surface source plot to the atmosphere. Atmos. Environ. 16, 1–7.

Plant and Soil 100, 71–97 (1987).

Nitrogen budgets in flooded soils used for rice production

D.S. MIKKELSEN
Agronomy and Range Science, University of California, Davis, CA 95616, USA

Key words Ammonium fixation Ammonia volatilization Dentrification Immobilization Run off losses N use efficiency Rice

Summary The flooded soil-rice plant ecosystem is extremely complex and final N budgets are the products of many N transformations modified by physical, chemical and biological agents, to a large extent controlled by O_2 fluxes, but interacting with each other over time. Topics reviewed include mineralization-immobilization, nitrification-denitrification, NH_4^+ fixation, NH_3 volatilization, leaching and run-off lossess. Nitrogen budgets constructed for water sown rice using temperate climate data clearly show that the major mechanisms by which fertilizer N is removed is crop absorption, nitrification-dentrification and NH_3 volatilization. Proper fertilizer management can reduce losses and desirably increase fertilizer use efficiency. Nitrogen budgets have proven useful in describing gains and losses from the various N transformation processes, all of which are environment and management dependent.

Introduction

The development of practices to improve the efficiency of fertilizer nitrogen (N) for rice requires that agronomists have a knowledge of the fate of the applied N and its effect on crop production. Identification of the various transformation processes, quantification of the size and transport rate of the various N pools, the establishment of interrelationships between the various biological processes and ultimately the effects on crop growth and yield are all essential components of a N budget. Adequate N nutrition is the major nutritional constraint to rice production in the world and the plant nutrient most difficult to manage in the rice ecosystem. Flooded soils are particularly unique in their chemical, physiochemical and biological properties and difficult to characterize because of the complex transformations which are involved. Soil N budgets must consider such processes as nitrification-denitrification, ammonia (NH_3) volatilization, NH_4^+-fixation immobilization-mineralization (ammonification) processes, leaching, floodwater losses, and ultimately the growing crop. The use of ^{15}N-labeled fertilizers has greatly increased the sensitivity for tracing fertilizer N, but quantitative measurement of the various complex N pools and agreement on interpretation of the data is yet to be fully achieved.

The literature on N budgets in submerged soils is very extensive. Investigators have studied the effects on N rates, sources, methods and times of N application on all aspects of N gains and losses, plant absorption, plant growth, crop yields and quality. These studies have been conducted over a wide range of soils, climates and management systems and are too extensive for detailed review. This paper will deal primarily with an overview of the nature of the submerged soil, the N transformation processes and temperate zone N budgets. The budget process ultimately seeks to improve N use efficiency in rice production, the socio-economic conditions of both rural and urban people and to limit unnecessary contamination of the environment.

Effects of soil submergence on soil properties

Several excellent reviews deal with the chemistry of submerged soils[37,38,39,40,41,47]. A brief review of the salient points dealing with N transformations is presented below:

Effects on physical properties

The immediate effect of submergence is interruption of the normal processes of gaseous exchange between the soil and the atmosphere. The pore space becomes saturated with water and the structural aggregates tend to break down. Water covering the soil acts as a barrier and diffusion of dissolved oxygen (O_2) into the interstitial water is reduced by four orders of magnitude slower than through a porous medium. Thus in a submerged soil, the chemical and the microbial demand for O_2 greatly exceeds the supply. As a consequence, O_2 levels in the soil drop quickly and within 6 to 8 hours of submergence the soil is virtually O_2-free except for a thin layer of soil at the soil-water interface. This thin layer of surface soil contains O_2 and is a few millimeters to about a centimeter thick.

Effects on physicochemical properties

Oxygen dissolved in the flooded water, from the atmosphere or from the photosynthetic activity of various hydrophytes, diffuses below the water layer into the thin oxidized soil layer. This layer supports microorganisms carrying on aerobic biological processes and the various mineral species are typically in oxidized forms such as SO_4^{-2}, NO_3^-, Fe^{+3} and Mn^{+4} compounds. The soil color in this zone is similar to wet aerobic soils.

Effects on biochemical properties

Immediately below the thin oxidized layer the O_2 content drops sharply and approaches zero within a very short distance. The depth of O_2 penetration into the soil depends on the balance between diffusion and

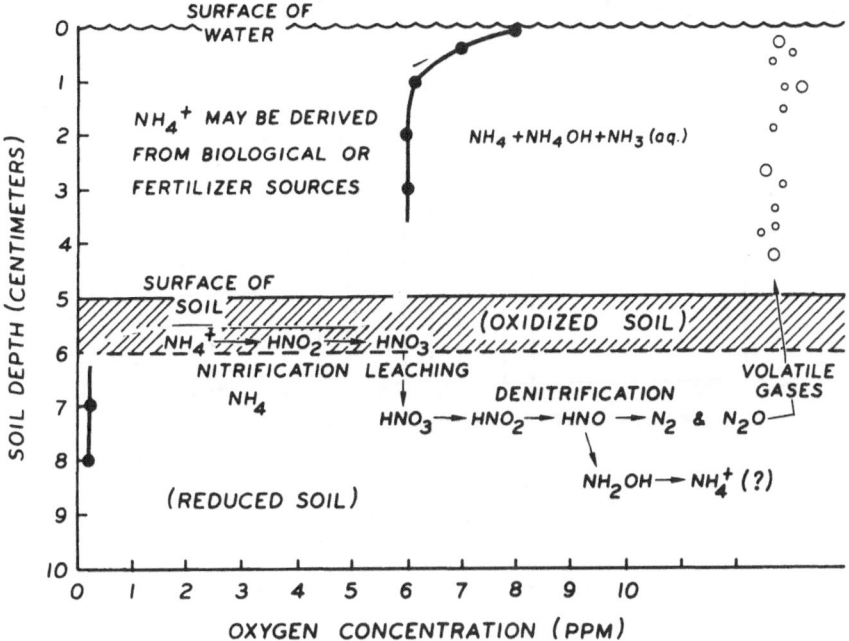

Fig. 1. Schematic representation of differentiated flooded soil profile.

consumption. In the absence of O_2, the aerobic microorganisms die or become quiescent and facultative or true anaerobes become active in the anaerobic zone. A schematic representation of the differentiated flooded soil profile is shown in Figure 1.

As a result of submergence, the redox potential of the reduced layer drops sharply as diagramatically represented by Patrick and Mahapatra[38] (Fig. 2). A well aerated soil is characterized by a redox potential of +400 millivolts (mV) or greater. If the reduction process is sufficiently intense, the soil may have a redox potential as low as −300 mV. The redox potential (Eh) of the oxidized zone of flooded soils may remain as high as +500 mV. The degree of oxidation and reduction of the redox systems — such as oxygen, nitrate, nitrite, manganese, iron and sulfur systems as well as various organic compounds determine the redox potential of soil. Free O_2 functions both physicochemically and biochemically to maintain these systems in an oxidized form.

In the reduced soil layer, anaerobic organisms utilize progressively weaker electron acceptors in place of O_2 for respiration. After O_2, the next strongest electron acceptor is NO_3^-. Nitrate is reduced to N_2 or N_2O gas at around +220 mV redox potential. This process, called denitrification, requires an energy source for the denitrifying bacteria. When O_2 and NO_3^- become exhausted redox potentials drop and Mn^{+3}, Mn^{+4} and

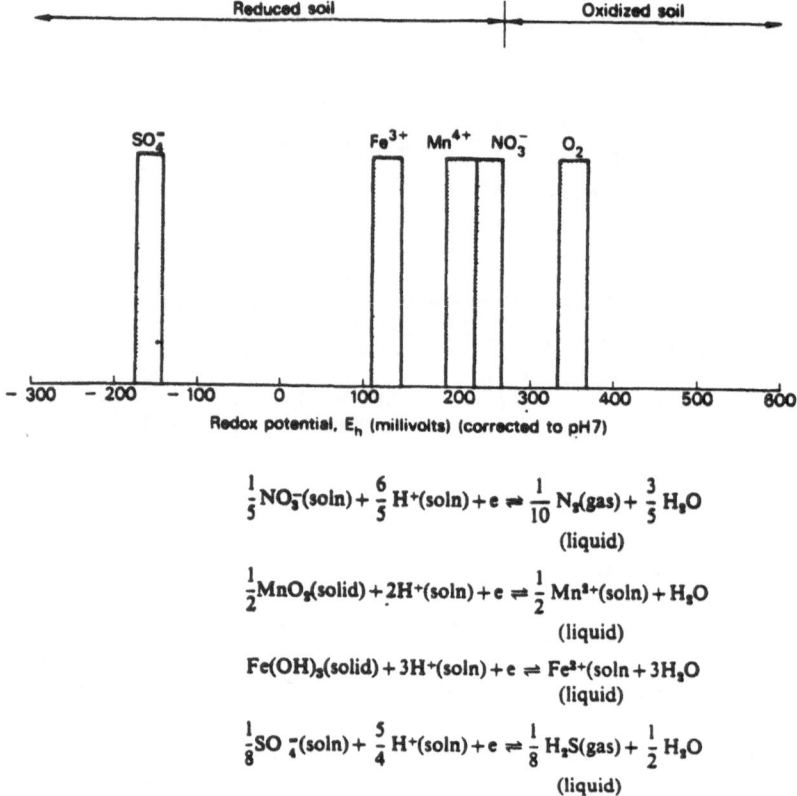

$$\frac{1}{5}NO_3^-(soln) + \frac{6}{5}H^+(soln) + e \rightleftharpoons \frac{1}{10}N_2(gas) + \frac{3}{5}H_2O$$
(liquid)

$$\frac{1}{2}MnO_2(solid) + 2H^+(soln) + e \rightleftharpoons \frac{1}{2}Mn^{2+}(soln) + H_2O$$
(liquid)

$$Fe(OH)_3(solid) + 3H^+(soln) + e \rightleftharpoons Fe^{2+}(soln + 3H_2O$$
(liquid)

$$\frac{1}{8}SO_4^-(soln) + \frac{5}{4}H^+(soln) + e \rightleftharpoons \frac{1}{8}H_2S(gas) + \frac{1}{2}H_2O$$
(liquid)

Fig. 2. Redox potential thresholds where oxidized mineral species become unstable[38].

Fe^{+3} hydroxides are reduced to Mn^{-2} (at $+200\,mV$) and Fe^{+2} (at $+120\,mV$), respectively. These reduced forms of Fe and Mn have higher solubilities than their oxidized forms. As a result, the availability of Fe and Mn increases under flooded condition. If the supply of electron acceptors is less than the rate at which electrons are made available, even stronger reducing conditions result and redox potential drops to around $-150\,mv$ and sulfate (SO_4^{-2}) is then reduced to S^{-2}. When SO_4^{-2} is exhausted, microorganisms use some of the energy stored in organic compounds by reducing H^+ and H_2 and by fermenting organic matter to CO_2, organic acids and alcohols. On further reduction CH_4 is produced from organic matter, usually at the values below -250 to $-300\,mV$. Soils tend to maintain Eh values in a specific range until the oxidized soil components are exhausted. The Eh then drops further and reduction of still weaker electron acceptors take place. For example, a reduced soil will tend to maintain an Eh around $-220\,mV$ as long as NO_3^- is present. When NO_3^- is exhausted, Eh drops and reduction of the next strongest electron acceptor occurs.

On submergence the solubility of soil P also increases. There is an accumulation of NH_4-N and disappearance of preexisting NO_3-N. Ammonia, amines, mercaptans and sulfides are produced from protein decomposition in submerged soils. Ammonification is positively correlated with the organic C and N percent in soil but negatively correlated with C:N ratio. Ammonium accumulation is greatly accentuated in a flooded soil system.

The pH of most soils after submergence tends to approach neutrality. The pH adjustment has several favorable and adverse effects on plant growth. Due to pH adjustment, the adverse effects of extremely high or low pH are minimized which reduces Al, Fe and Mn toxicities and increases the availability of P and Si. On flooding, the partial pressure of carbon dioxide (pCO_2) in the soil increases sharply and is the dominant gaseous product of anaerobic decomposition. The increased pCO_2 has a profound effect on soil pH. The change in pH on flooding is also affected by several other factors such as valence change from Fe^{+3} to Fe^{+2}, accumulation of NH_4^+-N and transformation of sulfate (SO_4^{-2}) to sulfide (S^{-2}). Ponnamperuma[41] concluded that the decrease in pH on submergence of alkaline soils is regulated by the Na_2CO_3-CO_2-H_2O system for sodic soils, $CaCO_3$-CO_2-H_2O system in calcareous soils while increase in pH in acid ferruginous soils is regulated by the $Fe(OH)_3$-Fe^{+2} system.

As a result of submergence, the ionic strength of the soil solution increases, reaches a maximum value during the peak soil reduction period, then decreases. In acid or slightly acid soils, the reduction of relatively insoluble Fe^{+3} and possibly Mn^{+4} to more soluble forms accounts for much of the increase in the ionic strength. In neutral to alkaline soils, Ca^{+2} and Mg^{+2} also make a contribution to ionic strength. Organic matter enhances the solubilities of Fe, Ca and Mg. If the soil is initially high in NO_3-N, the ionic strength of the soil may decrease on submergence due to the loss of NO_3^- by denitrification.

Reduction of soil is purely a biochemical process and microorganisms are essential for the changes. Reduction of minerals does not occur in sterile soils. Rice plants also affect the degree of reduction of soil due to O_2 secretion from the roots. A narrow zone of soil around the actively growing roots, however, may be oxidized while the bulk of the soil is reduced. During the active vegetative growth of rice the redox potentials of cropped-flooded soils are usually higher than fallow-flooded soils. Biochemical transformations of nutrients in the oxidized rhizosphere have not been widely studied, but their behavior is probably similar to aerated systems.

Effects of submergence on rice plants

Unlike many plant species, rice has unique qualities that allow it to survive and reproduce under upland, lowland and deep water conditions.

Although an aquatic medium is favored for rice growth and yield, root growth requires a supply of O_2 and an escape mechanism for CO_2 liberated during respiration. Due to a unique air-carrying channel system (aerenchyma) from the leaf blades to the root cortex, roots can aerate without the need of O_2 from the soil.

Physical and physicochemical processes of nitrogen transformation

Nitrogen fertilization is one of the most important factors affecting rice production in the United States and the world. In the temperate zone N fertilization accounts for 40–50% of the annual rice production. Nitrogen is one of the most difficult plant nutrients to manage because of the large number of potential transformation pathways. Figure 3 illustrates the complex interactions that exist in submerged soils where N losses may occur in the oxidized and reduced soil layers, from the floodwater, by outflow and leaching, absorption of nitrogen by plants and its loss by several mechanisms. The individual processes described in Figure 4 are discussed below.

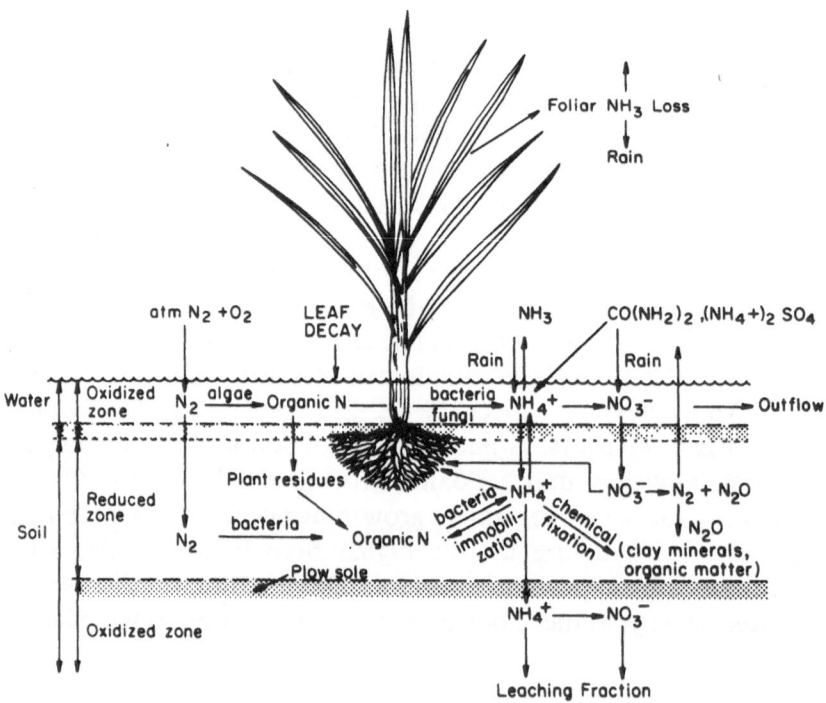

Fig. 3. Schematic representation of nitrogen transformations in a lowland rice ecosystem.

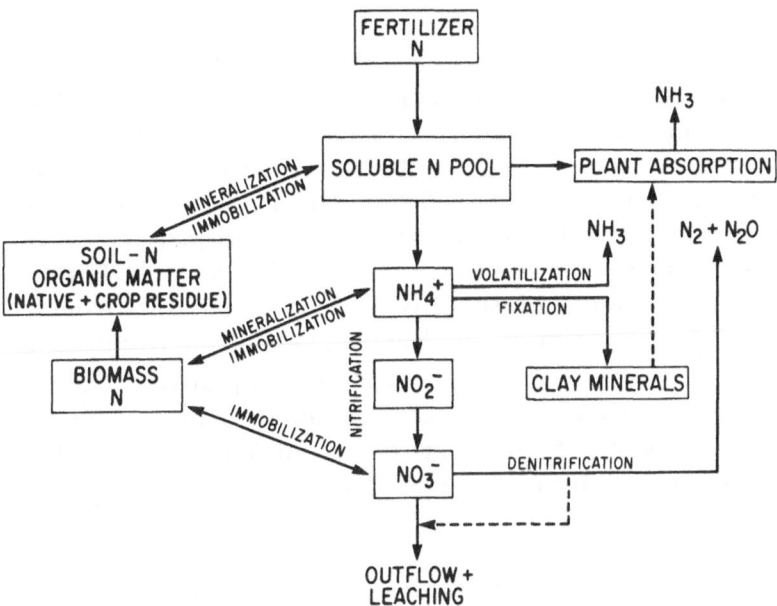

Fig. 4. Schematic representation of nitrogen transformation pathways in a flooded soil system.

Nitrogen transport in submerged soil

The movement of N in soils plays an important role in determining its availability for plant growth. Two main processes are involved in N transport, namely (i) convection of substances dissolved in the soil solution due to the mass-flow, and (ii) molecular or ionic diffusion due to concentration gradients[16]. Another possible mechanism includes ionic movement in an electrical field. Movement of N species, such as NH_3 (aq), NH_4^+, NO_3^- and urea occurs in soil by mass-flow, diffusion or both.

A factor which often determines N movement is the degree of interaction between the soil solution and the soil particle surface. A strongly adsorbed cation such as NH_4^+ will move less readily than an anion like NO_3^- or urea which are lightly adsorbed by most soils[16,36]. The quantity of NH_4^+-N transferred by diffusion per unit area per unit time is proportional to the diffusion coefficient and the concentration gradient. The apparent diffusion coefficient of NH_4^+-N is $0.216 \, cm^2/day$ as compared with $1.33 \, cm^2/day$ for NO_3^-, suggesting NO_3^- moves about 6 times faster than NH_4^+-N[42]. The transport of NH_4^+-N by ionic diffusion from the anaerobic to aerobic layer of the flooded soil is facilitated by large amounts of reduced Fe^{+2} and Mn^{+2}, lower CEC and high moisture content of the flooded soil. The general movement of NH_4^+-N is downward > lateral > upward[47]. Leaching and run-off processes also affect the movement and loss of N in these systems and are discussed below.

Leaching and run-off losses

Continuous submergence leads to losses of soluble soil and fertilizer N through leaching and run-off processes. Ammonium-N is much less subject to leaching than NO_3^- because of its adsorption on the cation exchange complex. Loss of NH_4^+ by leaching, however, is greater in waterlogged soil than in well drained soil. This is because NH_4^+ accumulates in waterlogged soil and reduced Fe and Mn displace NH_4^+ from the exchange complex and under a constant head of water NH_4^+ moves downward as percolation[38]. Losses of N by leaching may vary from 1 to 70% of applied N, while run-off and leaching losses from rice fields can range from 4 to 16 kg N/ha and 5 to 25 kg N/ha, respectively. Leaching may occur rapidly in coarse textured wetland soils or in soils with appreciable amounts of hydrous oxides of Fe and Al due to low CEC[47]. Leaching losses are less from slow-release fertilizers such as SCU and IBDU.

Chemical and biological transformations of nitrogen under submerged soil conditions

Nitrogen undergoes several chemical and biological transformation processes in the soil[4,5,37,38,39,47]. The chemical transformations include clay fixation of NH_4^+ and direct NH_3-volatilization. Biological transformation of nitrogen include mineralization-immobilization, biological N_2-fixation, nitrification-denitrification, and plant absorption.

The voluminous literature on nitrogen transformations in soil is reviewed in *Nitrogen and Rice*[5,27,32,37,38], *Nitrogen in Agricultural Soils*[3,21,48,56] and *Nitrogen Economy of Flooded Soils*[10]. Much of the research on N transformations has been done on upland crops. Biochemical transformations of N, especially in the oxidized rice rhizosphere, have not been widely studied. A recent study[45] extrapolating mass balance data suggests that nitrification-denitrification reactions do occur in the rice rhizosphere.

The systems of rice culture involving continuous and alternate flooding affects the behavior of both native and applied N. The special condition prevailing under the waterlogged soil environment accelerates the normal ammonification process and completely suppresses nitrification when O_2 is not present. On flooding, NH_4^+-N accumulates in the soil and NO_3^--N disappears. Large losses of NH_4^+-N, either applied as fertilizer at the soil-water interface or mineralized during the decomposition of organic matter, occurs as a result of waterlogging[46]. Nitrogen losses are usually more pronounced in the absence of crop residues[5,51]. The presence or absence of a growing crop also affects N transformation.

Nitrogen fertilizer sources for rice can be divided into two groups;

namely, organic crop residues and inorganic-N fertilizers. Organic crop residues must first undergo biological degradation before participating in other transformations as were described earlier. Inorganic N sources could be either soluble conventional type such as ammonium sulfate and urea or controlled (slow) release fertilizers such as SCU and IBDU. Urea and organic-N sources undergo enzymatic hydrolysis and are converted to NH_4^+-N. Addition of energy-rich organic material to the soil stimulates soil organic matter transformation (priming action) either positively or negatively[5,17,19]. The addition of N fertilizer has several actual and apparent effects on soil N transformations and plant uptake. Crop residue additions can affect the transformation and distribution of fertilizer N into different soil N fractions[8]. Biologically fixed-N is also subjected to these transformation processes[20].

Mineralization and immobilization

Tracer experiments have shown that there is a simultaneous synthesis (immobilization or tie-up) and degradation (mineralization or release) of organic compounds in the soil. These are due to microbial activity which leads to a continuous interchange between organic and inorganic N forms[5,14,21]. Thus, only the net mineralization or net immobilization of the mineral N is usually measured during the mineralization process[51]. It is the net balance of the two opposing processes that exerts a large influence over the availability of N to plants and the conversion of N in the soil as organic or fixed forms.

A dynamic equilibrium exists between the available (soluble plus exchangeable) and unavailable (organic and clay-fixed) forms of N in the soil. The individual changes are difficult to assess, since we can only measure the end result of these processes at a particular time. By using labeled-^{15}N materials, however, it is possible to measure and account for the dynamic changes of mineralization-immobilization processes.

Results of such studies indicate that added sources of inorganic N undergo rapid immobilization initially followed by a decrease in mineralization rate[4,5,7,8]. The magnitude and duration of inorganic-organic N equilibrium and transformation processes vary with conditions[4,9,47]. Net immobilization of N is less in flooded soil than aerobic, well-drained soils[7]. Yet, 20–80% of the added fertilizer N can be immobilized depending upon the conditions[6,9,13]. To achieve efficient use of N from the soil as biologically fixed-N, crop residues and fertilizer, it is necessary to consider the time-rate aspects of mineralization relative to the N requirement of rice. In the short-term, the supply of N to rice is governed by the rate of mineralization of organic-N to NH_4^+-N. Amino acid N is more susceptible to mineralization than other fractions of soil N.

Net mineralization does not occur until a wide C:N ratio of soil organic matter is narrowed to 20:1 or less[4,5]. Ammonification patterns can be rectilinear for air-dried soil brought under flooding, linear under continuous flooding or sigmoidal in very fine-textured soil under flooding[47]. Ammonification patterns of recently immobilized N show deviation from that of the native soil N in that the immobilized N slowly undergoes remineralization. Some N may progressively be stabilized and may remain for long periods, possibly decades, showing resistance to mineralization. Broadbent and Nakashima[7] and Hauck[18] have reported that mineralization of immobilized N can range from 2–10% during the growing season and about 1–3% per year thereafter. According to Ito and Watanabe[20], the mineralization rate can be about 5 times greater for biologically fixed-N (23.4% mineralized) as compared to native soil-N (4.6% mineralization). The amount of N mineralized in a soil during the growing season of a crop varies with soil and environmental conditions as well as the techniques used for measurement, and can range between 5–1166 ppm.

Non-exchangeable NH_4^+

The capacity of submerged soils to bind NH_4^+ and K^+ in a non-exchangeable form is determined largely by the amount and kinds of 2:1 type clay minerals present. Entrapment occurs between silica sheets through interlayer bonding accompanied by contraction of interlayer spacing[33]. Fixed-NH_4^+ is not removed by extraction with usual soil extractants which traditionally defines the fixed portion as being unavailable to plants and microorganisms. Lamm and Nadafy[24] suggest that like K^+, NH_4^+ may exist in several release forms characterized by a dynamic equilibrium such as:

$$\text{soluble } NH_4^+ \underset{}{\overset{fast}{\rightleftharpoons}} \text{exchangeable } NH_4^+ \underset{}{\overset{slow}{\rightleftharpoons}} \text{intermediate } NH_4^+ \underset{}{\overset{very\ slow}{\rightleftharpoons}} \text{fixed } NH_4^+$$

In this characterization 'intermediate NH_4^+' may be considered to occupy interlayer sites on the clay lattice which is exchangeable with K^+ and H^+ and in which defixation increases when the clay lattice is expanded. A distinction between 'native' and 'culturally induced' NH_4^+ fixation is made by some investigators who note a difference in their plant availability. 'Culturally induced' fixed-NH_4^+ is more available to crops and may be influenced by such factors as soil K^+ and NH_4^+ status, degree of lattice weathering, soil moisture status, particle size, competing ions and confined root masses[33]. Bajwa[1] reported that soils dominant by vermiculite and montmorillonite fix the largest amounts of applied NH_4^+ (94% and 91%, respectively), followed by beidelite (72%) and amorphous clays (45–64%). Fixation is negligible (10%) in clays with hydrous micas, halloysite and chlorite.

The clay-fixed (non-exchangeable) NH_4^+ can be slowly replaced and released by Na^+, Ca^{+2} and Mg^{+2}. Clay fixation of NH_4^+ increases with NH_4^+ concentration, depth of soil, alternate drying, high pH and liming, freezing and thawing, addition of nitrification inhibitors, and depletion of K^+ from the soil. The release of fixed-NH_4^+ increases when the situations mentioned above are reversed. Ammonium ions are also released when the exchangeable NH_4^+ content decreases below the equilibrium value as a result of plant uptake and leaching. Keerthisinghe *et al.*[22] showed that non-exchangeable NH_4^+ was released for plant uptake in submerged soil. As much as 2.2% of the total plant N was derived from the labeled fixed NH_4-N. The amount of ^{15}N-label taken up by rice was inversely related to the N application rate.

Soluble plus exchangeable nitrogen

The soluble plus exchangeable (available) N is the most important fraction in crop nutrition. The main sources of available N are added fertilizer and mineralized-N. When the N fertilizer sources are added to flooded soils, a rapid decline in the available N occurs with time, usually 30 to 45 days. Crop uptake of fertilizer N virtually stops early in the growing season. The exchangeable NH_4^+-N content in flooded rice soils may increase, however, due to mineralization of soil organic matter and the release of clay-fixed NH_4^+-N after a few days of flooding. Soil incorporation of straw decreases available N levels in the early season because of immobilization and increases in the late season due to its release. The available N is subjected to crop uptake, biological and non-biological fixation, nitrification-denitrification and volatilization[56].

Biological N_2-fixation

Nitrogen balance studies with rice often show an excess of N recovery over soil supply. This is assumed to be due to biological N fixation and these contributions vary from 15 to 50 kg N/ha per crop[23]. More N is biologically fixed in the presence of the rice plant, especially in the wet tropical seasons. Submerged soils with bluegreen algae and N-fixing bacteria fix N but a part of this may not be available for the current crop[44]. Reddy[42] has estimated the extent of biological N_2-fixation to be 0.41–0.74 mg N per kg dry soil per day.

Ammonia volatilization

Ammonia volatilization is an important pathway of N loss, especially in fertilized cropping systems. During recent years a number of studies

Table 1. Ammonia volatilization losses from wetland rice soils determined by different techniques

Kind of soil	Kind of study	Fertilizer	N fertilizer rate (kg/ha)	N loss (%)	Remarks	Reference
Flooded soils, pH 8.4	Laboratory	Ammonium sulfate Urea Surface applied	66	22	Laboratory incubation	17
Flooded soils, pH 8.4	Laboratory	Ammonium sulfate Urea Incorporated-broadcast	200	1.7–16.7	Laboratory incubation	25
Crowley silt, pH (8.0) flooded	Laboratory	Ammonium sulfate	112,224, and 448	0.04–0.10	Boric acid used as trap with air train	11
Flooded soils	Laboratory	Ammonium sulfate	50 and 200	0.50–7.0	Sulfuric acid used as trap for NH_3 without air train	26
Maahas clay	Greenhouse	Ammonium sulfate Incorporated broadcast	100	1–19.0	–do–	51
(Flooded)	Field	Ammonium sulfate Urea broadcast	100	3.3–4.0	–do–	51
Maahas clay (pH 7.0–8.4)	Field	Ammonium sulfate Broadcast	60	17–40	Open-dish system	2
Maahas clay (Flooded)	Field	Ammonium sulfate Broadcast	40–60	30–60	Open-closed systems	2
Maahas and Luisiana clays (pH 7–9.5)	Greenhouse	Ammonium sulfate Placed-broadcast	30–90	0.01–5.8	Sulfuric acid used as trap for NH_3 with air train	34
	Field	Ammonium sulfate Urea Placed-broadcast	90 30–90	0.25–6.8 0.25–5.8 1.0–20	–do–	27

Soil		Fertilizer	50–100	0.8–12.4	Method	28
Flooded soil (pH 7.0–7.5) Thailand	Field	Ammonium sulfate Incorporated-broadcast	50–100	0.8–12.4	–do–	28
Mahaas Clay	Field	Ammonium sulfate Broadcast-incorporated Top-Dresser-PI	80 / 40	5 / 11	Micro-meteorological / Micro-meteorological	15 / 15
Sacramento clay	Field	Ammonium sulfate Broadcast-incorporated Banded	80 / 80	5.6 / 1.6	Air train and ammonia Trapping	a
Mahaas clay	Field	Urea-incorporated	60	13	Micro-meteorological	13
Clay loam (7.6)		Urea-surface	100	21	Air Train-Acid Trap	49
Mahaas clay						
Maligaya si cl.	Field	Urea-broadcast	80	27–47	Micro-meteorological	12
Maligaya si cl.	Field	Urea-broadcast Ammonium sulfate-broadcast	58	36% / 38%	Micro-meteorological	12

a Mikkelsen et al, unpublished

have been concerned with NH_3 volatilization losses from flooded soils[27,53]. Under conditions where NH_4-N fertilizers are broadcast directly onto soil or water without adequate incorporation NH_3 volatilization losses can be appreciable (10–50%). Where appropriate technology is used to manage fertilizer application by placement either by banding, mud-ball or supergranule placement, NH_3 losses are very minimal (< 5.0%). Poor fertilizer management practices contribute to large N losses from flooded rice systems and may be a factor in reduced rice yields.

A wide range of NH_3 volatilization losses are reported from flooded soil systems[27]. Inconsistencies in reported losses exist because of imperfect systems of measurement and the complex range of cultural practices, floodwater, soil and atmospheric variables that make measurements very site specific. Forced-air exchange methods using enclosures and acid-traps are frequently used, although they have many limitations. Micrometeorological methods using energy balance and aerodynamic techniques measure NH_3 in the air but losses vary with wind speed, air and water temperature, heat fluxes and net radiation at a specific site. This methodology is perhaps most precise but is extremely labor intensive and require extensive instrumentation and analytical accuracy. The method lacks capability of comparing various ecosystems in close proximity.

A number of water, soil, air and fertilizer management factors affect the kinetics of NH_3 volatilization. The water parameters include NH_4^+-N concentration, diurnal pH, pCO_2, total alkalinity, water buffering capacity, depth, temperature, turbulence, transport fluxes and biotic activity. Dominant soil factors affecting volatilization are soil pH and pe, pCO_2 and carbonate chemistry, cation exchange characteristics and microbial activity. Atmospheric conditions of wind velocity, PNH_3, air temperature and radiation directly influence NH_3 loss. Crop and fertilizer management directly affect the loss patterns depending on N source, timing and method of application, cultural practices, transplanting, direct seeding, or seed drill, water management factors, field layout and plant canopy status.

Ammonia volatilization losses from flooded rice are highly dependent on the particular ecosystem involved. The range of losses measured by various researchers is found in Table 1. From these data it is evident that rate, source and method of N fertilizer application influence the concentration of NH_4^+ in the floodwater. Where urea is applied, urease activity as well as immobilization of N in the aquatic biota exchange reactions and water transport affect the quantity of NH_4^+ available for NH_3 volatilization. In addition to floodwater NH_4^+ concentration, volatilization depends to a large extent on floodwater pH, water depth, temperature and wind velocity. Aqueous NH_3 increases by a factor of 10 in the pH range of 7.5–9.5 in floodwater and increases in a linear fashion with increasing wind speed and volatilization temperature. Diurnal pH fluctuations occur with maximum values about 2 pm and decreasing during the evening, a pattern synchronized with cyclic photosynthesis and respiration of the aquatic biota. Water alkalinity derived chiefly from HCO_3-sources or from urea hydrolysis acts as a pH buffer which is essential if NH_3 volatilization is to be sustained over a prolonged period. The movement of NH_4^+ and urea-N in flooded soils has not received attention as a factor influencing NH_3 volatilization. Nitrogen movement in the soil can readily take place by liquid phase diffusion, solid phase diffusion and mass flow. Ammonium and urea-N spacial distribution, leaching and percolation losses are closely linked and may influence volatilization, denitrification, clay fixation, outflow and leaching losses.

Nitrification and denitrification losses

The biological oxidation of NH_4^+-N to NO_3^--N (nitrification) results in the conversion of the relatively immobile cation NH_4^+ into a more mobile anionic (NO_3^-) form, which in turn is susceptible to denitrification. The submerged soil is an ideal environment for denitrification since it possesses a thin oxidized surface underlain by a thick reduced layer. The oxidized layer supports nitrification and the reduced zone, deficient in oxygen and providing decomposable organic matter to energize the reduction of oxidized forms of nitrogen, supports denitrification. The existence of aerobic and anaerobic zones in close proximity in submerged soil, aided in part by the rice plant which transports oxygen to the rhizosphere, facilitates nitrification-denitrification. These reactions likely occur simultaneously. The anaerobic nature of the submerged soil causes the instability of NO_3^-, NO_2 and N_2O, which are used as terminal electron acceptors in the anaerobic respiration of various heterotrophic microorganisms lead to N_2 and N_2O loss[14].

The loss of N by denitrification may vary from 0 to 70% of the applied N fertilizer. The estimated average fertilizer N deficits range between 25 and 35% but field data are lacking to accurately characterize these losses. Diffusion of NH_4^+ from anaerobic to aerobic soil layers may account for half of the total loss. Where alternate soil submergence and drainage occur, conditions are highly favorable to nitrification and subsequent denitrification[43,46]. Denitrification rates depend on available soil C, temperature, pH, O_2 supply, redox potential, NO_3^--N concentrations and the activity of denitrifiers.

Denitrification is widely recognized as a major cause of N loss in lowland rice culture, but documentation of the absolute quantities involved are lacking. The low recovery of fertilizer N in rice is largely attributed to nitrification and subsequent denitrification. Such losses unaccounted for in plant absorption and soil retention are attributed to 'apparent denitrification loss.' Tracer studies, using ^{15}N enriched fertilizers in a N-budget approach, are needed in the field over long periods of time to statistically characterize denitrification losses. To date limitations due to costs, equipment and analytical sensitivity have precluded rapid advances in quantifying denitrification losses.

Some control of denitrification losses are possible which can improve N use efficiency. Proper placement of N in the soil reducing layer and timing of applications to meet the needs of growing crops are probably the most cost-effective means of reducing denitrification losses. The use of nitrification and urease inhibitors and controlled release N fertilizers have some promise to reduce N loss by retarding the soil nitrification process.

Nitrogen losses from plants

Recognition that significant N losses can occur from plant tops of annual and perennial crop species has been reviewed by Wetselaar and Farquhar[54]. Evidence that the absolute amounts of N in the aboveground parts decrease before harvest has been demonstrated for a number of crops, including rice. These losses, heretofore overlooked, represent real losses from the soil-water-plant system and need assessing in developing accurate N budget information. Typical N loss from plant tops observed in Australia and California are shown in Figure 5.

Tanaka and Navasero[50] reported N losses from rice crops grown under high levels of N application. The losses appear greatest about 3 weeks prior to flowering and maturity under high N inputs and somewhat later with lower N inputs. Other data support the observation that the N content of plant tops decline from the onset of flowering to maturity and may reflect significant plant losses both as direct NH_3 and amine loss from foliar parts but also possible excretions from roots. Foster and Stutte[14] provide data that N volatilization does occur from many plant species, including rice. Glutamine synthetase appears to be a main pathway of ammonia assimilation and catalyzing the refixation of NH_3 released during photorespiration. Preliminary data indicates that N released during photorespiration may escape as NH_3 and amines *via* plant foliage. Losses vary with increasing temperature in the range of 28 to 35°C. Various mechanisms of N loss from plants including gaseous losses, require further evaluation and need to be considered in developing N-budgets for rice and other crops.

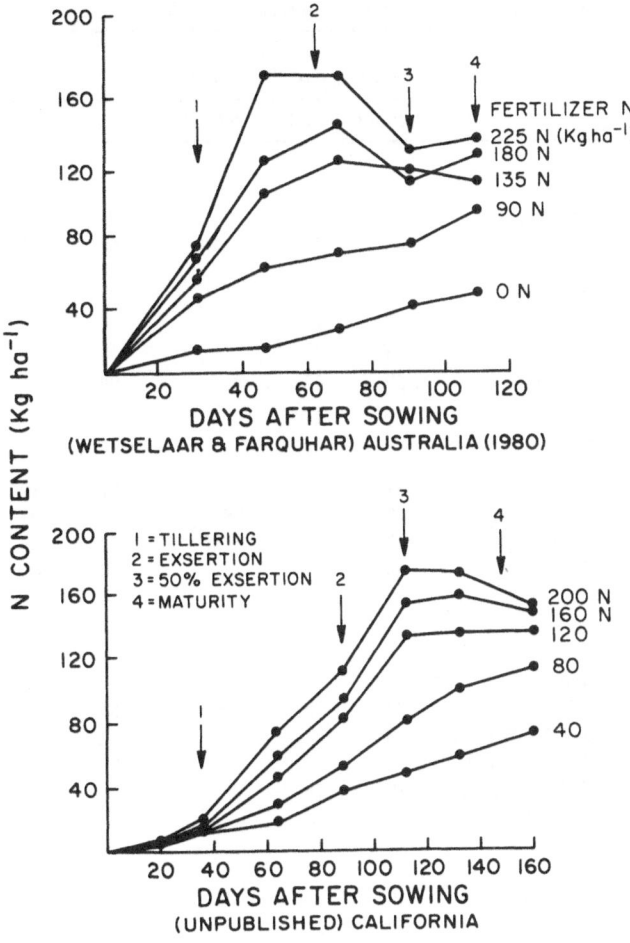

Fig. 5. Losses of N from plant tops during the reproductive period of rice in Australia and California.

Nitrogen budgets in rice

Nitrogen budgets have been used widely in recent years to expand knowledge of the N cycle for various ecosystems. Their major use has been to estimate the net N losses, or unaccountable N losses from specific cropping systems. Until labeled-^{15}N fertilizer materials became available, accurate accountability was difficult and the interrelated biological processes of mineralization, immobilization, NH_3 volatilization, clay fixation, nitrification-denitrification, outflow losses and plant absorption difficult to characterize. The methodology of ^{15}N uses in soils research, the advantages and limitations have been described by several authors[3,18,25]. These investigators emphasize that N-budget studies are

Table 2. Fertilizer ^{15}N management studies — water sown — direct seeded rice

Soil	Fertilizer treatment	Recovery of ^{15}N-fertilizer (%)				
		Plant	Soil	NH$_3$-VOL	Total	Loss
Montmorillonitic mollisol	100 kg N-field					
California	AS-BDCST-dry	15	34	–	49	51
	U-BDCST-dry	34	31	–	65	35
	AS-BDCST/INC-dry	22	39	–	56	44
	U-BDCST/INC-dry	29	32	–	61	39
	AS-band-dry	23	41	–	64	36
	U-band-dry	24	32	–	56	47
	Aq NH$_3$-band-dry	36	33	–	68	32
Montmorillonitic mollisol	90 kg N-field					
California	U-band-10 cm-dry	31	48	1.3	81	19
	U-BDCST surface-dry	17	22	35.0	73	27
	U-BDCST/INC-dry	22	39	1.6	62	38
	U 1/2 band 1/2 TD	34	25	–	60	40
Kaolinitic typic haploxeralfs	100 kg N-field					
California	U-band-10 cm-dry	26	34	–	60	40
	U-BDCST/INC-dry	22	32	–	54	46
	U 2/3 band, 1/3 TD (MT)	23	28	–	51	49
	U 2/3 band, 1/3 TD (PI)	25	27	–	52	48
	U 1/3 BDCST + 2/3 band	18	25	–	43	57
Montmorillonitic vertisol	90 kg N-field					
California	AS-band-10 cm-dry	48	27	4	79	21
	AS-2/3 band, 1/3 TD	57	23	8	88	12
	AS-TD 1/3 × 3	23	26	15	64	36
	120 kg N-field					
	AS-band-10 cm-dry	63	25	1	89	11
	AS-TD 1/3 × 3	24	29	18	71	29
Montmorillonitic mollisol	124/248 ppm N pot					
California	AS-BDCST surface-dry	29/34	57/48	–	86/82	14/18
	AS-banded-10 cm-dry	40/44	52/50	–	92/94	8/6
	AS-BDCST/INC-dry	29/33	54/49	–	83/82	37/28
	AS-TD — 1/2 × 2	36/40	49/48	–	85/88	15/12
	AS-split band + TD	39/47	50/40	–	89/87	11/13
	124/269 ppm N pot					
	U-BDCST surface-day	26/27	56/52	–	82/79	18/21
	U-banded-10 cm-dry	37/37	56/46	–	93/83	7/17
	U-BDCST/INC-dry	28/28		–	76/66	24/34
	U-TD — 1/2 × 2	32/33	48/42	–	80/75	31/34
	U-split band + TD	33/38	42/27	–	75/65	25/35

Table 2 cont.

Montmorillonitic mollisol	100 ppm N-pot		Broadbent and Mikkelsen (1968)		
California	AS-cropped	9.7	76		86	14
	AS-uncropped	–	57		57	43
Montmorillonitic mollisol	90 kg N		Manguiat and Broadbent (1977)		
California	AS-band 10 cm dry	59	25	2	86	14
	AS-BDCST/INC dry	29	33	10	72	28
	AS-2/3 band 1/3 TD	55	22	2	79	21
	AS-1/3 band, 2/3 TD	28	32	6	66	34

empirical, usually estimate events occurring over a single season, and that soil characteristics, environmental factors and crop management affect results. The 'priming effect' whereby fertilizer additions frequently increases soil N mineralization and crop recovery is not always considered.

Nitrogen transformation patterns and budgets with rice have been described by Vlek and Craswell[53] (53) for transplanted rice. They have shown that 40 to 60% of ^{15}N-labeled ammoniacal fertilizer N was lost when broadcast to floodwater when applied 2–4 weeks after transplanting and that broadcast and incorporation before transplanting resulted in a general reduction in fertilizer losses. Crop recovery of fertilizer N at harvest varied from 17.4 to 54.2%.

Nitrogen budget studies of mechanized rice production in the United States, where ammoniacal-N sources are usually applied to dry soil prior to direct sowing into water, have not been reported and represent a fertilizer use pattern of increasing importance in the world.

A summary of fertilizer budgets obtained from dry soil pre-flood fertilizer-^{15}N applications recorded from different soils in California appears in Table 2.

Rice ^{15}N budgets have been conducted with agronomic evaluations made with conventional fertilizer materials in randomized macro-plots. Application of ammoniacal N fertilizer by deep placement has been shown to be superior to surface application in several early studies[6,29,30]. Using ^{15}N-enriched ammonium sulfate and urea in a greenhouse experiment measuring the effects of N-source, timing placement and rate (30 vs 60 mg N/kg soil), data were presented showing that overall N losses were greater with urea than ammonium sulfate, especially at the higher N rate. Urea lost about 14% of the fertilizer N at the 30N rate and 23% at a higher rate. Actual uptake of ^{15}N-fertilizer ranged from 26 to

47%. Plants recovered 44 and 37% of banded ammonium sulfate, respectively, from band applications, and 29 and 26% of broadcast ^{15}N, respectively. Split N application recovery by the crop was intermediate. Fertilizer applications influenced the release of soil N and produced a dividend of crop N equivalent to 20–25% of the amount of fertilizer applied[6]. Reddy and Patrick[43] showed that when ammonium sulfate was deep placed, 49% of the fertilizer N was recovered by the crop, 26% was recovered in the soil and 25% was lost from the system. The same authors[43] report crop recovery of ^{15}N-labeled ammonium sulfate of 49 and 64% with multiple surface applications showing superior crop recovery. Residual ^{15}N measurements show that 24–26% remained in the soil organic fraction, from which about 6% was utilized by a succeeding rice crop.

In California field studies, ^{15}N-labeled fertilizers have been widely used to develop N budgets for the various pathways by which N is either utilized or lost and to develop strategies for fertilizer use which are consistent with the best crop yields (Table 2 and Figure 6). Labeled fertilizer N recovery by rice at harvest has varied from 10 to 59%. The magnitude of plant recovery from applied fertilizer has been poorly

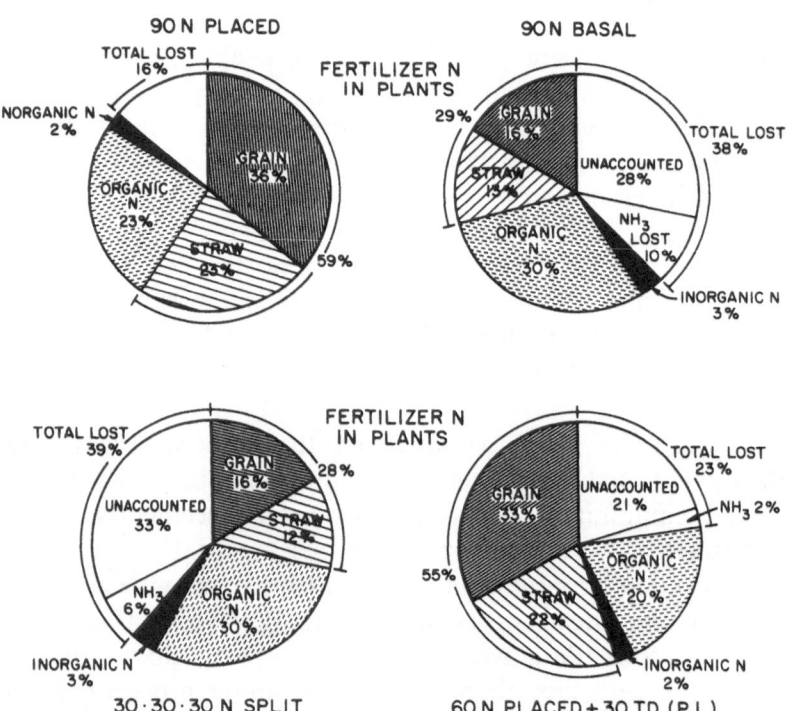

Fig. 6. Fate of fertilizer nitrogen in a direct-seeded rice ecosystem (California).

correlated with yield, although excellent correlations are observed with total N in the harvested crop and plant analysis. In general, however, high plant recovery of fertilizer N is correlated with low losses. It is evident that the application of fertilizer N has enhanced total uptake of N by the crop. In most cases where fertilizer has been applied, uptake of soil N has increased from 20–30%, in part due to the 'priming effect,' biological interchange, more vigorous plants and enhanced root development.

Soil retention of applied ^{15}N fertilizer has ranged from 22 to 57% with no consistent relationship to method or timing of N application. Banded fertilizer N usually provides the highest rate of fertilizer retention followed by broadcast-incorporated N and the least retention in surface or water broadcast applications. Labeled fertilizer N is found ultimately in the soil organic matter fraction representing from 19 to 34% of the total N applied. In typical clay soils used for rice production, possessing impervious clay layers, 90–96% of the residual nitrogen is retained in the 0–30 cm depth with the major portion in the 0–15 cm depth. These data indicate that relatively little movement of fertilizer N occurs in these typical soils with 25 to 40% clay. Clay fixation of NH_4^+-N accounts for 3 to 9% of fertilizer N retained in the soil. With cropping histories of 30–50 years continous rice, receiving an average of 100 kg N/ha annually and soils rich in K minerals, non-exchangeable NH_4^+ does not constitute a large portion of retained fertilizer N.

Total recovery of fertilizer N from crop and soil ranges between 51 and 93%, the highest values being recorded from large container experiments. Recovery of N from a urea source is usually slightly less than from ammonium sulfate, each applied in the same manner. With equal amounts of total N applied, ammonium sulfate gives an increased grain yield of about 5–10% more than urea. Losses of urea-N from the flooded soil system is usually significantly greater than for applied ammonium sulfate. Field experiments record that method of irrigation, flow rates and factors determining direction of water flow greatly influences the distribution of urea-N in the soil flow layer. Banded ammonium sulfate of aqua-NH_3 distribution are much less affected by the water transport system.

Ammonia volatilization measurements indicate that deep placement of ammoniacal-N, 10 cm deep in submerged soils, will reduce NH_3 volatilization to very low levels. One to 2% NH_3 volatilization losses are typically observed from deep placed applications, measured by micrometeorological and forced-air trapping methods. In contrast, fertilizer N applied broadcast on soil or into floodwater have shown volatilization losses as high as 35% of the total N applied. Urea-N consistently

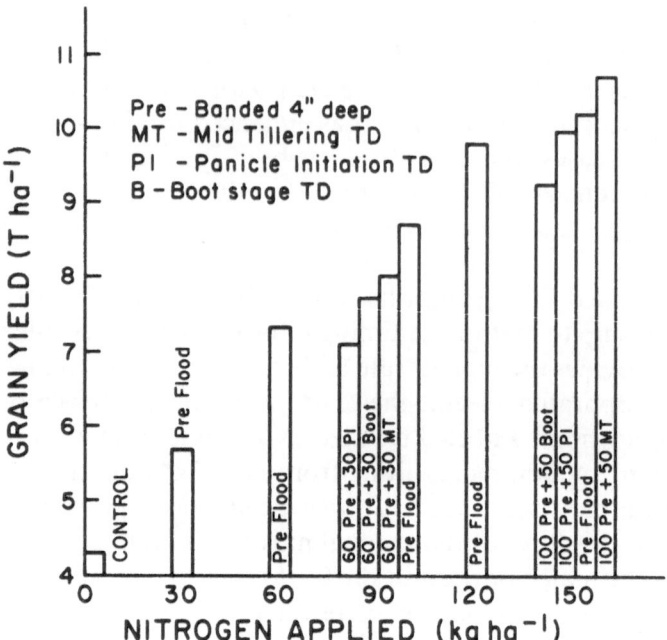

Fig. 7. Response of direct-seeded rice to rates and methods of nitrogen fertilization.

maintains higher volatilization losses applied broadcast than does ammonium sulfate.

Crop yields as affected by method of N application in 17 fertilizer experiments conducted under a wide range of soil conditions with water-sown rice show a highly significant difference with method of N application (Fig. 7). Deep placement of N gives statistically significant increases in grain yield over split applications, 2/3 N applied banded by pre-flood placement and 1/3 N applied at mid-tillering, at panicle initiation or in the boot stage of development. Split N applications applied at mid-tillering have been superior or equal to applications top-dressed at the panicle initiation stage.

Losses of plant N have been observed progressively from the time of panicle exsertion in rice until harvest. The seasonal loss from plant tops is sometimes of the order of 5% of the total plant N uptake. Further investigations are needed to more adequately quantify the magnitude of this loss and to characterize the conditions under which they occur (Fig. 5).

Nitrogen use efficiency

Efficient fertilizer use is one benefit desired from the development of N budgets in rice. It is apparent that ^{15}N-crop recovery values reveal only part of the effect of fertilizer N on crop yields. Valuable comparisons are provided on the benefits and losses likely to occur with specific N transformation processes, but they are very site specific, have value only within the context of specific soil-water-cropping systems, and do not provide useful indices of fertilizer benefits to increase grain production.

The effectiveness of fertilizer N treatments can be assayed by measuring the ratio of grain yield to N application rate. This is called the 'agronomic efficiency (AE)' of the fertilizer and is best measured as the increase in grain yield due to a given treatment. AE is defined as:

$$\text{AE} = \frac{\text{kg grain}}{\text{kg N applied}} = \left(\frac{\text{kg grain}}{\text{kg N absorbed}}\right)\left(\frac{\text{kg N absorbed}}{\text{kg N applied}}\right)$$

The ratio of grain yield to N absorbed is termed its 'physiological efficiency,' while the ratio of N absorbed to N applied is called the 'recovery fraction (RF).' This 'recovery fraction' is defined as:

$$\text{RF} = \left(\frac{\text{kg N absorbed}}{\text{kg N supply}}\right)\left(\frac{\text{kg N supply}}{\text{kg N applied}}\right)$$

It can be seen that RF is a measure of the effectiveness of a given treatment in increasing plant N uptake. The ratio of N absorbed to the N supply is termed 'uptake efficiency' and is a function of the properties of the root-surface area, distribution, and uptake per unit area. These factors are dependent upon soil physical and chemical status, and health and growth characteristics of the plant.

Moore et al.[31] applied this concept in describing the efficiency of fertilizer use in flooded rice. Available (soluble and exchangeable) N was measured with time in fertilized and unfertilized plots with concommitant measurements of N uptake by plants. They found that if 'the difference in plant N between fertilized plants and controls was plotted as a function of the parallel differences in soil N' for the first four weeks after application, a high correlation ($r^2 = 0.96$) resulted, with a slope of -0.55. This can be interpreted as a rate constant for the uptake of effective fertilizer N per unit change (decrease) of supply, or uptake efficiency. Calculation by the standard difference method gave a value of 63 percent for fertilizer N effectiveness. The difference between these two methods is the (kg N supply/kg N applied) term in the above equation for RF (actually fertilizer effectiveness). Values for AE will vary widely with soil fertility, rates of N applied, crop variety, and a wide range of management factors affecting yield. Values calculated from reported data fall in the range of 32.3–108 for field trials with much higher values calculated from pot trials[6]. Factors affecting AE generally parallel those affecting N uptake, yield, and fertilizer uptake efficiency. The shortcoming of this value is that it includes the grain yield due to soil N uptake which is undefined at a zero rate of application. For this reason it would perhaps be more useful to use the yield increase due to a given application rate as the numerator in the equation. This point is underscored by the findings of Reddy and Patrick[44] that show the uptake of fertilizer N to be more highly correlated with relative yield increase than with absolute yield.

The physiological efficiency (kg grain/kg N absorbed) portion of the AE equation is a function of the distribution of N in the plant at harvest and the percent of N in the grain. Murayama[32] reports that rice crops take up from 19–21 kg of N to produce a ton of brown rice and that this value is 'nearly constant.' This implies a physiological efficiency of about 50. Koyama et al.[23] call this value the productive efficiency and measured it for different times of top-dressing, finding that it varied with time of top-dressing and variety. Yoshida[57] suggests that a combination of parameters be used in determining efficiency of fertilizer use namely,

$$\begin{array}{l} \text{Efficiency of fertilizer N} \\ \text{(kg rice/kg applied N)} \end{array} = \begin{array}{l} \text{Percentage N recovered} \\ \text{(kg absorbed N/kg applied N)} \end{array} \times$$

$$\begin{array}{l} \text{Efficiency of utilization} \\ \text{(kg rice} \times \text{ kg absorbed N)} \end{array}$$

using values for recovery percentage and utilization efficiency for tropical transplanted rice. Yoshida[57] reports values of fertilizer use efficiency of 15 to 25 kg rice/kg applied N[57]. Nitrogen budgets have provided a large amount of research information about the N transformation processes in submerged soils. There remain many unanswered questions, however, that must be studied to fully integrate fertilizer management to the highest level of efficiency in the complete environment of wetland rice. Increased understanding of N budgets will lead to better fertilizer management practices, increased N use efficiency and improved socio-economic status of farmers.

Acknowledgement   Appreciation is expressed to G. P. Deo, A Abshahi, and A. Hafez in the review of literature and for technical assistance.

References

1   Bajwa M I 1982 Soil-clay mineralogies in relation to fertility management. Effect of clay mineral types on ammonium fixation under wetland rice culture. Agron. J. 74, 143–144.
2   Bouldin D R and Alimagno B V 1976 NH$_3$ volatilization from IRRI paddies following applications of fertilizer nitrogen. Unpublished report. 51 p. IRRI.

3 Bremner J M and Hauck R D 1982 Advances in methodology for research on nitrogen transformations in soils. *In* Nitrogen in Agricultural Soils Agron. No. 22, pp 467–502. American Society of Agronomy.

4 Broadbent F E 1953 The soil organic fraction. *In*. Adv. Agron. 5, 153–183. Academic Press New York.

5 Broadbent F E 1979 Mineralization of organic nitrogen in paddy soils. *In* Nitrogen and Rice. pp 105–118. Int. Rice Res. Inst. Los Banos, Philippines.

6 Broadbent F E and Mikkelsen D S 1968 Influence of placement on uptake of tagged nitrogen by rice. Agron. J. 60, 674–677.

7 Broadbent F E and Nakashima T 1970 Nitrogen immobilization in flooded soils. Soil Sci. Soc. Am. Proc. 34, 218–221.

8 Chou C H, Chiang Y C, Cheng H H and Farrow F D 1982 Transformations of ^{15}N-enriched fertilizer nitrogen during rice straw decomposition in submerged soil. Bot. Bull. Acad. Sinica 23(2), 119–133.

9 DeDatta S K 1978 Fertilizer management for efficient use in wetland rice soils. *In* Soils and Rice. pp. 671–701. Int. Rice Res. Inst. Los Banos, Philippines.

10 DeDatta S K and Patrick W H Jr. (Ed) 1986 Nitrogen Economy of Flooded Rice Soils Developments in Plant and Soil Sciences. Martinus Nyhoff Publishers, Dordrecht, The Netherlands.

11 Delaune R D and Patrick W H Jr. 1970 Urea conversion to ammonia in waterlogged soils. Soil Sci. Soc. Am. Proc. 34, 603–607.

12 Fillery I R P and DeDatta S K 1986 Ammonia volatilization from nitrogen sources applied to rice fields. I. Methodology, ammonia fluxes and nitrogen 15 loss. Soil Sci. Soc. Am. Proc. 50, 80–85.

13 Fillery I R P, Simpson J R and DeDatta S K 1984 Influence of field environment and fertilizer management on ammonia loss from flooded rice. Soil Sci. Soc. Am. J. 48, 914–920.

14 Foster E F and Stuttle C A 1986 Glutamine synthetase activity and foliar nitrogen volatilization in response to temperature and inhibitor chemicals. Annals Bot. 57, 305–308.

15 Freney J R, Denmead O T, Watanabe I and Craswell E T 1981 Ammonia and nitrous oxide losses following applications of ammonium sulfate to flooded rice. Aust. J. Agric. Res. 32, 37–45.

16 Gardner W R 1965 Movement of nitrogen in soil. *In* Soil Nitrogen. Agron Mono 10, 550–572. American Society of Agronomy.

17 Gupta S P 1955 Loss of nitrogen in the form of ammonia from waterlogged paddy soil. J. Indian Soc. Soil Sci. 3, 29–32.

18 Hauck R D 1981 Nitrogen fertilizer effects on nitrogen cycle. *In* Terrestrial Nitrogen Cycles Ecol bull (Stockholm), 33, 551–562.

19 Hauck R D and Bremner J M 1976 Use of tracers for soil nitrogen research. *In* Adv. Agron. 28, 219–266. Academic Press New York.

20 Ito O and Watanabe I 1981 Immobilization, mineralization and availability to rice plants of nitrogen derived from heterotropic nitrogen fixation in flooded soil. Soil Sci. Plant Nutr. 27, 169–176.

21 Jansson S L and Persson J 1982 Mineralization and immobilization of soil nitrogen. *In* Nitrogen and Agricultural Soils. Agron Mono 22, 229–252. American Society of Agronomy.

22 Keerthisinghe G, Mengel and DeDatta S K 1984 The release of nonexchangeable ammonium (^{15}N labeled) in wetland rice soil. Soil Sci. Soc. Am. J. 48, 291–294.

23 Koyama T and App A 1979 Nitrogen balance in flooded soils. *In* Nitrogen and Rice. pp 95–104. Int. Rice Res. Inst. Los Banos, Philippines.

24 Lamm C G and Nafady M H 1973 Plant nutrient availability in soils. III. Studies on potassium in Danish soils. Agrochimica 17, 435–444.

25 Legg J O and Meisinger J J 1982 Soil nitrogen budgets. *In* Nitrogen in Agricultural Soils. Agron No. 22, 503–566. American Society of Agronomy.

26 MacRae I C and Ancajas R 1970 Volatilization of ammonia from submerged tropical soils. Plant and Soil 33, 97–103.

27   Mikkelsen D S and DeDatta S K 1979 Ammonia volatilization from wetland rice soils. *In* Nitrogen and Rice. pp 135–156. Int. Rice Res. Inst., Los Banos, Philippines.

28   Mikkelsen D S, DeDatta S K and Obcemea 1978 Ammonia volatilization losses from flooded rice soils. Soil Sci. Soc. Am. J. 42, 725–730.

29   Mikkelsen D S and Finfrock D C 1957 Availability of ammoniacal nitrogen to lowland rice as influenced by fertilizer placement. Agron. J. 49, 296–300.

30   Mitsui S 1954 Inorganic Nutrition, Fertilization and Soil Amelioration for Lowland Rice. Yokendo Ltd. Tokyo 107 p.

31   Moore P A, Gilmour and Wells B R 1981 Seasonal patterns of growth and soil nitrogen uptake by rice. Soil Sci. Soc. Am. J. 45, 875–879.

32   Murayama N 1979 The importance of nitrogen for rice production. *In* Nitrogen and Rice, pp 5–23. Int. Rice Res. Inst. Los Banos, Philippines.

33   Nommik H and Vahtras K 1982 Retention and fixation of ammonium and ammonia in soils. *In* Nitrogen in Agricultural Soils. Agron No. 22, 123–171. American Society of Agronomy.

34   Obcemea W N, Mikkelsen D S and DeDatta S K 1977 Factors affecting ammonia volatilization losses from the flooded environment of rice. IRRI Saturday Seminar Mimeo. 45 p. Los Banos, Philippines.

35   Okuda, Takahasi E and Yoshida M 1960 On the volatilization of the ammonia transferred from urea applied under upland and waterlogged conditions. J. Sci. Soil Manure Japan, 31, 273–278.

36   Olson R A and Kurtz L T 1982 Crop nitrogen requirements, utilization and fertilization. *In* Nitrogen in Agricultural Soils. Agron Mono 22, pp 567–604. American Society of Agronomy.

37   Patrick W H Jr. 1982 Nitrogen transformations in submerged soils. *In* Nitrogen in Agricultural Soils. Agron Mono 22, pp 449–465. American Soceity of Agronomy.

38   Patrick W H Jr. and Mahapatra I C 1968 Transformations and availability to rice of nitrogen and phosphorus in waterlogged soils. *In* Adv. Agron. 20, 323–359. Academic Press New York.

39   Patrick W H Jr., Mikkelsen D S and Wells B R 1985 Plant nutrient behavior in flooded soil. *In* Fertilizer Technology and Use. 3rd Ed., pp 197–228. American Society of Agronomy.

40   Ponnamperuma F N 1972 The chemistry of submerged soils. *In* Adv. Agron. 24, 29–96. Academic Press New York.

41   Ponnamperuma F N 1977 Physiological properties of submerged soils in relation to fertility. Int. Rice Res. Inst. Paper Ser 5, 1–32.

42   Reddy K R 1982 Nitrogen cycling in a flooded-soil ecosystem planted to rice (*Oryza sativa*). Plant and Soil 67, 209–220.

43   Reddy K R and Patrick W H Jr. 1976 Yield and nitrogen utilization by rice as affected by method and time of application of labeled nitrogen. Agron. J. 68, 965–969.

44   Reddy K R and Patrick W H Jr. 1980 Uptake of fertilizer nitrogen and soil nitrogen by rice using ^{15}N-labeled nitrogen fertilizer. Plant and Soil 57, 375–381.

45   Reddy K R and Patrick W H Jr. 1986 Fate of fertilizer nitrogen in the rice root zone. Soil Sci. Soc. Am. J. 50, 649–651.

46   Sah R N and Mikkelsen D S 1983 Availability and utilization of fertilizer nitrogen by rice under alternate flooding. I. Kinetics of available nitrogen under rice culture. II. Effect on growth and nitrogen use efficiency. Plant and Soil 75, 221–234.

47   Savant N K and DeDatta S K 1982 Nitrogen transformations in wetland rice soils. *In* Adv. Agron. 35, 241–302. Academic Press New York.

48   Stevenson F J (Ed.) 1982 Nitrogen in Agricultural Soils. Agron No. 22, 1–940. American Society of Agronomy.

49   Stumpe J M, Vlek P L G and Lindsay 1984 Ammonia volatilization from urea and urea phosphates in calcareous soils. Soil Sci. Soc. Am. J. 48, 921–926.

50   Tanaka A and Navasero S A 1964 Loss of nitrogen from the rice plant through rain or dew. Soil Sci. Plant Nutr. 10, 36–39.

51   Tusneem M E and Patrick W H Jr. 1971 Nitrogen transformations in waterlogged soils. LSU Bull 657, 1–75.

52   Ventura W B and Yoshida T 1977 Ammonia volatilization from a flooded tropical soil. Plant and Soil 46, 521–531.

53   Vlek P L G and Craswell E T 1979 Effect of N source and management of ammonia volatilization losses from flooded rice-soil systems. Soil Sci. Soc. Am. J. 43, 352–358.

54   Wetselaar R and Farquhar G D 1980 Nitrogen losses from tops of plants. Adv. Agron. 33, 263–302 Academic Press New York.

55   Wetselaar R T, Shaw P, Firth P, Oupathum J and Thitepoca H 1977 Ammonia volatilization losses from variously placed ammonium sulfate under lowland rice field conditions in Central Thailand. pp 282–288. Proc. Int. Sem. Soil Env. Fert. Mgmt. Tokyo.

56   Young J L and Aldag R W 1982 Inorganic forms of nitrogen in soils. In Nitrogen in Agricultural Soils. Agron No. 22, 43–66. American Society of Agronomy.

57   Yoshida S 1981 Mineral nutrition of rice. Int. Rice. Res. Inst. pp 1–269. Los Banos, Philippines.

Plant and Soil 100, 99–112 (1987).
Ms. 100-08

The effects of low, regulated supplies of nitrate and ammonium nitrogen on the growth and composition of perennial ryegrass

S.C. JARVIS

The Animal and Grassland Research Institute, Hurley, Maidenhead, Berkshire SL6 5LR, UK

Key words Ammonium Exponential demand Flowing solution culture *Lolium perenne* L. Nitrate Ryegrass

Summary Perennial ryegrass was grown in flowing solution culture with nitrogen supplied in amounts that increased exponentially, *i.e.* in parallel with the rate of increase in growth. Nitrogen was supplied as either NO_3^- or NH_4^+, and the amounts to be added were calculated on the basis of extrapolated values for dry weights obtained from fitted curves. There were two rates of addition for each form of N aimed at providing adequate (5.0 per cent) and less than adequate (2.75 per cent) contents in the plants in each case. Measured plant weights and N concentrations were in close agreement with predicted values over a four week experimental period. There was no effect of N-form at high N, and these plants produced 46 per cent more dry matter than the plants at low N. Only minor differences in overall growth occurred with NO_3^- or NH_4^+ plants at low N, but the NH_4^+ plants had a greater shoot:root ratio. The absorption rate (m mol N g^- root d^{-1}) for NH_4^+-N was therefore greater than for NO_3^--N. The cation/anion composition of the plants was affected in a predictable way, and to a greater or lesser extent at high or low N, respectively, in NO_3^- or NH_4^+ plants. The major changes in cation composition came through effects on potassium absorption. Plants with low NO_3^- appeared to be under greater N stress than those with low NH_4^+ because of the lower shoot:root ratio and the greater C:N ratio in the shoots.

Introduction

Major differences in the composition of plants result from the changes in the flux of ions across roots which occur when either the rate or the form of nitrogen (N) supply is altered[13,14,17]. Changes in composition result from (i) the direct effects of uptake as either the cation, NH_4^+, or the anion, NO_3^-, and (ii) the cellular regulatory processes which follow the assimilation of these ions[17]. These changes may not only have important consequences for plant growth and productivity[5,16] but, in the case of forage species they may also affect the quality of fresh, and conserved, material to be consumed by ruminant animals.

Detailed studies of the effects of the form of N can most usefully be carried out in solution culture where control can be exerted over both the form and the amount of N presented to the root. Often such studies are carried out in static solution, but, because of the relatively high initial concentrations that are used, imbalances in the plant may result even to the extent that toxicity and reduced growth result[1,9]. However, systems of flowing solution culture have also been used which permit the concentrations of ions at the root/solution interface to be held constant and at realistically low levels[3,4,22].

It has also recently been suggested that traditional experimental techniques with treatments involving ranges of concentration have produced serious artifacts[11]. It is argued that rather than being determined by a fixed concentration at the root surface, uptake is determined by demand[2]. Plant requirements should therefore be met by the addition of nutrients in amounts that increase exponentially, *i.e.* in parallel with the rate of increase in growth[11]. Plants grown in this way will not contain unrealistic concentrations and imbalances of nutrients. Detailed studies which relate the rate of supply of nutrients to growth have been largely restricted to slow growing dicotyledenous species[10,11,12] and little attention has been paid to the relevance of the technique for faster growing graminaceous plants.

In the present studies, therefore, perennial ryegrass (*Lolium perenne* L.) was grown in flowing solution culture with N supplied as NO_3^- or NH_4^+ at two rates calculated on the basis of regular measurements of growth to provide (i) less than adequate and (ii) luxury concentrations in the plant. Subsequent effects on growth and composition were examined over a 28-day period.

Methods

Plant culture

Perennial ryegrass (*Lolium perenne* L. cv. Melle) was sown during May in four units of the system of flowing solution culture described by Clement et al.[3]. Five plants were grown in each of the six growth tubes in the 24 vessels in each unit. The units, which hold 300 l solution, were housed in a glasshouse in which day/night aerial temperatures were maintained at 25°C/15°C: solution temperature was maintained at 18C°. The initial concentration of nutrients was as described previously[15], but with N (as NO_3^-) and K at 0.1 mg l^{-1} (*i.e.* 7.14 and 2.56 μM, respectively). The pH of the solution was adjusted to, and automatically[8] maintained at 5.75 ($\pm$ 0.06). During the pretreatment period additional N and K were added to each unit on the basis of increases in dry weight of harvested plants; in total 21 mg of NO_3^--N and of K were supplied to each vessel in each unit during the first 40 days after sowing.

Plant harvest

Plants from a single vessel in each unit were harvested on days 21, 28, 35 and 40 after sowing. During the following treatment period, five culture vessels from each unit were harvested on days 47, 54, 61 and 68. At harvest, the plants were divided into shoots and roots and those from the pretreatment period were oven dried at 80°C to constant weight; one replicate of the treated plants was oven dried for rapid dry weight determinations and the remainder were stored at −20°C and then freeze dried for analysis.

Treatments

Four treatments were imposed from day 40, *i.e.* high and low N as ammonium (NH_4^+) or nitrate (NO_3^-) to provide plants that contained less than adequate (*i.e. c.* 1.96 mmol g^{-1}; 2.75%) or adequate (*i.e. c.* 3.57 mmol g^{-1}; 5.0%) concentrations of N in their dry matter from both forms of N. Mean total dry weights (shoots plus roots) per vessel on day 40 and on previous harvests were fitted to an exponential curve. A value was then extrapolated from the curve for dry weight after a further seven days. From this extrapolated value it was possible to calculate the amounts of N to be supplied over the next seven days in order to maintain the desired concentration in the plants. New growth curves were then fitted at each subsequent harvest (using data from that and all previous harvests) in order to calculate new rates of N supply for the next seven-day period. On each occasion, data for plants in the units at high N, and those from the two units at low N were each combined to provide two fitted curves.

The calculated amounts of N required varied for each seven-day period and were supplied continuously in solution as either (NH$_4$)$_2$SO$_4$ or Ca(NO$_3$)$_2$ with peristaltic pumps. Over any given period during a particular seven-day harvest interval, therefore, a constant amount of N was added to a particular culture unit.

The pH of the culture solution was maintained automatically at 6.00 ($\pm$ 0.06) from day 40 and the concentration of K was held at 10 mg l^{-1} (256 μM) by analysis and addition on three occasions per week. All other nutrient ions (except those of N) were added at the same time and at an appropriate ratio to K for healthy growth.

Analysis

Plant materials from a single culture vessel from each unit on day 40, and from two vessels from each unit at each subsequent harvest were finely ground and analysed for the following constituents: (1) Total N, by an automated micro-Dumas method (Carlo Erba Instruments 1400 N analyser). (2) Water soluble NO$_3^-$, by shaking 0.2 g dried plant material for 30 min in 20 ml deionised water, filtering and determining NO$_3^-$ in the filtrate with a specific ion electrode. (3) Water soluble K, by flame photometry of the filtrate used for the determination of water soluble NO$_3^-$-N. (4) Ca, Mg and P were measured after digestion of samples of plant material in a 5:1 mixture of 16M HNO$_3$ and 11.6M HClO$_4$. After making up to volume, Ca (in the presence of 10,000 μg ml^{-1} La) and Mg were determined by atomic absorption spectrophotometry using an air/acetylene flame and at wavelengths 422.6 and 285.2 nm, respectively. After appropriate dilution, P was determined with an automated vanadate/molydate colorimetric method. (5) S was determined by a turbidimetric auto-analyser method after plant material had been digested in 16 M HNO$_3$ (total-S), or extracted with a mixture of CH$_3$COOH, H$_3$PO$_4$ and HCl (SO$_4^{2-}$-S)[19]. (6) Ash alkalinity was measured after heating plant material slowly to 400°C in a muffle furnace and then at 500°C for 1 h. The resultant ash was treated with 1M HCl and this was then back-titrated against 0.5M NaOH using an autotitrator. (7) Total C was determined with an induction furnace (Leco Corporation HF 10 furnace).

Results

Overall growth

The growth of the whole plants during the treatment phase closely followed an exponential pattern (Fig. 1). On all occasions that data were transformed and curves were constructed, more than 98.4 per cent of the variation with time in the measured total dry weight could be accounted for by fitting to an exponential equation. Furthermore, there was good agreement between the estimated dry weight obtained by extrapolation and that measured at harvest. Shoot and root weights, when considered separately were also well described by exponential equations (on average, r = 0.996 and 0.992, respectively, for the transformed data).

When all the harvest data are considered, the equations for the plants with low N were similar with both forms of N. Thus over the 28 day

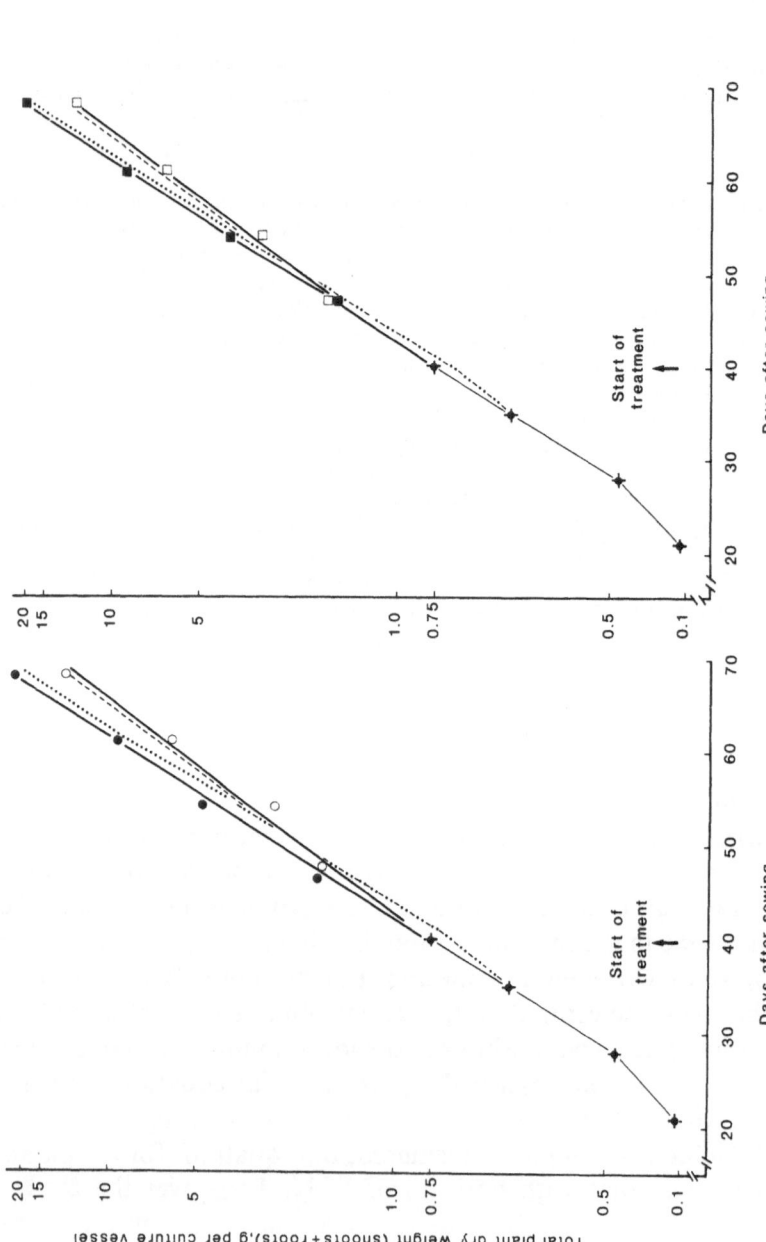

Fig. 1. Total dry weight (shoot plus root), (g per vessel) of perennial ryegrass during pretreatment (✦) and grown with low NO_3^- (○), high NO_3^- (●), low NH_4^+ (□) or high NH_4^+ (■): Solid lines are fitted for all measured weights after day 40, and broken lines are plotted for the estimated weights at the end of each seven-day period and as obtained by extrapolation from fitted exponential curves at each successive harvest with low N (– – –) or high N (· · · ·).

treatment period:

$$y = 0.0137e^{0.1010x} \quad (r = 0.992 \text{ for transformed data})$$

and $y = 0.0139e^{0.1009x}$ ($r = 0.999$ for transformed data) for NO$_3^-$ and NH$_4^+$, respectively, where y = dry weight (g) and x = time (d): a single fitted line is shown for these equations in Fig. 1.

At high N the values of the intercepts of the fitted lines were lower and differed slightly for the plants supplied with NO$_3^-$ or NH$_4^+$; there was no effect of N form on the slope. The equations for high NO$_3^-$ and high NH$_4^+$ were, respectively:

$$y = 0.0070e^{0.1188x} \quad (r = 0.999 \text{ for transformed data})$$

and $y = 0.0064e^{0.1187x}$ ($r = 0.999$ for transformed data).

Shoot:root dry weight ratio

There was no effect of the form of N on shoot growth at either high or low N and dry weight increased from, on average, 0.45 to 15.99 and 8.63 g per vessel, respectively, i.e. 46 per cent less growth at low N. In all treatments there was an overall trend for shoot:root ratio to increase with time over the treatment period (Fig. 2). With high N, the form of N had no consistent effect on the distribution of dry matter between shoots and roots. However, with low N there was a developing trend for

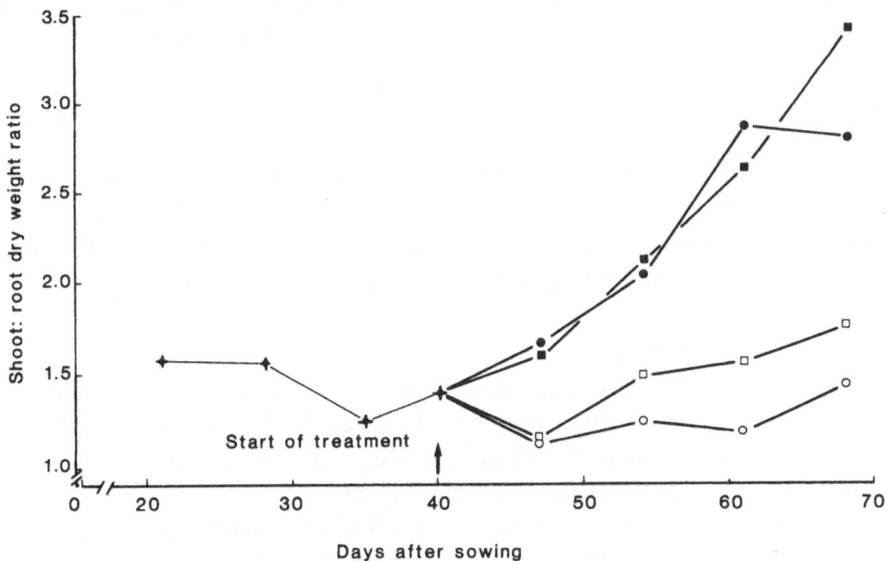

Fig. 2. Ratio of shoot:root dry weights (g per culture vessel) in perennial ryegrass during pre-treatment (+), and grown with low NO$_3^-$ (○), high NO$_3^-$ (●), low NH$_4^+$ (□) or high NH$_4^+$ (■).

plants with NH_4^+ to have a lower yield of roots and hence a larger (*i.e.* by 25 per cent over the final three harvests) shoot:root ratio than plants with NO_3^-.

Nitrogen absorption and distribution

The calculated rate of supply of N needed to sustain the target concentrations was 2.5 greater in the high N than in the low N treatments. On average, 83, 92, 95 and 94 per cent, respectively, of the N supplied in the low NO_3^-, low NH_4^+, high NO_3^- and high NH_4^+ treatments could be accounted for in the harvested plants (Table 1). The proportion of the added N that was absorbed by the low NO_3^- plants was generally lower than in the other plants.

Rates of N absorption were calculated using the formula adapted from that used to calculate the net assimilation rate of plants[21]; these represent net influx rates. Thus, absorption ($m\,mol\,N\,g^{-1}\,root\,d^{-1}$)

$$= \frac{(\log_e W_2 - \log_e W_1)(N_2 - N_1)}{(t_2 - t_1)(W_2 - W_1)}$$

where W_2 and W_1 are fresh or dry root weights (g) and N_2 and N_1 are total plant N contents (m mol per vessel) at times t_2 and t_1 (d). Expression on a fresh or dry weight basis had no effect on the relative differences in absorption (Table 1). In all treatments there was a general trend for the rates of absorption to increase with time; the rates were greater at high N than at low N by factors of 3.4 and 2.7 with NO_3^- and NH_4^+ treatments, respectively. The absorption rate was greater with NH_4^+ than with NO_3^- plants by, on average, factors of 1.5 and 1.1 with low and high N, respectively.

There was no effect of the form of N on distribution between shoots and roots, but the proportions in the shoots were slightly greater with high than with low N; on average, 69 and 76 per cent, of the N absorbed was in the shoots of high and low N plants, respectively).

Nitrogen concentrations

The average total N concentrations in the whole plants (Tables 2 and 3) were, over the final three harvests, 1.72 (2.41), 1.97 (2.76), 2.88 (4.03) and 3.10 (4.33) $m\,mol\,g^{-1}$ (%) at low NO_3^-, low NH_4^+, high NO_3^- and high NH_4^+, respectively; these values represent 88, 100, 81 and 87 per cent of the target concentrations. In all the treatments, there was little overall change in concentration in the shoots with time from day 54 and the mean values were 2.10, 2.39, 3.15 and 3.25 $m\,mol\,g^{-1}$ for low NO_3^-,

Table 1. Effects of high and low nitrate or ammonium N supply on N absorption and distribution by perennial ryegrass

Treatment	Plant age, days	Total N supplied, mmol vessel⁻¹		Total N content, mmol vessel⁻¹		Rate of absorption, mmol g⁻¹ root d⁻¹		Proportion (%) of N supplied present in plants	Proportion (%) of total plant content in shoots
		During preceeding period	Cumulative	Shoots	Roots	(fresh weight basis)	(dry weight basis)		
—	40	1.53	1.53	1.02	0.40	—	—	93	72
Low NO₃⁻	47	1.14	2.66	1.66	0.90	0.014	0.295	96	65
	54	4.02	6.68	3.22	1.64	0.026	0.322	73	66
	61	6.71	13.39	7.34	3.14	0.036	0.266	78	70
	68	13.64	27.02	16.23	6.56	0.033	0.422	84	71
Low NH₄⁺	47	1.14	2.66	1.62	0.87	0.026	0.289	94	65
	54	4.02	6.68	4.25	1.64	0.043	0.483	88	67
	61	6.71	13.39	8.63	3.09	0.042	0.468	87	74
	68	13.64	27.02	21.81	7.28	0.057	0.697	100	75
High NO₃⁻	47	3.26	4.78	2.69	1.27	0.065	0.750	84	68
	54	7.04	11.83	9.52	3.01	0.096	1.135	106	76
	61	19.36	31.89	22.84	5.78	0.100	1.160	92	80
	68	36.89	68.07	51.71	13.39	0.103	1.352	96	79
High NH₄⁺	47	3.26	4.78	3.22	1.20	0.076	0.940	92	73
	54	7.04	11.83	8.57	2.84	0.086	1.109	97	75
	61	19.36	31.19	20.27	7.26	0.115	1.284	87	74
	68	36.89	68.07	54.18	14.04	0.141	1.671	100	79

Table 2. Effects of high and low nitrate or ammonium supply on elemental composition of perennial ryegrass shoots (mmol g⁻¹)

Treatment	Plant age, days	Element (mmol g⁻¹)									
		C	N Total	NO$_3^-$	S Total	SO$_4^{2-}$**	P	K	Ca	Mg	COO$^-$
—	40	—	2.36	—	0.18	—	0.21	0.76	0.31	0.14	—
Low NO$_3^-$	47	32.4	1.70	0.01	—	—	0.19	1.06	0.25	0.10	1.33
	54	34.7	2.21	0.04	0.17	—	0.19	1.10	0.27	0.09	1.39
	61	33.7	2.18	0.03	0.14	0.11	0.18	1.16	0.22	0.07	1.47
	68	33.4	1.91	0.02	0.13	0.10	0.16	1.16	0.21	0.07	1.37
Low NH$_4^+$	47	34.9	1.76	0.01	—	—	0.21	0.98	0.24	0.10	1.09
	54	33.9	2.34	0.01	0.26	—	0.19	0.91	0.22	0.08	1.02
	61	33.9	2.34	0.01	0.20	0.10	0.18	0.94	0.19	0.06	0.85
	68	33.4	2.49	0.01	0.17	0.06	0.16	1.00	0.17	0.05	1.03
High NO$_3^-$	47	33.7	2.98	0.07	—	—	0.22	1.26	0.30	0.12	1.77
	54	32.7	2.96	0.06	0.15	0.10	0.17	1.11	0.26	0.09	1.60
	61	32.7	3.24	0.08	0.12	0.07	0.16	1.29	0.25	0.09	1.62
	68	32.7	3.24	0.08	0.13	0.09	0.18	1.35	0.24	0.08	1.66
High NH$_4^+$	47	—	3.21	0.01	—	—	0.23	1.02	0.22	0.09	0.91
	54	34.4	3.27	0.01	0.23	—	0.20	0.94	0.20	0.07	1.15
	61	34.9	3.10	0.01	0.24	—	0.19	1.00	0.17	0.06	0.98
	68	34.8	3.38	0.01	0.23	0.19	0.19	0.98	0.19	0.05	0.99
S.E. (df)		0.52 (15)	0.170 (16)	0.006 (8)**	0.005 (12)		0.008 (16)	0.005 (16)	0.016 (16)	0.008 (16)	0.071 (16)

*single sample only
**excludes data for low and high NH$_4^+$

Table 3. Effects of high and low nitrate or ammonium supply on elemental composition of perennial ryegrass roots (m mol g^{-1})

Treatment	Plant age, days	Element (m mol g^{-1})									
		C	N		S		P	K	Ca	Mg	COO$^-$
			Total	NO$_3^-$	Total	SO$_4^{2-}$*					
—	40*	—	1.26	—	—	—	0.22	0.67	0.07	0.15	—
Low NO$_3^-$	47	36.3	1.02	0.01	—	—	0.19	1.04	0.05	0.06	1.09
	54	34.0	1.41	0.03	0.17	0.14	0.25	0.98	0.07	0.05	1.10
	61	33.7	1.11	0.02	0.12	0.09	0.22	1.04	0.05	0.03	1.24
	68	32.7	1.11	0.01	0.13	0.09	0.21	1.37	0.06	0.03	1.13
Low NH$_4^+$	47	35.7	1.06	0.01	—	—	0.21	0.96	0.04	0.06	0.95
	54	32.5	1.76	0.01	0.16	0.11	0.27	0.77	0.04	0.03	0.95
	61	32.5	1.26	0.01	0.17	0.09	0.26	0.77	0.04	0.03	0.66
	68	36.7	1.59	0.01	0.18	0.11	0.25	0.98	0.05	0.02	0.84
High NO$_3^-$	47	32.5	1.81	0.07	—	—	0.19	1.10	0.04	0.07	1.38
	54	33.1	1.92	0.08	0.15	0.07	0.18	1.06	0.05	0.04	1.34
	61	32.4	2.35	0.10	0.12	0.09	0.18	1.10	0.05	0.04	1.10
	68	32.7	2.35	0.11	0.09	0.09	0.23	1.29	0.05	0.03	1.14
High NH$_4^+$	47	32.2	1.91	0.01	—	—	0.25	1.00	0.04	0.04	0.61
	54	33.1	2.29	0.01	0.17	0.13	0.26	0.89	0.05	0.02	1.07
	61	33.4	2.91	0.01	0.13	0.11	0.24	0.73	0.04	0.02	0.78
	68	33.2	2.99	0.01	0.15	0.10	0.24	0.75	0.05	0.01	0.77
S.E. (df)		0.61	0.126	0.005	0.008	—	0.009	0.005	0.004	0.003	0.091
		(16)	(16)	(8)**	(12)		(16)	(16)	(16)	(16)	(16)

*single sample only

**excludes data for low and high NH$_4^+$

low NH_4^+, high NO_3^- and high NH_4^+, respectively (Table 2). In the roots, concentrations (Table 3) were lower but followed similar trends to those in the associated shoots. Plants grown with NH_4^+ at both rates of supply had higher concentrations of total N than those with NO_3^- in both plant parts. Only minor proportions of either the shoot or the root N contents were present as soluble NO_3^- even in those plants grown with high NO_3^-.

Other plant components

The total carbon (C) contents (Tables 2 and 3) were similar in both shoots and roots. However, at high N the shoots of plants grown with NH_4^+ had a higher C content than those grown with NO_3^-. The mean C:N ratio in the shoots ranged from 17 in low N plants to 10 in high N plants, and although it was greater with NO_3^- than with NH_4^+ plants at low N (*i.e.* 17.1 and 15.7, respectively), there was little effect at high N supply. Carbon concentrations in the roots were greater at low than at high N and there was no effect of the form of N. C:N ratios therefore decreased with increasing N supply and, because of differences in N content, were also greater with NO_3^- than with NH_4^+ plants; values ranged from, on average, 13 (high NH_4^+) to 30 (low NO_3^-).

Sulphur concentrations were greater with NH_4^+ plants in both shoots and roots. There was little effect of treatment on the form of S in the

Table 4. Effects of high and low nitrate or ammonium N supply on absorption and distribution of potassium by perennial ryegrass

Treatment	Plant age, days	Total K, m mol vessel^{-1}		Rate of absorption, m mol g^{-1} dry root d^{-1}	Proportion of K in shoots, % total content
		Shoots	Roots		
—	40	0.34	0.21	—	62
Low NO_3^-	47	1.12	0.91	0.391	55
	54	1.63	1.14	0.105	59
	61	3.91	2.93	0.308	57
	68	9.86	8.06	0.379	55
Low NH_4^+	47	0.90	0.79	0.288	53
	54	1.67	0.94	0.208	64
	61	3.61	1.91	0.128	65
	68	8.76	4.80	0.322	65
High NO_3^-	47	1.48	0.77	0.507	66
	54	3.67	1.66	0.408	69
	61	9.09	2.70	0.464	77
	68	21.55	7.32	0.633	75
High NH_4^+	47	1.02	0.63	0.321	62
	54	2.45	1.11	0.300	69
	61	6.52	1.82	0.380	78
	68	15.71	3.53	0.447	82

plants, much of it being present as SO_4^{2-}. The effects on total P concentrations were similar to those for S, but were more marked in roots than in shoots. Carboxyl concentrations (as assessed by ash alkalinity) were, as were those of the other anions, constant with time and increased in the order low NH_4^+ = high NH_4^+ < low NO_3^- < high NO_3^-.

With the cations, all concentrations were reduced when NH_4^+ was supplied and the biggest effects on composition came through changes in K (Tables 2 and 3). At high N the total contents of K in the plants grown with NH_4^+ was only 66 per cent of that with plants with NO_3^- (Table 4). A similar but smaller effect was apparent at low N. These differences are also reflected in the rates of absorption of K; thus, despite their smaller root system, plants grown with low NH_4^+ had an absorption rate that was only 72 per cent of that of plants grown with low NO_3^-. Greater proportions of the absorbed K were transported to shoots at high N; thus 75 and 61 per cent of the total K was present in the shoots of plants with high and low N, respectively, over the final three harvests. With low N a greater proportion of the absorbed K was transported to shoots in plants with NH_4^+.

Discussion

There is an accumulating amount of evidence from studies in flowing solution culture with concentrations held constant[3,4,22] and adequate flow rates past the roots[7] which indicates that nutrient demand by plants can be met by much lower concentrations than previously thought. The concentrations of N in solution in the present studies would always have been low. If it is assumed that absorption of N occurred rapidly, and calculating on the basis on the amounts added in one hour, the concentrations at high N range from 1.25 (day 40-47) to 3.75 (day 54-68) μM, i.e. 17.5 to 52.5 $\mu g l^{-1}$. At low N the comparable concentrations ranged from 0.40 to 1.30 μM, i.e. 5.6 to 18.2 $\mu g l^{-1}$. In fact, concentrations at any one time were likely to have been much lower because (i) additions were as a continuous slow input and (ii) removal from solution was rapid as indicated by the close agreement between plant contents and the amounts added to the solution.

The rate of supply at low N did not provide sufficient N for maximum growth. It has been suggested that a critical concentration for NO_3^--N in grasses is 0.14 per cent or, alternatively, that the critical organic N concentration is 3.5 per cent[20]. Nevertheless, despite less growth, the plants at low N looked healthy and sustained an exponential form of growth. In other studies with birch[10,12], the N deficiency symptoms which developed when the internal N concentration decreased rapidly, disappeared when the N status and relative growth rate stabilised although the plants continued to grow less well than plants with a higher rate of N supply. These effects are comparable to many field situations, especially in natural habitats, where plants attain a steady state and adjust growth in response to the nutritional resources of the site. The changing amounts of N given to the present plants are equivalent to the increasing amounts that reach roots in the field through an exponentially expanding root system.

Once steady state conditions are achieved, strong relationships between addition rate, internal N concentrations and growth rates exist[11]. The traditional means of assessing response curves may not be accurate and often reflects further reductions of growth which is already below optimum. Similarly, there may be an over-estimate of growth as plants become stressed because measurements are made of an integrated effect of a decreasing and not a stable N status. It is of interest to note the low content of NO_3^--N in the plants at high NO_3^- supply. In other studies with ryegrass grown at very low, constant concentrations of NO_3^--N $(14 \mu M)$, much higher proportions of the N in the plants were present as NO_3^-[4]. The low NO_3^- contents of the present plants with a comparable total N status indicates a rapid reduction of adsorbed NO_3^-, *i.e.* a different pattern of utilisation when supply is calculated on the basis of growth under steady state conditions and not regulated by maintained external concentrations.

The principle of an overall target concentration in the tissues of the whole plant is obviously an approximation. Future models should take account of (i) the differences in concentration between shoots and roots and (ii) the changing distribution of N that occurs with the change in dry matter distribution with a reduced N supply.

In general the effects of the different forms of N on growth were slight. Differences in the absorption and utilisation of N as NO_3^- or NH_4^+ often reflect inadequately controlled conditions, which may result in widely fluctuating pH at the root surface as well as the possibility of a toxic imbalance of ions. Nitrogen stress often results in the relative enlargement of root systems, but it has been found with slow growing birch seedlings that the shoot:root ratio was altered only whilst the plants were reaching equilibrium with the new rate of supply[12]. With the present ryegrass plants there was an effect of N form under low N conditions and with stabilised internal concentrations. The reason for the developing trend for the greater reduction in the ratio with low NO_3^- than with low NH_4^+ is not clear, but may be related to different sites of transformation in the plant. In the field, adaptation of root distribution and morphology is particularly important with restricted NH_4^+ supply because of the low mobility of NH_4^+ compared with NO_3^- [18]. Where mobility is not restricted, as in the present system, this aspect has little relevance and other processes and mechanisms become dominant and roots become more efficient absorbers with low NH_4^+ than with low NO_3^-.

Not only may growth responses *per se* be masked in studies involving concentration ranges, but there may also be other effects. Thus, the required external concentration for one nutrient may apparently be increased by high concentrations of others. This would create an artificially high demand and an unrealistic content and possible imbalance. However, the present data confirm many previous observations that plant composition is considerably influenced when N is supplied in different forms. Graminaceous plants, including Lolium, generally take up a large excess of nutrient anions over cations and this is associated with a high rate of OH^- extrusion[6]. When NH_4^+ is the sole source of nitrogen this pattern is reversed and H^+ is extruded, and K absorption is considerably influenced, as with the present plants. Various models have been put forward to describe the possible mechanisms[13,17] involved in maintaining intracellular electroneutrality and keeping the tissue pH between narrow limits. No attempt has been made to construct an ionic balance for the present ryegrass plants, but the expected effects were quickly demonstrated, and to a greater or lesser degree when plants were grown with high or low N, respectively. There was, however, little change in the balance over the four week period. This contrasts with the onto-genetic changes demonstrated recently in a six week study with Ricinus[1], where significant changes in chemical composition could be related to plant age and not just to N source. Although it is possible that such changes would be less likely in plants remaining in a vegetative state, the results with Ricinus may also reflect the constraints imposed when plants are grown in static solution culture.

It was also demonstrated with Ricinus that the C:N ratio for NO_3^- grown plants was slightly, but significantly, higher than that for NH_4^+-grown plants[1]. In the present ryegrass shoots grown with high N there was no different in the C:N ratio which was, on average for days 54-68, 10.4 and 10.7 for NO_3^- and NH_4^+ plants, respectively. At low N, the comparable ratios were 16.1 and 14.1, respectively. The higher C:N ratio in shoots of low NO_3^- plants is indicative of the greater N-stress of these plants and may have important implications for plants growing in soil with restricted supplies of N. This difference is related to the greater ability of the present plants to absorb NH_4^+ (on average, 10 per cent more NH_4^+ than NO_3^-) when they are presented with low supplies related to their growth rate.

Acknowledgements I am grateful to D J Hatch for expert assistance and advice in the culture of the plants and to M J Henwood for assistance with chemical analysis. The Animal and Grassland Research Institute is funded through the Agricultural and Food Research Council, London.

References

1   Allen S, Raven J A and Thomas G E 1985 Ontogenetic changes in the chemical composition of *Ricinus communis* grown with NO$_3^-$ or NH$_4^+$ as N source. J. Exp. Bot. 36, 413–425.

2   Clarkson D T 1981 Nutrient demand as the pacemaker for nutrient absorption and root development. *In* Aspects of Crop Growth Agronomy Conference, A.D.A.S. Eds. J Davies and F E Shotton. pp 1–14. HMSO London.

3   Clement C R, Hopper M J, Canaway R J and Jones L H P 1974 A system for measuring the uptake of ions by plants from flowing solutions of controlled composition. J. Exp. Bot. 25, 81–99.

4   Clement C R, Hopper M J and Jones L H P 1978 The uptake of nitrate by *Lolium perenne* from flowing nutrient solution. I. Effect of NO$_3^-$ concentration. J. Exp. Bot. 29, 453–464.

5   Cox W J and Reisenauer H M 1973 Growth and ion uptake by wheat supplied nitrogen as nitrate, or ammonium, or both. Plant and Soil 38, 363–380.

6   Dijkshoorn W 1958 Nitrate accumulation, nitrogen balance and cation-anion ratio during the regrowth of perennial ryegrass. Neth. J. Agric. Sci. 6, 211–221.

7   Edwards D G and Asher C J 1974 The significance of solution flow rate in flowing culture experiments. Plant and Soil 41, 161–175.

8   Hatch D J and Canaway R J 1984 The control of pH in a system of flowing solution culture using a microcomputer. J. Exp. Bot. 35, 1860–1868.

9   Haynes R J and Goh K M 1978 Ammonium and nitrate nutrition of plants. Biol. Rev. 53, 465–510.

10   Ingestad T 1979 Nitrogen stress in birch seedlings. II N, K, P, Ca and Mg nutrition. Physiol. Plant. 45, 149–157.

11   Ingestad T 1982 Relative addition rate and external concentration; driving variables used in plant nutrition research. Plant, Cell Environ. 5, 443–453.

12   Ingestad T and Lund A B 1979 Nitrogen stress in birch seedlings. I Growth technique and growth. Physiol. Plant. 45, 137–148.

13   Israel D W and Jackson W A 1978 The influence of nitrogen nutrition on ion uptake and translocation by leguminous plants. *In* Mineral Nutrition of Legumes in Tropical and Subtropical Soils. Eds. C S Andrew and E J Kamprath. pp 113–129. CSIRO, Melbourne.

14   Jarvis S C and Robson A D 1983 The effects of nitrogen nutrition of plants on the development of acidity in Western Australian soils. II Effects of differences in cation/anion balance between plant species grown under non-leaching conditions. Aust. J. Agric. Res. 34, 355–365.

15   Jarvis S C and Hatch D J 1985 Rates of hydrogen ion efflux by nodulated legumes grown in flowing solution culture with continuous pH monitoring and adjustment. Ann. Bot. 55, 41–51.

16   Lycklama J C 1963 The absorption of ammonium and nitrate by perennial ryegrass. Act. Bot. Neerl. 12, 361–423.

17   Raven J A and Smith F A 1976 Nitrogen assimilation and transport in vascular land plants in relation to intracellular pH regulation. New Phytol. 76, 415–431.

18   Robinson D and Rorison I H 1983 Relationships between root morphology and nitrogen availability in a recent theoretical model describing nitrogen uptake from soil. Plant Cell Environ. 6, 641–647.

19   Sinclair A G 1974 An Auto-Analyser method for determination of extractable sulphate in plant material. Plant and Soil 40, 693–697.

20   Whitehead D C 1966 Nutrient minerals in grassland herbage. CAB Mimeo Publ. 1/1966. CAB, Farnham Royal.

21   Williams R F 1948 The effects of phosphorus on the rates of intake of phosphorus and nitrogen and upon certain aspects of phosphorus metabolism in gramineous plants. Aust. J. Sci. Res. (Series B) 31, 331–361.

22   Woodhouse P J, Wild A and Clement C R 1978 Rate of uptake of potassium by three crop species in relation to growth. J. Exp. Bot. 29, 885–894.

Plant and Soil 100, 113–126 (1987).   Ms. 100-10

Changes in P forms and availability as influenced by management practices*

I.P. O'HALLORAN, J.W.B STEWART and E. DE JONG
Department of Soil Science, University of Saskatchewan, Saskatoon, Saskatchewan, Canada

Key words   [137]Cs activity   Erosion   Inorganic P   Organic P fractionation   Soil texture

Summary   The objective of this study was to evaluate the changes in soil phosphorus (P) forms and potential availability that occurred in a grassland soil when it was cultivated and cropped to winter wheat (*Triticum aestivum* L.) in a crop-fallow rotation using three tillage systems: no-till, stubble mulch, and plow (bare fallow). The experiment was located in western Nebraska on a Duroc loam (fine silty, mixed, mesic Pachic Haplustolls). After 14 years, significant differences had developed in the form of soil P as determined by a sequential fractionation procedures between the original control (grass-sod) and the bare fallow treatments. Most of the differences were associated with increases in sand content caused by erosion, accompanied on the bare fallow plots by a mixing of sand from lower horizons throughout the surface 0–15 cm.

Introduction

Most of the soils in the cultivated area of the Northern Great Plains of N. America developed under a grassland ecosystem and are classified as mollisols in the USDA classification. Under native prairie conditions, these soils have a high base exchange capacity, are dominated by calcium and magnesium and generally have a high (4–8%) organic matter content. Cultivation and cropping to cereal crops using summerfallow has change the carbon equilibrium in these soils with the result that organic matter contents now are approximately 40–70% of original values after 50–75 years of cultivation[2,27]. The extent and consequence of this loss of organic matter varies with farming practice, soil type and climate. Wheat-fallow management also has reduced nutrient supplying capabilities, promoted loss of tilth, and increased susceptibility to erosion in parts of the northern plains[24]. Much of the initial work on organic matter quality in these soils concentrated on differences between carbon (C), nitrogen (N), phosphorus (P) and sulfur (S) levels of a soil under permanent pasture or virgin vegetation as compared to cultivated soils[2,11,27]. Recent management practices, which incorporate soil conservation measures such as stubble mulching and no-till practices, allow greater efficiency of water use and slow down or even eliminate soil organic matter changes[12,13,20]. Changes are slow and often become clear only after prolonged periods of cultivation. To date, more attention has been paid to C and N than to P and S, despite the fact that losses during cultivation

* Contribution R502 from the Saskatchewan Institue of Pedology, University of Saskatchewan, Saskatoon, SK, Canada S7N 0W0.

of all four elements appear to be related to changes in soil organic matter content.

Part of the reason for the neglect in measuring the change in soil P is due to the difficulties associated with separating inorganic and organic P fractions. However Hedley *et al.* use a sequential fractionation procedure to separate soil P into various inorganic and organic fractions based on chemical extracts, which differentiate between groups of P compounds according to their chemical properties. These extracts are in turn related to their availability to growing plants, and this method has been used to document the long and short term changes in soil P fractions as influenced by environmental gradients, cropping practices and management[18,19,21,28,29,30]. In an earlier study of soils developed on aeolian deposits overlying glacial till in the Canadian prairies we were able to separate changes in the distribution of soil P due to the influence of texture from those caused by management practices over the last 17 years period[18,19]. The purpose of the present study was to investigate the influence of management and cropping practices over a 15 year period on the forms and distribution of soil P at another site in the Northern Great Plains where the soil had been developed under different parent material and where slightly different tillage practices were used.

Materials and methods

The field site chosen was in the Northern Great Plains, at the High Plains Agriculture Laboratory located 8.3 km north of Sidney, Nebraska. The soil was a Duroc loam (fine silty, mixed mesic Pachic Haplustoll) with slopes of less than 1%. This site remained in native mixed prairie grass sod until 1970 when it was mold board plowed. In 1970, the recorded chemical properties in the 0–10 cm layer were pH 7.4, % C 2.33, % N 0.19, and $NaHCO_3$-P 27 mg kg^{-1}.

In 1970, a tillage experiment was initiated, comprising two major blocks (one for fallow and one for winter wheat in a wheat-fallow rotation). Four treatments were established: no-till, stubble mulch, bare fallow tillage systems and a grass sod control (Fig. 1). Weed control in the no-till system was achieved with contact herbicides and the only soil disturbance was from the wheat drill. A special no-till drill equipped with rolling coulters and slot openers was used to keep soil disturbance at a minimum. Stubble mulch tillage involved primary tillage with a 1.5 m V-blade sweep plow in the spring of the fallow year, followed by one or two operations with the V-blade sweep plow or with a chissel fitted with 18 cm sweep blades, depending on the amount of residue, weeds present and soil-water content. The final operations before fall seeding were with a bare rod weeder to firm the seed bed. Bare fallow tillage consists of mold board plowing to a 10 cm depth in the spring of the fallow year. This was followed by 2 or 3 spring-tooth operations to pack the soil and control weeds. The final two operations were similar to the stubble mulch treatment. The plot sizes were 8.5 m by 45.5 m. There were three replicates and four treatments in a randomized complete block design. Fertilizer treatments were not applied. Earlier work at this site on changes in C and N contents (Table 1) and N supply power with management are given in papers by Lamb *et al.*[12,13] and details of changes in soil physical properties are contained in studies of Mielke and co-workers[15,16].

We collected soil samples from these plots in the following manner by taking cores (7.0 cm inside diameter) to a depth of 15 cm on a grid sampling pattern with two transects per plot. The transects started 1.5 m from the end of the plot and samples were taken at 2.10 m intervals; transects were 2.1 m from the edge of the plots. With this sampling arrangement, 40 samples were taken per plot resulting 480 samples for the complete site.

Each soil sample was air-dried, crushed to pass through a 2-mm sieve, mixed and subsampled.

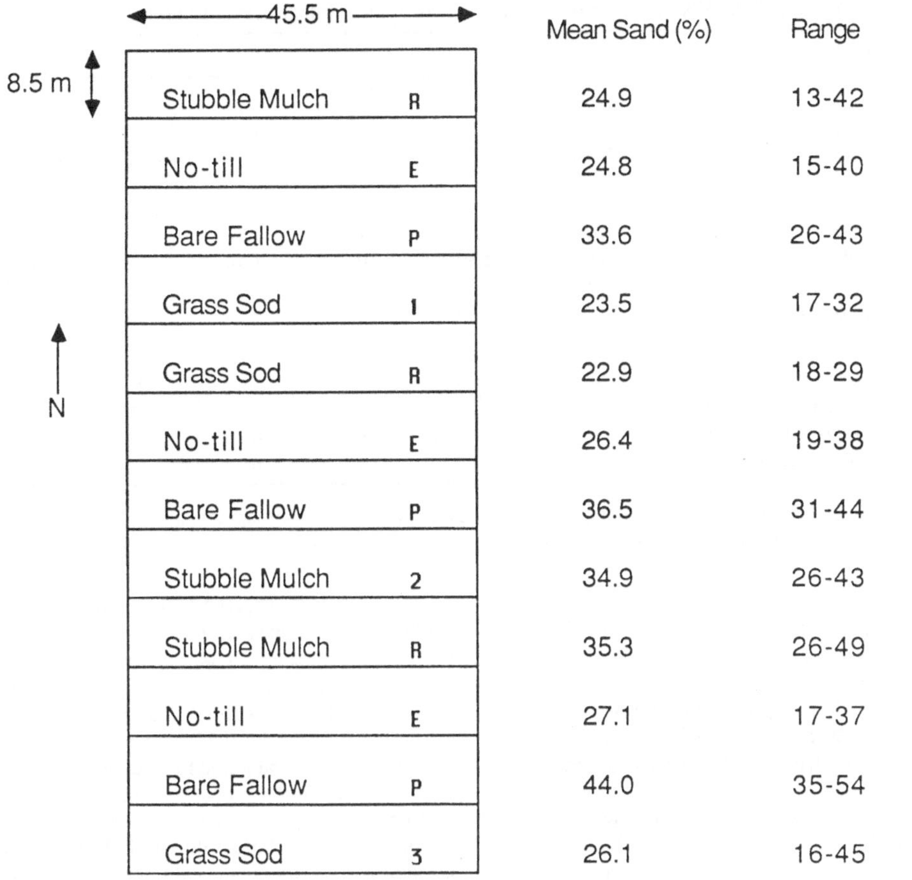

	Mean Sand (%)	Range
Stubble Mulch R	24.9	13-42
No-till E	24.8	15-40
Bare Fallow P	33.6	26-43
Grass Sod 1	23.5	17-32
Grass Sod R	22.9	18-29
No-till E	26.4	19-38
Bare Fallow P	36.5	31-44
Stubble Mulch 2	34.9	26-43
Stubble Mulch R	35.3	26-49
No-till E	27.1	17-37
Bare Fallow P	44.0	35-54
Grass Sod 3	26.1	16-45

← 45.5 m → ; 8.5 m ; N

Fig. 1. Plot design showing position of treatments in all three replicates plus the range and mean sand percentage in each treatment.

Particle size analysis was performed on the < 2 mm soil (pretreated for carbonates and organic matter) using the pipette method[14]. The soil subsamples were ground to pass through a 40 mesh sieve, and used for chemical analyses. Soil pH was determined in 0.01 M $CaCl_2$ (5:1 solution/soil ratio). Organic C was determined by the method of Tiessen et al.[26]. Bulk density measurements were carried out on the intact cores as outlined by Blake[3].

Soil P was fractionated by the method of Hedley et al.[10], with a few minor modifications (Fig. 2). In this sequential extraction procedure the extractants are selected such that relatively labile forms of inorganic P (Pi) and organic P (Po) are removed from the soil first, followed by the removal of more stable P forms by stronger chemical extractants. An anion exchange resin Dowex™ 1 × 8–50, > 30 mesh, in bicarbonate form) was used to extract Pi (resin-Pi) directly exchangeable with solution P[1]. Sodium bicarbonate (0.5 M, pH = 8.5) was then used to remove labile Pi and Po sorbed to the soil surface, plus a small amount of microbial P[4,5]. Sodium hydroxide (0.1 M) was used next to remove Pi more strongly bound to Fe and Al compounds[32,33] and Po (NaOH-Po) associated with humic compounds[5,9]. A portion of the bicarbonate and hydroxide extracts was acidified to precipitate extracted organic matter, and the Pi in the supernatant solutions (bicarb-Pi and NaOH-Pi) determined colorimetrically[17]. Total P (Pt) in the neutralized extracts was determined by acidified by acidified ammonium persulfate oxidation[6] and Po (bicarb-Po and NaOH-Po) calculated as the difference between Pt and Pi. The procedure of Hedley et al.[10] calls for a second extraction with sodium hydroxide after ultrasonification of the soil samples. However, preliminary analysis of these soils indicated that less that 5% of the total soil P was extracted by sodium hydroxide after

Table 1. Total C and N and statistical analysis as related to tillage systems and soil depth §

Tillage	Depth (cm)					
	0–10		10–20		20–30	
	C	N	C	N	C	N
	Total C and N (Mg·ha⁻¹)					
Bare fallow	13.1	1.29	11.0	1.16	10.5	1.14
Stubble mulch	14.7	1.48	11.0	1.18	10.1	1.04
No-till	18.4	1.83	10.7	1.14	10.6	1.09
Grass sod	17.6	1.76	13.8	1.40	10.0	1.03
LSD (0.05)	1.6	0.14	2.3	0.14	1.0	0.07
	Statistical analysis (Mg·ha⁻¹)					
Tillage	× ×	× ×	×	×	ns	ns
Grass sod *vs* cropped	× ×	× ×	× ×	× ×	ns	ns
No-till *vs* stubble mulch	× ×	× ×	ns	ns	ns	ns
No-till *vs* bare fallow	× ×	× ×	ns	ns	ns	ns

$^{×, × ×}$ Significant in the $P < 0.05$, 0.01 levels respectively, ns = not significant.
§ Data compiled from Lamb *et al.*[12]

sonification. Therefore, it was decided to eliminate this step and proceed to the $1 M$ HCl acid extraction (HCl-Pi), which was used to remove the relatively insoluble apatite-type minerals[34]. The residue containing insoluble Pi and more stable Po forms (residual-P) was digested with concentrated H_2SO_4 and H_2O_2[25]. This extraction procedure recovered an average of 102% of the total soil P. Total soil Po (soil-Po) was determined by the method of Saunders and Williams[22] as modified by Walker and Adams[31]. Phosphate was determined using the reagent of Murphy and Riley[17].

To calculate soil erosion losses on the cultivated sites, each was compared to a nearby uneroded control site. Two or more samples were selected at random from the original sampling transects. Later a second set of five samples was collected to 20 cm depth along the centre of the bare fallow and grass sod plots, and near the edges of the bare fallow-sod grass border on rep. 1. The samples were air-dried, weighed and analyzed for ^{137}Cs[7] using a high purity Ge-crystal coupled to a 4096 Multichannel Analyzer (MCA). The MCA identified the various gamma peaks of the sample and measured their peak area and percentage error to 90% confidence level. The samples contained in 1 L Marinelli beakers (model 130G, GA-MA and Associates Ltd., Miami, Florida) were placed over the crystal; sample and crystal were enclosed in a 10 cm thick lead castle which was lined with lucite. The activity of ^{137}Cs in Bq m^{-2} was calculated using the bulk density calculated from the weights and dimensions of the cores. It was assumed that the ^{137}Cs originally deposited on the cultivated site was the same as the average for all the sampling sites in the control. Gains of soil by deposition and losses of soil by erosion were calculated as described by de Jong[8].

Regression equations were calculated using the procedures outlined by Snedecor and Cochran[23].

Results

Table 2 presents data on the total P content in the soil and in each of the Hedley P fractions for each replicate and treatment expressed as P in $mg·kg^{-1}$ soil. Total P was highest in the grass sod and lowest in the bare fallow samples with significant changes between grass sod and no-till, and between no-till and both stubble mulch and bare fallow. Changes in P concentration within the fractions were significantly different among cultivations. All measured fractions were different between the grass sod

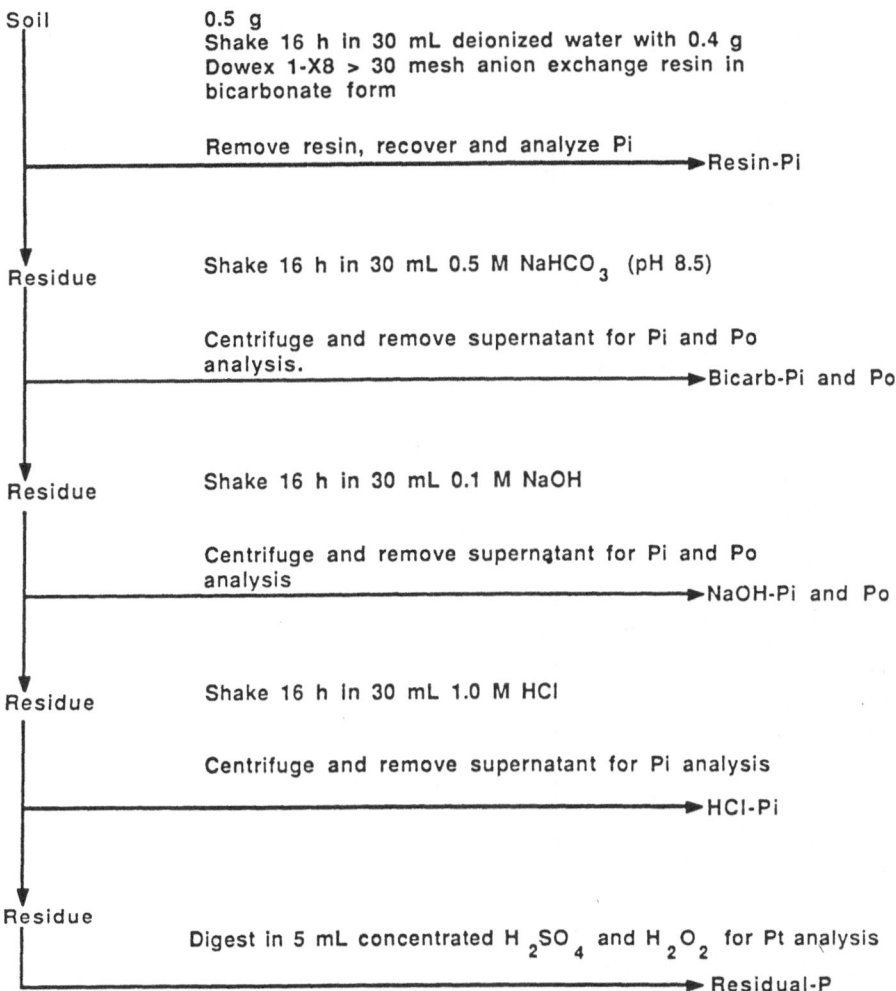

Fig. 2. Hedley sequential fractionation procedure for soil phosphorus (Pi, Po and Pt refer to inorganic, organic and total P, respectively).

and bare fallow plots and all measured fractions were significantly different between the grass sod and stubble mulch treatments with the exception of bicarbonate Pi and HCl Pi. Differences between grass sod and no-till were only significant in total P concentration and in resin Pi concentration.

If the P content in the various Hedley fractions is expressed as a percentage of the total P in the soil (Table 2), differences between tillage treatments largely disappear (statistical data not shown). Similarly, if the same data are expressed as P levels in $kg \cdot ha^{-1}$ to a depth of 15 cm (Table 3) using the concentration in $mg \cdot kg^{-1}$ and the measured bulk density (Table 2) there is no significant difference in total P between treatments.

Table 2. Mean phosphorus sand, clay and bulk density levels in 0–15 cm soil samples for all treatments

Treatment	Rep.	Hedley P fractions (mg kg^{-1})							Total-P*	Sand (%)	Clay (%)	Bulk density (g cm^{-3})
		Resin-Pi	Bic-Pi	NaOH-Pi	HCl-Pi	Bic-Po	NaOH-Po	Resid-P				
Bare fallow	All[¶]	45.1 b[§] (8.2)[†]	14.2 b (2.6)	25.6 c (4.7)	199.8 c (36.6)	12.8 b (2.4)	61.9 b (11.3)	187.6 c (34.3)	547.0 c	38.0	24.1	1.14
Stubble mulch	All	45.4 b (7.9)	16.0 ab (2.8)	27.7 bc (4.9)	204.5 ab (35.7)	12.7 b (2.2)	65.3 b (11.4)	200.6 bc (35.0)	572.3 c	31.7	26.0	1.11
No till	All	48.4 b (7.9)	17.5 a (2.9)	29.7 ab (4.9)	210.4 a (34.4)	13.9 a (2.3)	76.5 a (12.5)	210.4 a (35.4)	612.0 b	26.1	28.9	1.04
Grass sod	All	59.9 a (8.9)	17.9 a (2.7)	32.8 a (4.9)	211.3 a (31.8)	14.3 a (2.2)	86.7 a (13.0)	243.2 a (36.4)	666.2 a	24.2	30.3	0.95

* Sum of P in Hedley fractions.
[¶] Total of 120 samples per treatment.
[§] Means not followed by the same letter in any fraction differ at the 0.05 level of probability by Duncan's New Multiple Range Test.
[†] Numbers in parentheses represent the percentage of the total soil P found in each fraction.

Table 3. Overall mean phosphorus levels (expressed as P kg per ha to a depth of 15 cm) in Hedley P fractions for all treatments

Treatment	Replicate	Hedley P fractions (kg ha^{-1})						Residual-P	Total-P
		Resin-Pi	Bic-Pi	NaOH-Pi	HCl-Pi	Bic-Po	NaOH-Po		
Bare fallow	All[¶]	77.0 b[§]	24.2 a	43.9 a	341.8 a	22.2 a	105.7 b	320.9 a	935.6 a
Stubble mulch	All	75.7 b	26.7 a	46.2 a	341.7 a	21.3 ab	109.2 b	335.2 a	956.2 a
No-till	All	75.2 b	27.2 a	46.1 a	327.2 a	21.6 a	135.4 a	359.4 a	950.7 a
Grass sod	All	84.7 a	25.3 a	46.6 a	301.5 b	20.3 b	122.8 a	345.4 a	946.3 a

[¶] Total of 120 samples per treatment.
[§] Means not followed to the same letter in any fraction differ at the 0.05 level of probability by Duncan's New Multiple Range Test.

However, significant differences between the grass sod and cultivated treatments were found in four fractions (resin Pi, HCl Pi, bicarbonate Po and NaOH Po) reflecting differences in P distribution. Among the three tillage treatments (stubble mulch, bare sod and no-till) the only differences observed were between NaOH Po fractions.

Table 2 also presents data on mean sand, clay and bulk density values in the four treatments. Percentage sand was highest in the bare fallow plots and lowest in the grass sods plots, with the stubble mulch and no-till having intermediate values. Similarly, bulk density values were lowest in the grass sod plots and highest in the bare fallow plots. Sand content was found to be the most variable size fraction as well as being the fraction most closely correlated to total soil P and N.

Further examination of the differences in sand content were carried out. As the plots had been sampled on the grid pattern it was possible to examine the spatial variability in texture over the complete site. The spectral analysis and auto-correlation for surface sand content did not indicate a significant cyclic pattern within each plot. However, moving across the sites, significant periodicity in the sand content was found. The wave length indicated by the spectral analysis was in the order of approximately 34 m. Closer examination of the data indicated that periodicity of the data did not follow a sine wave function but rather a box function with peaks occurring every 34 m. Due to the initial randomization of the plots two treatments, no till and bare fallow, were always situated in the same relative position in each replicate (see Fig. 1) and therefore the replicates for these treatments were spaced approximately 34 m apart. The highest sand contents in each replicate were always found on the bare fallow plots suggesting that it was the equal spacing of the experimental design that was related to changes in sand content. It was concluded that the changes in sand content in the plots had to do with the cultural treatments. To test this hypothesis the relationship between the percentage sand and the quantity of soil P in various fractions and in total P was examined. Data presented in Fig. 3 and Table 4 show that as sand content increased in all treatments total P decreased. As demonstrated by the regression lines (Fig. 3) total P in the grass sod plot was always significantly higher than total P in the cultivated soils (bare fallow, no-till and stubble mulch) at any sand content. Total P values in the cultivated treatments (no-till, stubble mulch and bare fallow) were not significantly different from each other especially not at higher sand contents.

A similar pattern was seen in the resin Pi relationship with sand content (Fig. 4), with highest values of resin Pi in any treatment being observed in the grass sod plots (i.e. higher values at all percentage sand

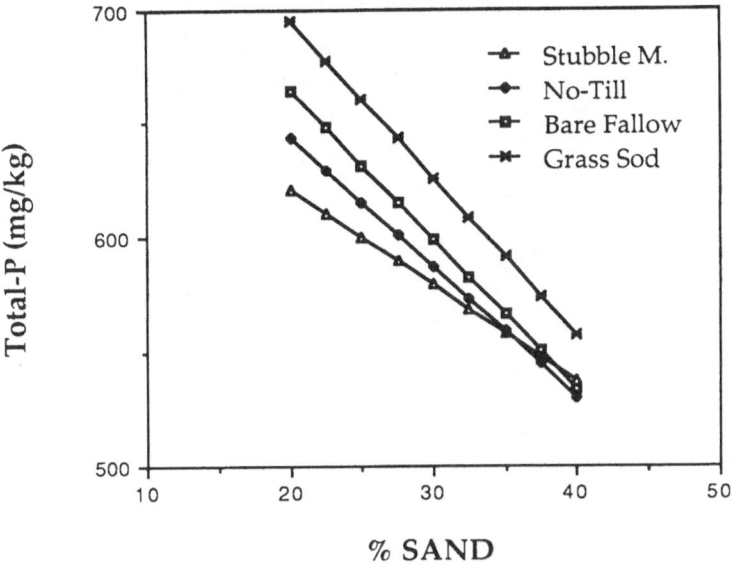

Fig. 3. Relationship between sand content and measured total P (in mg·kg^{-1}) in the four management systems. Graphs shown were derived from linear regression equations (see Table 4 for equations and significance).

content). There was no significant difference between the no-till and bare fallow plots although a similar pattern was shown in higher values at lower sand contents. Differences observed between treatments when expressed as mgkg^{-1} were related to sand content.

The differences between the average sand content in the bare fallow plot of approximately 38.0 ± 6% and that of a grass sod plot of approximately 24.2 ± 2% could be due to two factors. The first is that with repeated mold board plowing of these plots to a depth of 10 cm it

Table 4. Relationship between % sand and the quantity of soil P in the Resin-Pi fraction and in total P

Rotation	Equation[§]	Correlation coefficient[†] r^2
Bare fallow	Resin-Pi = 75.7677 − 0.8068 × (% sand)	0.604
Stubble mulch	Resin-Pi = 61.3115 − 0.5020 × (% sand)	0.440
No-till	Resin-Pi = 73.1965 − 0.9496 × (% sand)	0.439
Grass sod	Resin-Pi = 74.2313 − 0.5976 × (% sand)	0.058
Bare fallow	Total P = 794.727 − 6.5129 × (% sand)	0.906
Stubble mulch	Total P = 705.259 − 4.1926 × (% sand)	0.761
No-till	Total P = 756.774 − 5.6575 × (% sand)	0.702
Grass sod	Total P = 833.064 − 6.8977 × (% sand)	0.582

[§] Number of data points per regression equation is 120.
[†] All significant at $P < 0.05$ except Grass Sod Resin-Pi.

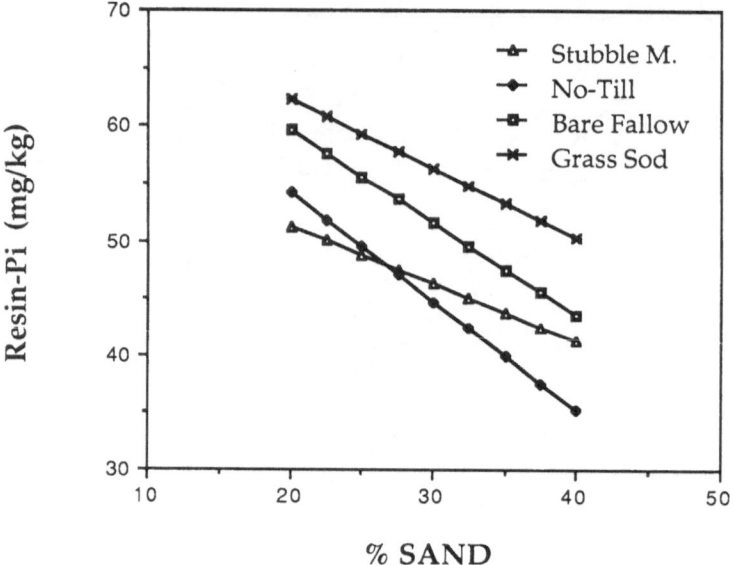

Fig. 4. Relationship between sand content and measured resin Pi (in mg·kg⁻¹) in the four management systems. Graphs shown were derived from linear regression equations (see Table 4 for
equations and significance).

was possible that with time and/or surface erosion the plowing went
below our sampling depth of 15 cm and subsoil material which was richer
in sand was brought to the surface. Other observations confirmed that
sand was evenly mixed to a depth of 15 cm in the bare fallow plots. The
sand % in the 15–30 cm layer of the cultivated plots averaged
53.8 ± 3.3% in the bare fallow plots, 51.9 ± 0.4% in the no-till and
stubble mulch plots (Elliott and Cole, personal communication*). If
plowing eventually mixed in material from this enriched sand layer
below 15 cm into the top layer, it can be calculated that the sand content
of the 15–30 cm layer would need to be ≈ 80% if mixing 5 cm into the
top 15 cm caused the observed differences. Mixing of subsoil into the top
20 cm can only partially explain the observed changes in sand content.

The second possibility was that differences in sand content represented
a loss of finer material by wind erosion from the unprotected surface of
the soil in the bare fallow treatment. To differentiate between these two
possibilities, it was decided to look at the cesium content of all the
samples that had been collected from the plots from the 0–15 cm layer.
De Jong et al.[7] have shown that in much of the Northern Great Plains,
the Cs concentration on deposits in any one area was relatively uniform
and, therefore, ^{137}Cs activity in the surface horizon can be used as a
measure of soil erosion. Data obtained from the 0–15 cm samples are

* Natural Resource Ecology Laboratory, Colorado State University, Fort Collins, CO, USA.

shown in Table 5 and it can be shown quite clearly that the Cs level as expressed in $Bq \cdot m^{-2}$ was significantly lower in the bare fallow treatments compared to the grass sod, no-till or stubble mulch. Grass sod had the highest Cs levels, followed by no-till, followed by stubble mulch. The differences between bare fallow and other treatments were significant at the $P < 0.02$.

To further decide between the differences in Cs level being due to a deeper plowing on the bare fallow treatment and a mixing of material from lower subsoil horizons, a second sampling was carried out on the bare fallow and grass sod plots. This time five samples, to a depth of at least 20 cm, were taken from a transect along the centre of each plot (*i.e.* 4.25 m from the border of the plot), the soil mixed and analyzed for Cs content. From Table 5 it can be seen that there is a significant difference ($p < 0.10$) between the Cs levels in the bare fallow and the grass sod treatment in the second sampling.

Table 5. ^{137}Cs Analyses on 0–15 cm and 0–20 cm depth samples ($Bq\ m^{-2}$)

	Analysis of 0–15 cm samples (Original sampling)			
	Rep 1	Rep 2	Rep 3	Mean
Bare fallow	2923	2849	2294	2664 b§
Stubble mulch	3737	3293	3848	3626 a
No-till	3552	4736	4218	4144 a
Grass sod	4514	4181	4551	4403 a
	Statistical analyses			
	dF	SS	MS	F value
Blocks	2	0.0124	0.051	0.036
Treatments	3	38.62	12.9	0.05*
Error	6	8.51	1.42	
Total	11	47.15		

§ Means not followed by the same letter are significantly different, as assessed by Fisher's New Protected LSD.
* Significant at $P < 0.025$.

	Analysis of 0–20 cm samples (Second sampling)			
	Rep 1	Rep 2	Rep 3	Mean
Bare fallow	3108	3826	2997	3244
Grass sod	3404	3922	4033	3786
	Statistical analyses			
	dF	SS	MS	F value
Reps	2	1.952	1.676	
Treatment	1	6.161	6.161	10.846*
Error	2	1.136		

* Significant at $p < 0.10$.

The first problem with the Cs data that had to be addressed was the differences in absolute Cs values observed between replicates in the bare fallow treatment and between the values obtained with 0–15 cm and 0–20 cm sampling. We found that highest values were obtained at the borders of the bare fallow plots. If the bare fallow plot bordered on a grass sod plot, samples taken near to the border at that plot were also higher in Cs values than samples taken in the middle of the grass sod plot. This confirmed the hypothesis of material eroding from the bare fallow plot although it is difficult to assess the exact quantity of eroded material. From the ^{137}Cs data it can be estimated that the range of erosion that had taken place ranged from 15–40% of the top soil on the bare fallow plot, 15–20% on the stubble mulch and < 5% on the no-till.

Using a value of 15% for erosion loss of top soil in the 0–15 cm layer of the bare fallow treatments, and assuming that this loss was finer material (all clay and silt), it can be calculated that the sand values found in the bare fallow treatments were the combined result of subsoil material (15–30 cm layer) being mixed with topsoil material and of erosion.

Discussion

It can therefore be concluded that the differences existing in P forms and availability were related to changes in sand content. These changes were due to the removal of clay-sized particles from the surface of the unprotected bare fallow plots and to mixing of lower horizon material into the 0–15 cm layer. The degree of erosion was found to be related to the amount of cover, bare fallow having the greatest amount of erosion, stubble mulch next followed by no-till and grass sod both with low amounts of erosion.

In contrast to documented changes in C and N content (Table 1) with tillage[12,13], changes in P were mainly related to changes in sand content brought about by erosion and by mixing due to plowing. Phosphorus exists in both organic and inorganic forms. Bringing sandier material to the surface does not change the total P in the soil on a kg · ha^{-1} but does change the amount of the P that is in HCl-Pi form (*i.e.* apatite-like materials).

These results contrast sharply with changes observed in earlier studies[18,19] where changes in sand content were due to uneven aeolean deposition in the glaciation of the Northern Great Plains. In the latter site[19], changes in P form and availability with decreasing sand content were found to closely resemble those changes expected to occur with a weathering sequence of soils, with the degree of weathering of the soil increasing with decreasing amounts of sand. In the present study changes in sand content were directly related to management practices and subsequent erosion.

Acknowledgements The authors wish to thank the Nebraska Agricultural Research Division, University of Nebraska for permission to sample the long-term rotation plots at the High Plains Agricultural Laboratory, Sidney, Nebraska and G A Peterson and C V Cole, Colorado State University for assistance in the collection of samples for cesium analyses. This work was supported by the Natural Sciences and Engineering Research Council of Canada. The technical assistance of J O Moir and H Trotter is gratefully acknowledged.

References

1   Amer F, Bouldin D R, Black C A and Duke F R 1955 Characterization of soil phosphorus by anion exchange resin adsorption and ^{32}P equilibration. Plant and Soil 6, 391–408.
2   Bettany J R, Saggar S and Stewart J W B 1980 Comparison of the amounts and forms of sulfur in soil organic matter fractions after 65 years of cultivation. Soil Sci. Soc. Am. J. 44, 70–75.
3   Blake G R 1965 Bulk density. pp 374–390. In Methods of Soil Analysis, Part 1. Ed C A Black, Agronomy 9, Am. Soc. of Agron, Madison WI.
4   Bowman R A and Cole C V 1978 Transformations of organic phosphorus substances in soil as evaluated by $NaHCO_3$ extraction. Soil Sci. 125, 49–54.
5   Bowman R A and Cole C V 1978 An exploratory method for fractionation of organic phosphorus from grassland soils. Soil Sci. 125, 95–101.
6   Environmental Protection Agency 1971 Methods of Chemical Analysis of Water and Wastes. Environmental Protection Agency, Cincinnati, Ohio.
7   De Jong E, Villar H and Bettany J R 1982 Preliminary investigations on the use of ^{137}Cs to estimate erosion in Saskatchewan. Can. J. Soil Sci. 62, 673–683.
8   De Jong E, Begg C B M and Kachanoski R G 1983 Estimates of soil erosion and deposition for some Saskatchewan soils. Can. J. Soil Sci. 63, 607–617.
9   Fares F, Fardeau J C and Jacquin F 1974 Etude quantitative du phosphore organique dans différents type de sols. Phosphore et Agriculture 63, 25–41.
10 Hedley M J, Stewart J W B and Chauhan B S 1982 Changes in inorganic and organic soil phosphorus induced by cultivation practices and laboratory incubations. Soil Sci. Soc. Am. J. 46, 970–976.
11 Keeney D R and Bremner J M 1964 Effect of cultivation on the nitrogen distribution in soils. Soil Sci. Soc. Am. Proc. 28, 653–656.
12 Lamb J A, Peterson G A and Fenster C R 1985 Wheat fallow tillage system's effect on a newly cultivated grassland soil's nitrogen budget. Soil Sci. Soc. Am. J. 49, 352–356.
13 Lamb J A, Peterson G A and Fenster C R 1985 Fallow nitrate accumulation in a wheat-fallow rotation as affected by tillage system. Soil Sci. Soc. Am. J. 49, 1441–1446.
14 McKeague J A (ed.) 1978 Manual on Soil Sampling and Methods of Analysis, 2nd Ed. Can. Soc. Soil Sci., Ottawa, Ontario.
15 Mielke L N, Doran J W and Richards K A 1986 Physical environment near the surface of plowed and no-tilled soils. Soils Tillage Res. 7, 355–366.
16 Mielke L N, Wilhelm W E, Richards K A and Fenster C R 1984 Soil physical characteristics of reduced tillage in a wheat-fallow system. Trans. Am. Soc. Agr. Eng. 27, 1724–1728.
17 Murphy J and Riley J P 1962 A modified single solution method for the determination of phosphate in natural waters. Anal. Chim. Acta 27, 31–36.
18 O'Halloran I P, Kachanoski R G and Stewart J W B 1985 Spatial variability of soil phosphorus as influenced by soil texture and management. Can. J. Soil. Sci. 65, 475–487.
19 O'Halloran I P, Stewart J W B and Kachanoski R G 1987 Influence of texture and management practices on the forms and distribution of soil phosphorus. Can. J. Soil. Sci. 67 (In press.)

20   Ridley A O and Hedlin R A 1968 Soil organic matter and crop yields as influenced by the frequency of summerfallowing. Can. J. Soil Sci. 48, 315–322.

21   Roberts T L, Stewart J W B and Bettany J R 1985 The influence of topography on the distribution of organic and inorganic soil phosphorus across a narrow environmental gradient. Can. J. Soil Sci. 65, 651–665.

22   Saunders W M H and Williams E G 1955 Observations on the determination of total organic phosphorus in soils. J. Soil Sci. 6, 254–267.

23   Snedecor G W and Cochran W G 1967 Statistical Methods, 6th ed. Iowa State University Press, Ames, Iowa, 593 p.

24   Stewart J W B 1985 Research needs and management strategies for stabilizing and improving organic matter and nutrient status of Chernozemic soils. Intecol Bull. 12, 57–67.

25   Thomas R L, Sheard R W and Moyer J R 1967 Comparison of conventional and automated procedures for nitrogen, phosphorus, and potassium analysis of plant materials using a single digestion. Agron. J. 59, 240–243.

26   Tiessen H, Bettany J R and Stewart J W B 1981 An improved method for the determination of carbon in soils and soil extracts by dry combustion. Commun. Soil Sci. Pl. Anal. 12, 211–218.

27   Tiessen H, Stewart J W B and Bettany J R 1982 Cultivation effects on the amounts and concentration of carbon, nitrogen and phosphorus in grassland soils. Agron. J. 74, 831–835.

28   Tiessen H, Stewart J W B and Moir J O 1983 Changes in organic and inorganic phosphorus composition of two grassland soils and their particle size fractions during 60–90 years of cultivation. J. Soil Sci. 34, 815–823.

29   Tiessen H, Stewart J W B and Cole C V 1984 Pathways of phosphorus transformation in soils of differing pedogenesis. Soil Sci. Soc. Am. J. 48, 853–858.

30   Wagar B I, Stewart J W B and Moir J O 1986 Changes with time in the form and availability of residual fertilizer phosphorus on Chernozemic soils. Can. J. Soil Sci. 66, 105–119.

31   Walker T W and Adams A F R 1958 Studies on organic matter. I. Influence of phosphorus content of parent materials on accumulations of carbon, nitrogen, sulfur and organic phosphorus in grassland soils. Soil Sci. 85, 307–318.

32   Williams J D H and Walker T W 1969 Fractionation of phosphate in a maturity sequence of New Zealand basaltic soil profiles: 1. Soil Sci. 107, 22–30.

33   Williams J D H and Walker T W 1969 Fractionation of phosphate in a maturity sequence of New Zealand basaltic soil profiles: 2. Soil Sci. 107, 213–219.

34   Williams J D H, Syers J K, Harris R F and Armstrong D E 1971 Fractionation of inorganic phosphate in calcareous lake sediments. Soil Sci. Soc. Am. J. 35, 250–255.

Plant and Soil 100, 127–134 (1987).

Ms. 100-11

The mechanism of rock phosphate solubilization in the rhizosphere

P.H. NYE and G.J.D. KIRK
Soil Science Laboratory, Department of Plant Sciences, University of Oxford, Oxford OX1 3PF, UK

Key words   Model   Rhizosphere   Rock phosphate   Solubilization

Summary   A simple model is described, and experimentally tested, for predicting the rate of dissolution of rock phosphates in soil, including the effect of solubilization by plant roots. The sensitivity of the model to its input parameters is assessed and it is seen that plants may significantly increase their P uptake by acid secretion. The model provides a rational basis for the selection of P solubilizing crops and inter-crops.

Introduction

Per unit of P, rock phosphate is about one third the price of simple superphosphate, so it is obviously very important, particularly for poor countries often gravely short of phosphate in their soils, to know in what conditions it may be profitably used. Yet it is surprising that in 1986 the value of rock phosphates is still assessed by an ammonium citrate extraction and correlation with field trials. Here is an inorganic crystalline compound, whose composition we can measure, being added to soils whose properties we can also measure. We ought to be able to predict how rapidly it will dissolve from first principles. A complete account of the development of a predictive model has been given in a series of papers[1,2,3,4,5,6] that are physico-chemically and numerically rigorous. In this paper, only the main ideas and results will be given.

The dissolution of a planar layer of DCPD

When tackling a problem of this kind it pays to start with as simple a system as possible, test each assumption, and explore more complicated ground from an assured base. Dicalcium phosphate dihydrate (DCPD) is the most soluble, and therefore experimentally most convenient, of the insoluble calcium phosphates that typify rock phosphates. We started by considering the dissolution of a planar source of DCPD in contact with one end of a long column of moist soil (Fig. 1), so that the concentration profiles of the constituents of the DCPD could be measured, in order to test the model. The dissolution reaction of DCPD in acid conditions is

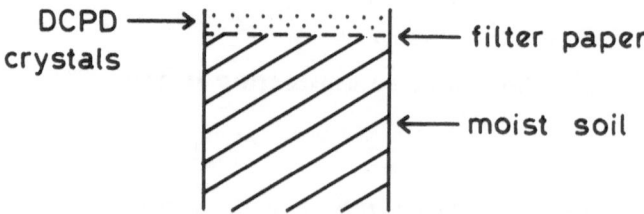

Fig. 1. Experimental system for measuring concentration profiles of phosphate diffusing into soil from a planar source of DCPD.

$$CaHPO_4 \cdot 2H_2O \; \rightleftharpoons \; Ca^{2+} + H_2PO_4^- + OH^- + H_2O$$

The rate of dissolution is limited by diffusion of the constituents from the interface into the soil. This rate will depend upon the concentrations of the constituents at the interface and the diffusion characteristics of the soil. At the interface we have five unknowns: the concentrations in the soil solution of phosphate, base and calcium, and the fluxes of phosphate and base. These may be determined by the following five relationships.

1. The concentrations of phosphate, base and calcium must satisfy the solubility product condition

$$[Ca^{2+}][H_2PO_4^-][OH^-] \; = \; K_{SP}$$

2. The solution must be electrically neutral so that

$$[Ca^{2+}] \; = \; ([H_2PO_4^-] + [OH^-])/2$$

3. The flux of phosphate across the interface is given by

$$F_P \; = \; ([H_2PO_4^-]_{surface} - [H_2PO_4^-]_{soil}) \left(\frac{D_P}{\pi t}\right)^{1/2} b_P$$

where D_P = diffusion coefficient of phosphate in soil
b_P = phosphate buffer capacity

4. The flux of base across the interface is given by

$$F_{Base} \; = \; (pH_{surface} - pH_{soil}) \left(\frac{D_{HS}}{\pi t}\right)^{1/2} b_{HS}$$

where D_{HS} = soil acidity diffusion coefficient
b_{HS} = pH buffer capacity

5. For continuous dissolution the fluxes of phosphate and base must be equal. For example, if phosphate diffused faster than base, the concentration of OH^- at the interface would rise until the flux of base again equalled that of phosphate. Thus

$$F_P \; = \; F_{Base}$$

Table 1. The experimentally determined total amounts of phosphate and base dissolved at different times, and the corresponding predicted amounts of DCPD dissolved

Time (h)	Experimental		Predicted
	Total P (mmol)	Total base (mmol)	Total DCPD (mmol)
99	0.51	0.52	0.52
186	0.60	0.65	0.63
432	0.93	1.00	0.97

With these conditions it was possible to predict the concentration profile of phosphate in the soil and the rate of dissolution. The experimental results shown in Table 1 and Fig. 2 were obtained by measuring the total inorganic phosphate in thin sections of the column of soil. The model predictions were obtained using independently measured soil parameters.

The dissolution of particles of DCPD

To predict the rate at which particles of DCPD dissolve, each particle was assigned a sphere of influence of radius

$$R = \left(\frac{3}{4\pi N}\right)^{1/3}$$

where N is the number of particles per unit volume of soil. The planar model was altered in three principal ways. The equations for the fluxes of phosphate and base were modified to allow for the spherical geometry. The restricted volume of soil available to each particle was allowed for. The decreasing size of each particle as it dissolved was also allowed for. To test the model, particles of DCPD labelled with ^{45}Ca were prepared,

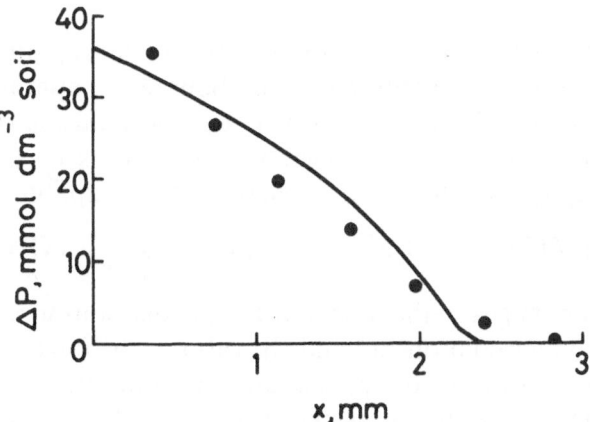

Fig. 2. Predicted and measured phosphate concentrations near a planar source of DCPD after 48 h.

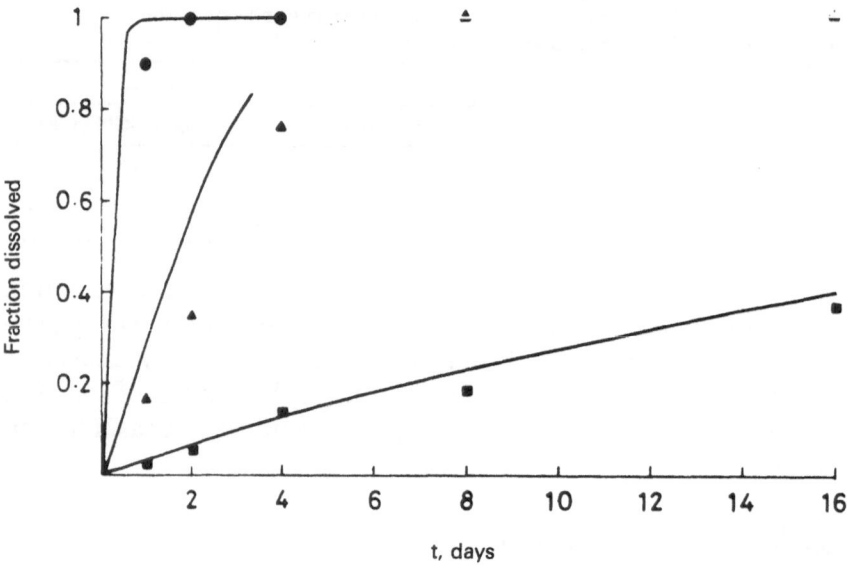

Fig. 3. The predicted and measured rate of dissolution of DCPD particles of different size at pH = 6.0. Values of the average particle radius (mm), (●) 0.021, (▲) 0.064, (■) 0.231

divided into size classes and mixed into moist soil. The difficulty of determining the amount dissolved at any given time was solved by recovering the ^{45}Ca from the soil while dissolving only a trace from the particles. Fig. 3. shows that the rate of dissolution of particles of the different size classes could be satisfactorily predicted from independently derived parameters; thus indicating that the underlying mechanisms are correctly modelled.

The dissolution of rock phosphates

The composition of rock phosphate ranges from fluorapatite, $Ca_5(PO_4)_3F$, to fully substituted carbonate apatite, $[Ca_{9.4}Na_{0.4}Mg_{0.2}(PO_4)_{4.5}(CO_3)_{1.5}F_{2.6}]^{0.5+}$. However, the mechanisms of dissolution are the same as for DCPD except that the ratios of the constituents are different, e.g. the dissolution reaction of fluorapatite is

$$Ca_5(PO_4)_3F + 6H_2O \rightleftharpoons 5Ca^{2+} + 3H_2PO_4^- + F^- + 6OH^-$$

Thus only three major changes to the model of DCPD dissolution are needed: the solubility product relationship and the ratio of the fluxes of P and base across the interface must be revised, and the additional ions dissolved must be included in the charge balance for electrical neutrality in solution. Fig. 4 shows the predictions of the revised model for the

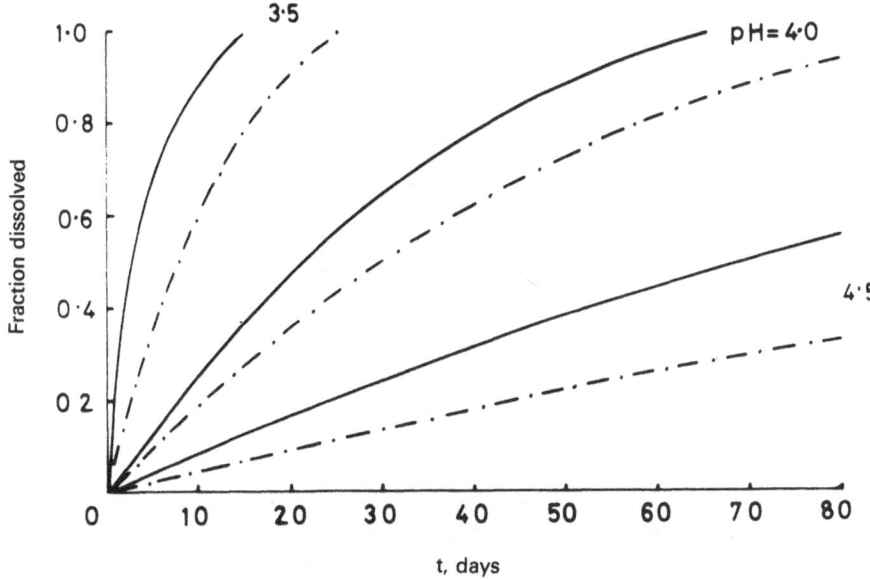

Fig. 4. The effect of soil pH on the rate of dissolution of fluorapatite and fully substituted carbonate apatite. $-\cdot-\cdot-$ fluorapatite, ——— carbonate apatite.

effect of soil pH on the dissolution of fluorapatite and the more soluble, fully substituted carbonate apatite.

The effect of plant roots on the dissolution and uptake of P from rock phosphates

Plant roots may increase the rate of dissolution of rock phosphate by lowering the concentration of P in the soil solution and in some instances by lowering the pH of the soil. Generally root uptake of Ca will not significantly lower the concentration of Ca in solution. Some, but not all of the dissolved P will be absorbed by the roots. The average effect of such roots on the pH of the soil may be calculated from its pH buffer capacity and the amount of acid the roots excrete, which in turn equals the excess of nutrient cations over anions that they absorb. Assuming that roots and rock phosphate particles are randomly interspersed in the soil it is easy to combine a model of rock phosphate dissolution with an established model of root uptake, since both models have the average concentration of P in the soil solution as a central controlling variable.

Rate of dissolution per unit volume of soil $=$

$$([H_2PO_4^-]_{surface} - \overline{[H_2PO_4^-]})\, \overline{b_p} \overline{D_p} 4\pi a_s \left(\frac{2R + a_s}{2R - 2a_s}\right) N$$

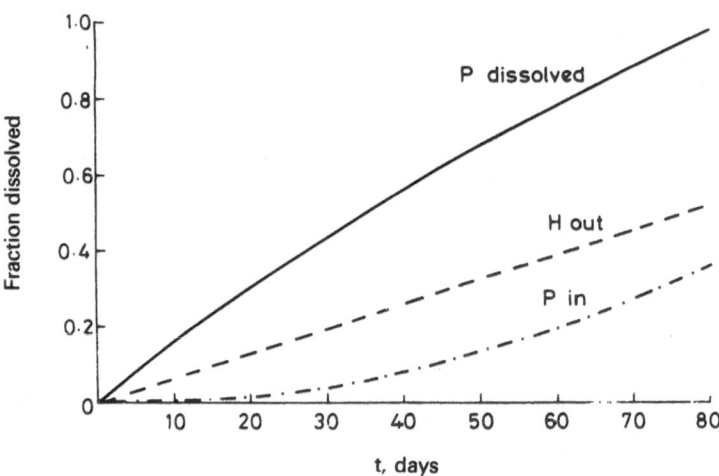

Fig. 5. The effect of acid excretion by roots on the rate of dissolution and the uptake of P from fully substituted carbonate apatite. The rate of acid excretion is 3×10^{-11} mol dm^{-2} root surface s^{-1}. $L_v = 1000$ dm m^{-3}, $a_r = 0.005$ dm. This rate is typical of a well nodulated legume root system. It would add about 10 mmol dm^{-3} of acid to the soil in 100 days, giving, if no apatite dissolved, a reduction in the average soil pH of about 0.14 units.
——— fraction of P applied that has dissolved
– – – – fraction of OH$^-$ applied that has dissolved
– · – · – fraction of P applied that has been taken up

Rate of uptake per unit volume of soil $=$

$$\frac{2\pi \overline{b_P}\, \overline{D_P}\, L_v}{\ln (x/1.65a)}\, \overline{[H_2PO_4^-]}$$

where: a_s = particle radius
  R = radius of sphere influenced by one particle
  N = number of particles per unit soil volume
  L_v = root length per unit volume by soil
  x = radius of soil cylinder exploited by one root

Fig. 5 shows the effect of roots on the dissolution of fluorapatite and fully substituted carbonate apatite when the roots are excreting acid at a rate typical of a well nodulated legume. The figure also shows the amount of P absorbed by the roots. Much of the released P is absorbed by the soil, especially when its P buffer capacity is high.

Discussion

The predictive power of the model is illustrated by the sensitivity diagrams shown in Fig. 6a and b which summarize, for a particular set of conditions, the effects of the most important variables. It will be seen

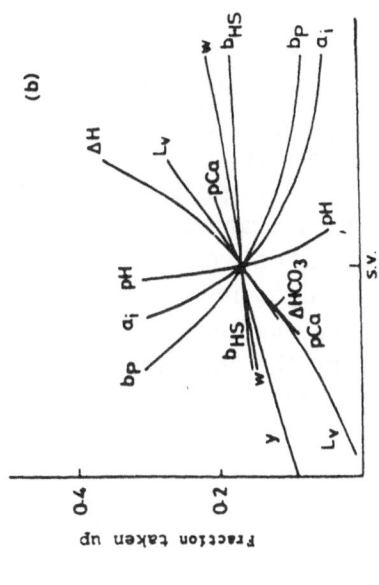

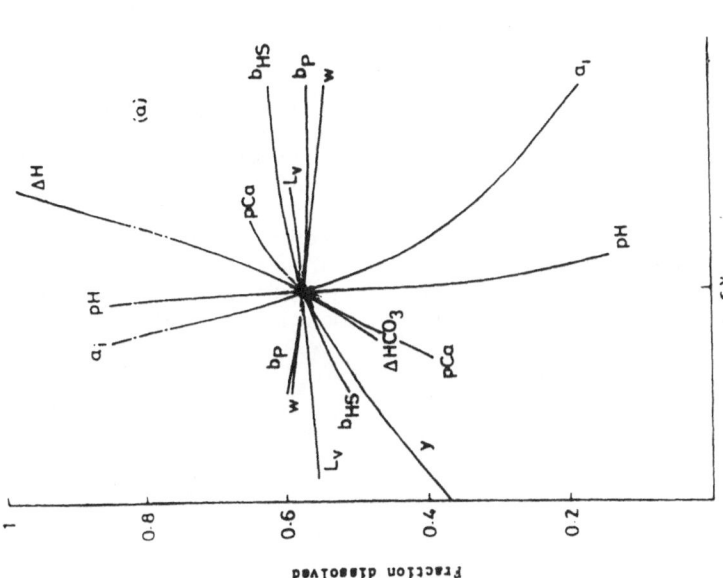

Fig. 6. The effects of the most important variables on the fraction of rock phosphate that has dissolved. The standard values (s.v.) are: initial particle radius $(a_i) = 0.1$ mm, application rate (w) $= 0.16$ kg P m^{-3} soil, $b_P = 70$ mol dm^{-3} soil, $b_{HS} = 0.07$ mol dm^{-3} soln, $b_{HS} = 0.07$ mol dm^{-3} soil/mol dm^3 soln, initial pH $= 4.5$, $\theta f = 0.05$, $L_v = 1000$ dm dm^{-3}. The ranges in values are: b_P 35–140 mol dm^{-3} soil/mol dm^3 soln, b_{HS} 0.035–0.14 mol dm^{-3} soil pH^{-1}, pH 4.25–5.25, pCa 2–4, w 0.08–0.32 kg P m^{-3} soil, a, 0.05–0.2 mm, L_v 100–1500 dm dm^{-3}, H$^+$ secreted by roots 0–10^{-9} mol dm^{-3} soil s^{-1}, HCO$_3^-$ secreted by roots 0–5 $\times$ 10^{-10} mol dm^{-3} soil s^{-1}, y (degree of CO$_3$ substitution in Ca$_{10-x}$[NaMg]$_k$(PO$_4$)$_{6-y}$(CO$_3$)$_y$F$_z$] 0–1.5.

that the fraction of the P applied that is dissolved is particularly sensitive to the soil pH, the rate of acid excretion by the roots, and the size of the fertilizer particles. The fraction of the P applied that is absorbed by the plant is also sensitive to the density of roots in the soil, although the net rate of dissolution is not. At a lower soil P buffer capacity, the effect of root uptake on the rate of dissolution should be greater. This illustrates the complexity of the system, and the range of agronomic responses to be expected. There is considerable scope for using rock phosphates on soils not usually thought suitable for it (i.e. other than strongly acid, low Ca soils) and for manipulation of soil and crop management practices to suit the use of rock phosphates.

The computer programme for the model is only 50 lines of FOR-TRAN and requires only one or two minutes of CPU time on a micro-computer, to simulate three months of dissolution. The model could therefore provide agricultural advisors with a realistic alternative to field trials for assessing the effectiveness of particular rock phosphates in given conditions, and a rational basis for the selection of P solubilizing crops and inter-crops.

References

1   Kirk G J D and Nye P H 1985 The dissolution and dispersion of dicalcium phosphate dihydrate in soils. I. A predictive model for a planar source. J. Soil Sci. 36, 445–459.
2   Kirk G J D and Nye P H 1985 The dissolution and dispersion of dicalcium dihydrate in soils. II. Experimental evaluation of the model. J. Soil Sci. 36, 461–468.
3   Kirk G J D and Nye P H 1986 The dissolution and dispersion of dicalcium dihydrate in soils. III. A predictive model for regularly distributed particles. J. Soil Sci. 37, 511–524.
4   Kirk G J D and Nye P H 1986 The dissolution and dispersion of dicalcium phosphate dihydrate in soils. IV. Experimental evaluation of the model for particles. J. Soil Sci. 37, 525–528.
5   Kirk G J D and Nye P H 1986 A simple model for predicting the rates of dissolution of sparingly soluble calcium phosphates in soil. I. The basic model. J. Soil Sci. 37, 529–540.
6   Kirk G J D and Nye P H 1986 A simple model for predicting the rates of dissolution of sparingly soluble calcium phosphates in soil. II. Applications of the model. J. Soil Sci. 37, 541–554.

Plant and Soil 100, 135–147 (1987).
Ms. 100-12

The rock-phosphate solubilizing capacity of *Pueraria javanica* as affected by soil pH, superphosphate priming effect and symbiotic N₂ fixation

P.H. DE SWART and A. VAN DIEST
Department of Soil Science and Plant Nutrition, Agricultural University, De Dreÿen 3, 6703 BC Wageningen, The Netherlands

Key words Biogas production Mixed farming N₂ fixation Nutrient uptake balance *Pueraria javanica* Rock phosphate mobilization Soil acidification

Summary When *Pueraria javanica* was grown on acid soil in a pot experiment, the legume's acidifying effect, originating from an uptake pattern in which on an equivalence basis more nutritive cations than -anions were absorbed, was sufficient to mobilize rock phosphate even when added as very insoluble material. In neutral soil, a small quantity of triple superphosphate proved necessary to set into motion a chain of reactions in which the priming action of the TSP enables nodulation to take place followed, in order, by N₂ fixation, soil acidification and rock phosphate mobilization.

More attention should be paid in tropical regions to mixed farming systems in which leguminous crops, like Pueraria, produce fodder for livestock whose faeces and urine, when properly collected, can be used for the manufacturing of biogas, after which the residues can serve as manure to food crops. Even when they prove unsuitable for beneficiation, many rock phosphates found in African and Latin American countries can be made useful as fertilizer for leguminous fodder crops.

Introduction

It is generally known that the form in which nitrogen is absorbed by a plant has an influence on the pattern of absorption of other nutrients, and also on the ratio in which nutritive cations and -anions are absorbed. In case of NH_4 nutrition, the combined equivalents of nutritive cations absorbed always exceed those of anions absorbed. In the case of NO_3 nutrition, the reverse is usually — but not always — the case: the combined equivalents of anions absorbed exceed those of cations absorbed.

In the former case, the plant maintains its electroneutrality presumably by extruding H^+ ions with a consequent pH-lowering effect on the rhizosphere. In the latter case electroneutrality is maintained presumably by an extrusion of OH^- ions with an ensuing pH-raising effect on the rhizosphere.

When leguminous plants make use of nitrogen fixed from the atmosphere, they exhibit a nutrient uptake pattern in which neither form of ionic nitrogen — NH_4^+ or NO_3^- — plays a major role. In such a situation, the combined equivalents of absorbed cations (Σ (K^+, Ca^{2+}, Mg^{2+}, Na^+)) usually exceed those of absorbed anions (Σ ($H_2PO_4^-$, SO_4^{2-}, Cl^-))

resulting for maintenance of electroneutrality in an extrusion of H^+ ions and in an acidification of the rhizosphere.

Several authors[1,2,5,14] reported on the acidifying effect of legumes growing in symbiosis with N_2 fixing Rhizobium strains. The extent of this acidification is expressed in moles of H^+ extruded either per mole of nitrogen fixed or per gram of dry matter produced.

For soybean, values in the range of 0.2–0.5 moles and for alfalfa in the range of 0.45–0.7 moles of H^+ extruded per mole of nitrogen fixed were found[1,12]. For field bean (*Vicia faba*) the values reported[4] were between 0.47–0.87 moles of H^+ extruded per mole of nitrogen fixed.

All abovementioned crops are legumes of subtropical and temperate regions growing optimally in neutral to alkaline soils. One of the objectives of the experiment to be discussed was to examine to what extent the tropical legume *Pueraria javanica* is capable of acidifying its rhizosphere.

Pueraria species are used in tropical and subtropical regions as cover crops and fodder crops. They are known to respond well to fertilization. With proper nutrition yields of up to 20 tons of dry matter per ha and per growing season have been reported[8]. It was further reported[9] that N_2 fixation in Pueraria proceeds optimally at pH 4, although other investigators claim[15] that growth and N_2 fixation are promoted by liming soils up to pH values of 6.5. The highest reported value of N_2 fixed was 200 kg in a 5-month growth period[13]. P contents of Pueraria reported vary widely from a low of 0.16% P to a high of 0.65% P in dry matter.

Hampered N_2 fixation in legumes of subtropical and temperate-region legumes growing on acid soils has been ascribed to Mo deficiency[15], to Al toxicity resulting in P deficiency[11], and to Ca deficiency[3,10,11].

As far as P deficiency is concerned, it is to be expected that in acid soils alkaline rock phosphates can set into motion a chain of reactions depicted in Figure 1. Prior to soil acidification brought about by the alkaline uptake pattern of legumes making use of N_2 fixed from the atmosphere, root nodules should be formed. It is known that in the period of nodule formation, the legumes show a rather high demand for phosphate. When the soil itself is unable to meet this demand for phosphate it might be reasoned that in acid soils phosphate released from the rock phosphate source can supply enough P to the legumes to set the depicted chain of reactions into motion. The resulting acidification of the rhizosphere will help to make additional phosphate from the rock phosphate source available to the legume so that the availability of phosphate will not be the factor limiting the rate of N_2 fixation and the rate of growth of the legume.

The possibility exists that even for a tropical legume like Pueraria, high soil acidity forms an obstacle to optimal N_2 fixation and crop growth. When, however, the soil pH would be raised or Pueraria would

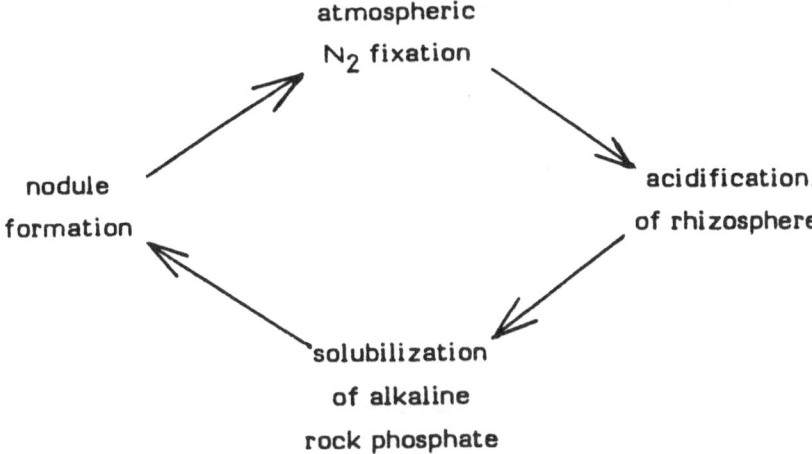

Fig. 1. Presentation of the cyclic interaction between rock phosphate solubilization and atmospheric N_2 fixation in legumes.

be grown on neutral soil low in native phosphate, it is to be expected that the initial availability of a rock phosphate source would be too low to meet the P requirement of a young legume in the phase of nodule formation. In such a situation, the cycle of reactions shown in Figure 1 would not be set into motion by the addition of alkaline rock phosphate.

The second and main objective of the present experiment was to examine whether or not the addition of a small quantity of superphosphate would bring about a priming effect through which the above chain of reactions could be set into motion under conditions enabling *Pueraria javanica* to serve optimally as host plant to Rhizobium fixing atmospheric nitrogen.

Materials and methods

The experiment was conducted in a greenhouse with the use of 7-liter enamel pots. The soil selected was a sandy loam of fluvio-glacial origin found at a depth of approximately 10 m in a sand pit. The soil was extremely low in available P which was a prerequisite for the experiment. Some relevant soil characteristics were: clay fraction ($< 2\,\mu$m): 15%, silt fraction (2–$50\,\mu$m): 15%, sand fraction ($> 50\,\mu$m): 70%; CEC: 6 me/100 g; pH-KCl: 4.4, pH-H_2O: 5.4; C content: 0.03%; N content: 0.01%; P-total content: 140 mg per kg soil, P extractable in ammonium lactate-acetic acid (P-AL): 4 mg per kg soil, P-sorption capacity: 500 mg P per kg soil.

The soil was placed into the pots in three layers as shown in Figure 2. In the top 1-kg layer which had not received any fertilizer, three *Pueraria javanica* seedlings were planted. This layer was separated from the underlying 1-kg layer through 1.6-mm mesh plastic screening material allowing roots to pass through. Halfway down this 1-kg soil layer 50 mg P fertilizer was placed in various forms, as explained below. This 1-kg soil layer was again separated from the underlying 5 kg of soil through insertion of another plastic screen. To this 5-kg quantity of soil an additional quantity of 450 mg fertilizer P, whenever needed, was applied.

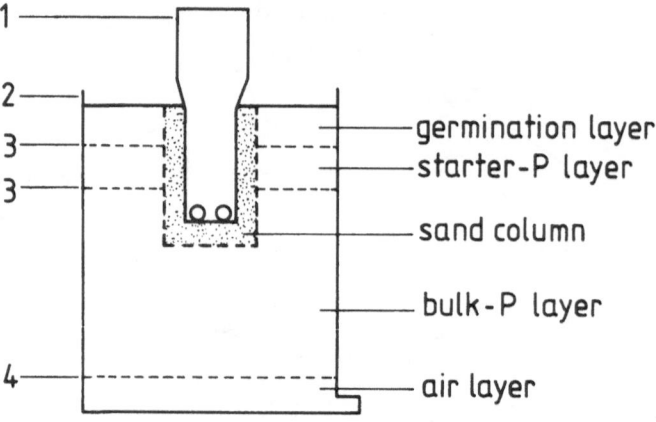

1. plastic tube for watering
2. enamel 7-l pot, diameter 20 cm, height 20 cm
3. plastic screening material
4. perforated PVC disc

Fig. 2. Placement of fertilizer P in the pots used.

All pots received the following quantities of fertilizer other than phosphate, uniformly mixed with the lower 6 kg soil in each pot: 3 g potassium magnesium sulphate (28% K_2O and 8% MgO) and 5 ml of a trace-element solution containing B, Mn, Cu, Zn, Mo and Co.

To provide the seedlings with nitrogen in the period prior to nodule development, 70 mg N in the form of $Ca(NO_3)_2$ was added to each 1-kg starter-P soil layer.

The three phosphate sources used in the experiment were alkaline rock phosphate from Khouribga, Morocco, alkaline rock phosphate from Tilemsi, Mali, and triple superphosphate. All phosphates had passed a 0.125-mm sieve. The Morocco phosphate contained 31.5% P_2O_5, 40% of which was soluble in a 2% citric acid solution. For Mali phosphate these values were 30.6% P_2O_5 and 26% of the P soluble in the citric acid solution.

The experimental set-up included 7 different phosphate treatments in three replicates. Of the total of 42 pots, 30 received 500 mg P of which 50 mg P was placed, as mentioned above, in the starter-P soil layer. In treatments without starter-TSP, a quantity of 50 mg P in the form of one of the rock phosphates was likewise placed in the 1-kg soil layer between the screens.

At the conclusion of the experiment, estimates of the contribution of the starter-TSP to the P nutrition of the plants were obtained by determining that part of the initially present TSP that could not be recovered in the starter-P soil layer. It was assumed that 1. the manner in which the P had been placed in the middle of the 1-kg soil layer, 2. the limited distance over which P diffuses in a soil, and 3. the high P-sorption capacity of the soil used had prevented any displacement of fertilizer P beyond the boundaries of the 1-kg soil layer.

The use of ^{32}P-labeled starter-TSP was initially contemplated, but was rejected in view of the considerable length of the growth period (approx. $5\frac{1}{2}$ months) and of the limited half-life of ^{32}P.

The P recovered in the plant material beyond that contributed by the starter-TSP was considered as withdrawn from the rock-phosphate and native soil P sources. The contribution of the native soil P was estimated from the P contained in plants that had grown on pots to which no P had been added at all. As will be shown in Table 2, the contribution of the native soil P to the P nutrition of the Pueraria plants was found to be extremely low.

Table 1. Materials used in the various treatments of the pot experiment

Treatment no.	Bulk phosphate	Starter phosphate (TSP)	Lime treatment	Legend
1	Mali rock phosphate	−	−	MaRP
2	Mali rock phosphate	+	−	MaRP + stP
3	Morocco rock phosphate	−	−	MoRP
4	Morocco rock phosphate	+	−	MoRP + stP
5	Triple superphosphate	+	−	TSP + stP
6	none	+	−	stP
7	none	−	−	−
8	Mali rock phosphate	−	+	MaRP + l
9	Mali rock phosphate	+	+	MaRP + stP + l
10	Morocco rock phosphate	−	+	MoRP + l
11	Morocco rock phosphate	+	+	MoRP + stP + l
12	Triple superphosphate	+	+	TSP + stP + l
13	none	+	+	stP + l
14	none	−	+	l

The last remaining P treatment to be discussed was one in which the 50 mg starter-P as TSP was added to the pertinent 1-kg soil layer without the additional 450 mg fertilizer P added to the bulk-P soil layer. This treatment was included to investigate to what extent Pueraria was capable of growing with the use of starter-TSP only.

Since the plants were to be grown at two pH-levels, one half of the pots contained soil at the original soil pH level (pH-H_2O 5.4), whereas the other half contained soil whose pH had been raised with the use of lime to a value of 6.4.

An overview of all treatments is presented in Table 1. The pots were placed in the greenhouse in a completely randomized design.

Approximately one week after planting, the three plants per pot were inoculated with the proper Rhizobium strain through injection of a bacterial suspension into the soil. Before the final harvest took place, the plants had been cut 4 times. After drying, weighing and grinding, the plant material of each cutting was analyzed for N, P, K, Ca, Mg, Na, S and Cl[16]. In the present paper, only the compounded data for the 4 cuttings and the final harvest will be reported. Only above-ground material was harvested. The root systems were inspected for nodule formation. In all treatments, except for the zero-P treatment, nodules were observed.

For calculation of the nutrient uptake pattern it was assumed that all 70 mg NO_3-N originally added as starter-N were taken up as such. This implies that the remaining N present in the plant was obtained through N_2 fixation, since the contribution of soil N was found to be negligible. This fixed N was therefore omitted from the calculation of the balance between the sum of cations and the sum of anions absorbed, resulting in estimates of the quantity of H^+ extruded by the plants.

Results and discussion

In Table 2, the various yield characteristics of *Pueraria javanica* are presented. The data in each column will be discussed separately.

Dry-matter yields

For both pH-levels, yields were highest when all P had been added as TSP. However, in the low-pH treatments, the yield obtained with Morocco phosphate was no lower, and that obtained with Mali phosphate was only slightly lower in the cases in which starter-TSP was used. The

Table 2. Yield characteristics of *Pueraria javanica*, as affected by variations in P fertilizer- and lime treatments

Treatment	Dry matter yield (g per pot)	N absorbed (mg per pot)	P absorbed (mg per pot)	N content (% of d.m.)	P content (% of d.m.)	Final soil pH(H₂O)
MaRP	47.9*b	1540 d	106 c	3.2 a	0.22 ab	4.7 ef
MaRP + stP	53.0 b	1790 c	126 bc	3.4 a	0.24 a	4.7 ef
MoRP	52.2 b	1620 d	135 b	3.1 a	0.26 a	4.8 de
MoRP + stP	56.3 b	1840 bc	147 b	3.3 a	0.26 a	4.8 de
TSP + stP	58.6 b	2060 b	183 a	3.5 a	0.31 a	4.5 f
stP	23.4 d	600 f	32 e	2.6 b	0.14 b	5.1 c
noP	3.6 f	60 h	2 f	1.7 c	0.06 c	5.3 c
MaRP + 1	11.9 e	350 g	27 e	3.0 a	0.23 a	6.2 a
MaRP + stP + 1	48.0 b	1390 de	102 c	2.9 a	0.21 ab	5.0 cd
MoRP + 1	33.1 c	1050 e	71 d	3.2 a	0.21 ab	5.5 b
MoRP + stP + 1	50.9 b	1670 d	131 b	3.3 a	0.26 a	5.1 e
TSP + stP + 1	71.3 a	2330 a	213 a	3.3 a	0.30 a	4.5 f
stP	22.6 d	520 f	24 e	2.3 b	0.11 bc	5.6 b
noP	2.0 f	40 h	1 f	2.5 b	0.05 c	6.2 a

* Values followed by a common letter are not significantly different from each other according to Duncan's multiple-range test.

considerably lower yields obtained with starter-TSP only make it clear that sizeable contribution to the growth of the plants were made by the rock phosphates.

For practical purposes it is important to note that the yields recorded for rock phosphates used without starter-TSP were not very much lower than those obtained with starter-TSP and that no differences in yield were found for the two rock phosphate sources which, as reported earlier, differ considerably in their solubilities in citric acid.

The situation is markedly different in the high-pH treatments. With unlimited availability of phosphate, as obtained with the use of TSP, the yields were higher than in any other treatment. With the use of starter-TSP, the yields obtained with the rock phosphates lagged considerably behind the yield obtained with TSP, but were at about the same levels as acquired in the low-pH treatment. Large differences, however, occurred as a result of the use of starter-TSP. Without such a starter, yields declined considerably, also in comparison with those obtained in the low-pH treatments. It is also evident that at the higher soil pH Mali rock phosphate without starter-TSP was considerably less useful than was Morocco rock phosphate.

Atmospheric nitrogen fixation

The data on N contained in the plant material follow the data on dry-matter production rather closely. The most noteworthy observation is that with the use of TSP as phosphate source N_2 fixation was higher in neutral than in acid soil. This finding supports the earlier reported observation[15] that, in spite of its ability to serve as a host plant to Rhizobium in acid soils, Pueraria still responds favorably in this respect to liming.

In contrast to dry-matter yields in acid soil, N_2 fixation was promoted by the use of starter-TSP in combination with rock phosphates, but the response to starter-TSP was considerably lower than in neutral soil. As was the case with dry-matter yield, in acid soil no difference was observed between Morocco and Mali rock phosphates in their effects on N_2 fixation. In contrast, the differences in neutral soil were large.

Finally, it can be observed that with the use of rock phosphates, with and without starter-TSP, atmospheric N_2 fixation was always higher in acid than in neutral soil. Since the reverse was found when TSP was used, it can be concluded that the positive influence of an improved availability of rock phosphate in acid soil overshadowed the negative influence of acidity on the effectiveness of the N_2 fixation mechanism.

Phosphate absorption

As expected, P absorption was highest when TSP was used as sole P source. In the acid soil, P absorption was not raised through the application of starter-TSP. Neither was there any difference in P absorption for the two rock phosphate materials used, which implies that in the acid soil the sparsely soluble Mali rock phosphate was just as valuable for Pueraria as was the more soluble Morocco rock phosphate, which conclusion could also be deduced from the dry-matter yields.

In the neutral soil, the situation was considerably different. Both with and without starter-TSP, Morocco rock phosphate proved to be a much better P source for Pueraria. In addition, the differences in quantity of P absorbed with and without the use of starter-TSP were substantial. In fact, the difference between the two values for Mali rock phosphate is larger than the quantity of P added as starter-TSP. This finding already shows that there must be a priming effect exerted by the starter-TSP on the solubilization of the rock phosphate. Table 3 supplies more quantitative information on this matter.

As explained in the previous section, estimates of difference in contribution of rock phosphate to the P nutrition of Pueraria with and without the use of starter-TSP were obtained by determining the quantity of starter-TSP that could not be recovered from the 1-kg starter-P soil layer at the end of the growth period. Both values obtained for the acid soil were not significantly different from zero, whereas for the neutral soil both values differed significantly from zero.

These findings make it clear that in acid soil the solubilization of even sparsely soluble rock phosphate, like the one from Mali, proceeded rapidly enough to supply sufficient P to the young Pueraria plants to allow for nodule formation, for N_2 fixation and for the ensuing acidification of the rhizosphere, resulting in further rock phosphate solubilization. In neutral soil, however, this sequence of processes will not be set into motion without the priming action of a small quantity of readily

Table 3. Phosphate in the above-ground portions of *Pueraria javanica*, absorbed from rock phosphate sources

Treatment	Rock phosphate-P absorbed (mg pot^{-1})	Difference
MaRP	106 b	12
MaRP + stP	118 b	
MoRP	135 a	19
MoRP + stP	154 c	
MaRP + 1	27 d	74
MaRP + stP + 1	101 b	
MoRP + 1	71 c	47
MoRP + stP + 1	118 b	

soluble TSP. The data of Table 3 also show that even with the use of starter-TSP, the contribution of Morocco rock phosphate to Pueraria's P nutrition in neutral soil lagged behind that obtained in acid soil, whereas in this respect the Mali rock phosphate showed no difference.

Nitrogen contents

The nitrogen contents of Pueraria showed a great deal of similarity for treatments in which 500 mg P per pot was supplied, irrespective of P source. Even when the phosphate material was highly unavailable, like in the case of Mali rock phosphate without starter-TSP in the neutral soil, did the plants maintain a N content of 3%.

N contents dropped to much lower values when P supply per pot was limited to 50 mg starter-P or when no P was supplied at all.

Phosphate contents

A high degree of correlation is to be seen between P- and N contents. Although values of 0.30% and 0.31% P are not significantly higher than many other values obtained, it seems that the former values approach the level of luxury consumption. It was found earlier[1] that luxury consumption of P in legumes occurs readily when the P source used is highly available and when in cases of successful nodulation and N_2 fixation the plants grow rapidly without absorbing much NO_3. When in such situations NO_3 is not competing with H_2PO_4 for carrier sites, P can be absorbed in excessive quantities. The high P-sorption capacity of the soil used in the present experiment probably prevented such excessive P uptake, but a slight tendency toward luxury consumption of P is noticeable.

Like in the case of N, it can be observed that the use of rock phosphate with and without starter-TSP always resulted in similar P contents, and that lower P contents were found only when P application was limited to 50 mg P or was absent. It can be speculated that only with a uniform distribution of P throughout the whole soil column circumstances are favourable for N_2 fixation resulting in localized soil acidification enabling solubilization of rock phosphate needed for a build-up of a normal P content in Pueraria.

Soil pH

In both acid and neutral soil, the highest pH values were found with the lowest yields obtained. These values were not much different from the initial values of pH 5.4 and 6.4, respectively. Likewise, in both acid and neutral soil, the lowest final pH values were found with the highest yield values and highest N_2 fixation values acquired. These findings make it clear that soil acidification is a function of the rate of growth of legumes and of the rate of N_2 fixation taking place in the nodules of these legumes. Further observation of the pertinent data also reveals that a close correlation exists between growth rate of Pueraria and rate of soil acidification. In the neutral soil, with both rock phosphates used the application of starter-TSP led to increased soil acidification. The effect of the starter-TSP is an indirect one in that it promotes growth and N_2 fixation and thus gives rise to circumstances favoring soil acidification.

Nutrient uptake balances

It must of course be realized that the abovementioned soil pH decreases were obtained under unnatural circumstances of a relatively dense root system developing in a limited quantity of soil contained in a pot. To transform the findings obtained to information useful under practical circumstances, the quantities of H^+ extruded as a consequence of differences in nutritive cation and -anion uptake observed were expressed per unit of nitrogen fixed from the atmosphere and per unit of above-ground dry matter produced, as discussed in the Introduction section (Table 4).

For treatments with 500 mg P applied per pot, the values in both columns show a great deal of similarity, in each case with one exception, namely 'Mali rock phosphate plus lime'. In this treatment, little P was available to plants and, consequently, nodulation and N_2 fixation never developed well. As a result, the 70 mg NO_3-N applied as starter-N con-

Table 4. Acid excreted by *Pueraria javanica* roots per unit N_2 fixed from the atmosphere, and per g above-ground dry matter produced

Treatment	Acid excreted per unit N_2 fixed (mM/mM)	Acid excreted per g d.m. produced (mM/g)
No lime added		
MaRP	0.47	1.09
MaRP + stP	0.48	1.05
MoRP	0.49	1.10
MoRP + stP	0.49	1.06
TSP + stP	0.45	1.06
stP	0.52	0.83
0	–	–
Lime added		
MaRP + l	0.51	1.01
MaRP + stP + l	0.42	0.64
MoRP + l	0.49	1.10
MoRP + stP + l	0.49	1.04
TSP + stP + l	0.49	1.10
stP + l	0.64	0.86
l	–	–

stituted a relatively large portion of total-N accumulated in the crop. This anionic N exerted a suppressive influence on the $(C_a–A_a)$ value thus accounting for the low $(C_a–A_a)/N$ and $(C_a–A_a)/$dry matter values.

The $(C_a–A_a)/N$ values ranging from 0.42–0.51 for treatments with 500 mg P per pot applied compare well with values mentioned earlier for legumes of temperate and tropical regions[12,1,4].

For practical purposes, values of $(C_a–A_a)/$g dry matter are more useful. In the present experiment, these values ranged from 0.64 to 1.10 mmol/g. When it is assumed that Pueraria can produce 10 tons of dry matter per ha per growing season, and that per g dry matter 1.0 mmol H^+ is produced, it follows that per ha and per growing season the quantity of acid produced by a Pueraria crop can amount to 10 kmoles H^+ per ha. The quantity of pure calcitic limestone needed to neutralize this quantity of acid produced will be 500 kg per ha.

With an average $(C_a–A_a)/N$ value of 0.48, as obtained in the present experiment, and with the estimate of 10 tons of Pueraria dry matter produced per growing season, it can be calculated that the amount of atmospheric nitrogen fixed per ha and per growing season will be 292 kg N. This value, applying to a $5\frac{1}{2}$-month growing period, compares well with the earlier mentioned value of 200 kg N fixed in a 5-month growing period[13], although in this latter case no mention was made of the quantity of dry matter produced.

Conclusions

Roughly speaking, in tropical agriculture two distinctly different situations can be observed with respect to deficiencies of major nutrients. In lowland rice culture, nitrogen is usually the nutrient whose limited availability is determining crop yields, whereas in upland crop culture yield levels are often determined by the level of availability of phosphate.

Among these upland crops, many belong to the large group of leguminous plant species. In the present experiment, it was shown that in acid soils even a highly insoluble rock phosphate, such as the one from Tilemsi in Mali, can be useful to Pueraria, and that with the use of a limited quantity of superphosphate as starter the same holds for neutral soils.

It can be held against a crop like Pueraria that as a fodder crop it cannot be used for direct human consumption. It may, however, be reasoned that more than has been the case up till now, crops like Pueraria should be looked upon as sources of fodder and, indirectly, as sources of energy.

In a continent like Africa, the pressing need for energy material, *e.g.* for the preparation of meals, could be filled to a large extent when more use would be made of the possibility to produce biogas. For the manu-facturing of such biogas, manure produced by livestock is one of the most valuable basic materials. If this livestock would be raised in feed lots, useful for a convenient collection of faeces and urine, it could feed on fodder consisting of leguminous biomass. For the production of leguminous fodder crops, phosphate is often the nutrient limiting yields most.

The results of the present experiment make it clear that phosphate as key nutrient for the growth of fodder crops can be supplied in the form of sparsely soluble rock phosphates.

The food crops needed for human consumption in farming systems as described above, could be grown with the residues available after the production of biogas. For a recycling of nutrients other than N and P within such farming systems, it would be needed to apply the residues partially to the leguminous fodder crops to be grown.

Thus, in two ways the nutrient phosphate can be viewed as a key factor in tropical farming systems. In the first place, as was shown in the present paper, it is needed to initiate the development of nodules essen-tial for the growth of legumes in farming systems without fertilizer-N inputs. When the legumes are grown in acid soils, sparsely soluble rock phosphates, as they are known to be present in many African and Latin American countries, can be used without any need for a beneficiated source of phosphate to serve as priming material. In neutral soils, the priming action of a limited quantity of superphosphate is needed to set into motion a series of processes in which the rock phosphate is being mobilized.

In a broader scope, phosphate can perform a key role in mixed farming systems in which legumes are grown to supply fodder for cattle whose excrements can serve to produce biogas and manure. In tropical Africa and Latin America more attention should be paid to the develop-ment of such mixed farming systems.

References

1   Aguilar Santelises A 1981 Rock-phosphate mobilization induced by the alkaline uptake pattern of legumes utilizing symbiotically fixed nitrogen. Doct. Dissertation, Agric. Univ. Wageningen, Netherlands.
2   Aguilar Santelises A and van Diest A 1981 Rock-phosphate mobilization induced by the alkaline uptake pattern of legumes utilizing symbiotically fixed nitrogen. Plant and Soil 61, 27–42.
3   Andrew C S 1976 Effect of calcium, pH and nitrogen on the growth and chemical composition of some tropical and temperate pasture legumes 1. Nodulation and growth. Aust. J. Agric. Res. 27, 611–623.
4   Bekele T, Cino B J, Ehlert P A J, van der Maas A A and van Diest A 1983 An evaluation of plant-borne factors promoting the solubilization of alkaline rock phosphates. Plant and Soil 75, 361–378.

5   Beusichem M L van 1981 Nutrient absorption by pea plants during dinitrogen fixation. 1. Comparison with nitrate nutrition. Neth. J. Agric. Sci. 29, 259–272.
6   Blasco L M and Bohorguez A N 1968 Pasture species in the Amazone region (in Colombia). 1. Analysis of some chemical components. Agricultura Tropical 24, 175–177.
7   Dirven J G P and Ehrencron V K R 1963 Een bemestingsproef bij kudzu (*Pueraria phaseoloides*). Surinaamse Landbouw 11, 2, 39–45.
8   Febles G and Padilla C 1970 The effect of inoculation and foliar urea on kudzu (*Pueraria phaseoloides*) and pigeon pea (*Cajanus cajan*). Revta. Cub. Cienc. Agric. 4, 148–151.
9   Loustalot A J and Telford E A 1948 Physiological experiments with tropical kudzu. J. Am. Soc. Agron. 40, 503–511.
10 Lowther W L and Loneragan J F 1970 Calcium in the nodulation of legumes. Proc. 11th Int. Grassland Congress, Queensland, Australia, 446–450.
11 Munns D N 1965 Soil acidity and growth of a legume. 2. Reactions of Al and P in solution and effects of Al, P, Ca and pH on *Medicago sativa* and *Trifolium subterraneum* in solution culture. Aust. J. Agric. Res. 16, 743–755.
12 Nyatsanga T and Pierre W H 1973 Effect of nitrogen fixation by legumes on soil acidity. Agron. J. 65, 936–940.
13 Rajaratnam J A and Ang Poo Guan 1972 Nitrogen fixation by *Pueraria phaseoloides* in Malaysia. Malay. Agric. J. 1, 2, 92–97.
14 Raven J A and Smith F A 1976 Cytoplasmic pH regulation and electrogenic H^+ extrusion. Curr. Adv. Plant Sci. 24, 649–660.
15 Smith J H and Chandler J V 1961 Tropical kudzu moves into Puerto Rico. Crops and Soils, March, 12–14.
16 Swart P H de 1985 Invloeden van bodem-pH, verschillende ruwe fosfaten en het gebruik van geringe hoeveelheden 'starter-fosfaat' op het luchtstikstofbindingsproces bij *Pueraria javanica*. Internal Research Report, Dept. of Soil Science and Plant Nutrition, Agric. Univ. Wageningen, Netherlands.

Plant and Soil 100, 149–156 (1987).
Ms. 100-13
© 1987 *Martinus Nijhoff Publishers, Dordrecht.*

Relative efficiency of nitrogen fixation and the occurrence of hydrogenase in pea root nodules

R.O.D. DIXON
Botany Department, The King's Buildings, Mayfield Road, Edinburgh EH9 3JH, UK

Key words Ammonia Hydrogenase Hydrogen diffusion Pea Relative efficiency

Summary The apparent $K_{m(hydrogen)}$ for uptake of hydrogen by pea root nodules was determined. This enabled the concentration gradient necessary for the evolution of hydrogen to be calculated for nodules with no hydrogenase activity. This indicated that hydrogen inhibition of nitrogenase is not likely to be the cause of the low relative efficiency of legume root nodules. The factors that affect electron allocation between protons and nitrogen in nitrogenase are reviewed and it is concluded that there must be some as yet unknown factor that affects electron distribution in *Rhizobium* nitrogenase. One possibility is put forward and considered. A strain of *Rhizobium* was used that was found to possess hydrogenase activity in combination with pea variety Feltham First but not with variety Meteor. The control of this enzyme is briefly discussed.

Introduction

Nitrogen fixation is an expensive process in terms of ATP and reductant consumption. Part of this energy expended is used in the reduction of protons to hydrogen. The minimum amount of hydrogen produced is one molecule of hydrogen for each molecule of nitrogen reduced. It was with some concern then that it was found by Schubert and Evans[15] that legume root nodules evolved a greater proportion of hydrogen than 25%, which is the minimum. This showed a lack of efficiency by these nodules as energy was being channelled into hydrogen production rather than the reduction of nitrogen. In order to quantify this, Schubert and Evans put forward the concept of Relative Efficiency (RE) which is given by:

$$RE = 1 - \frac{\text{rate of } H_2 \text{ evolution in air}}{\text{rate of } H_2 \text{ evolution in 20\% } O_2, 80\% \text{ Ar}}$$

(or rate of C_2H_2 reduction)

Maximum RE will be 0.75 whereas some root nodules had an RE as low as 0.4. The reasons for this lack of efficiency have not been fully explored. One theory put forward[6,7] was that the efficiency was low because of the inhibition of nitrogenase by hydrogen. The purpose of the work given here was to further test this theory. It has not been possible to obtain the definitive test, which would be to find the concentration of hydrogen within the nodule, as this is not, at the moment, technically possible.

However, with the publication of the K_m of hydrogenase[1], it has become possible to determine the ease of diffusion of hydrogen into the nodule when hydrogenase is present. As the diffusion pathways and resistances are the same, into and out of the nodule, it is possible to calculate the approximate internal concentrations of hydrogen necessary for hydrogen to diffuse out of the nodule at the rate at which it is formed.

The activity of uptake hydrogenase in legume root nodules, although a bacterial enzyme, was shown to be controlled by the host plant as *Rhizobium leguminosarum* strain ONA 311 showed high hydrogenase levels in peas, var Meteor, but no detectable activity in nodules of broad bean, *Vicia faba*[5]. A further extension of this is that *Rhizobium* strain 128C53 possessed high hydrogenase activity in pea varieties Alaska and JI1205 but very little activity in the variety Feltham First[2]. The *Rhizobium* strain, used in the work described here, however had a high hydrogenase activity in pea var Feltham First and no detectable activity in var Meteor. This finding is discussed.

Materials and methods

Rhizobium leguminosarum strain E100 was used throughout this work. Peas, *Pisum sativum*, var Feltham First and Meteor were obtained from local suppliers. Because of the dependence of the results on the pea variety used, three different accessions of seed were used and the results, with regard to the presence or absence of detectable hydrogenase activity, were the same in each case.

Hydrogen was assayed by gas chromatography using a Pye 104 gas chromatograph with a katherometer detector and a column of molecular sieve 5A at a flow rate of argon carrier gas of 40 mls a minute. The temperature was 60°C. Determinations of hydrogen evolution and uptake were carried out on about 300 mg nodules attached to short pieces of root placed in flasks with 10 ml gas volume. Samples of 0.5 ml were withdrawn through a 'Subaseal' cap for analysis. For hydrogen uptake assays the flasks were briefly evacuated and then filled with the desired gas mixture through a side arm. Hydrogen evolution was measured with air as the gas phase. Determinations were done in triplicate. The results presented for the effect of hydrogen concentration are from 3 separate experiments on the same batch of plants.

Results and discussion

Relative efficiency

The low RE values obtained for a large number of legume root nodules indicates that a greater proportion of electrons are allocated to the reduction of protons than the minimum necessary for the nitrogen reduction reactions. Assays of nitrogenase, *in vitro*, have shown that a number of factors can affect electron allocation between nitrogen and protons and it has been considered that any one of these, dependent upon the conditions, may be responsible for high hydrogen evolution in legume root nodules. The factors that affect electron allocation are given

Table 1. Factors which affect the allocation of electrons in nitrogenase

ATP:ADP[18]	pH[8]
Protein 1:protein 2[20]	Hydrogen concentration[21]
Nitrogen concentration[9]	Temperature[14]
Electron Flux[10]	

in Table 1. It was noticed, however, that low RE values are generally associated with high activity of nitrogenase in the root nodule. The factors that affect electron allocation are all associated with an inhibition of electron flux[10], and therefore activity of nitrogenase, with the exception of hydrogen inhibition and nitrogen concentration. It has been calculated that sufficient nitrogen can diffuse to the nitrogenase site to prevent a shortage of this substrate[14] and experiments in which extra nitrogen was supplied made no difference to the RE[7]. It was therefore concluded, by a process of elimination, that the only factor remaining was hydrogen inhibition. Examination of the diffusion pathways, intercellular space, showed that in peas, which generally have a low RE the intercellular spaces were more limited than in soya bean, which has in general a higher RE[6]. This gave support to the hydrogen inhibition hypothesis and to date no evidence that was inconsistent with hydrogen inhibition has been produced. The only fact that controverted the hypothesis was the high diffusivity of hydrogen. Evidence to support, or disprove, the hydrogen inhibition hypothesis would be an assay of actual hydrogen concentrations within the nodule or a determination of the resistance to hydrogen diffusion within the nodule.

The publication of the K_m of hydrogenase has enabled a measure of the resistance to hydrogen diffusion within the nodule to be obtained. This is because when the enzyme is working at half maximal activity the concentration of substrate must equal the K_m value. In the case of hydrogenase this will be $1-2\,\mu M$. Thus the concentration of hydrogen outside the nodule necessary to maintain the internal concentration to $1-2\,\mu M$ can be determined. Accordingly root nodules with hydrogenase activity were provided with hydrogen at different concentrations and the rate of uptake determined. The results of such experiments are given in Figure 1. It can be seen that the concentration of hydrogen that gives half maximal velocity of hydrogen uptake is about 2.5%. The rate of hydrogen uptake at 2.5% hydrogen is of the same order as the rate of hydrogen evolution from nodules that lack hydrogenase. As the pathways are the same the internal concentration of hydrogen will be around 2.5%, or less for those nodules which evolve hydrogen at a lower rate. Thus the concentration of hydrogen is the root nodule will not accumulate to sufficiently high levels to inhibit nitrogen fixation.

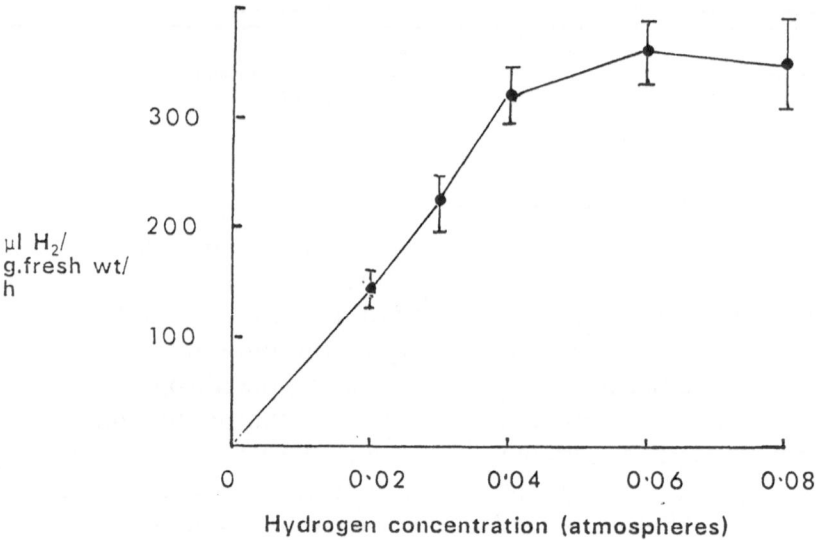

Fig. 1. Rates of hydrogen uptake by pea root nodules in response to different concentrations of hydrogen supplied. Each point is an average of 9 determinations and the *bars* indicate the standard error of the mean.

With the realisation of the fact that low REs are not due to hydrogen inhibition it becomes more difficult to understand the role of hydrogenase in legume root nodules and also the increases in efficiency obtained by some workers who have compared Hup$^+$, hydrogenase-containing, strains with Hup$^-$, no hydrogenase strains, and obtained increases in yield and nitrogen fixation with Hup$^+$ strains[13]. This was explicable if hydrogenase relieved hydrogen inhibition and improved the RE in which case the savings in energy could be substantial. If the RE is not affected then the savings in energy by the reutilisation of the hydrogen produced would only be marginal. It is not possible to detect a low RE in nodules which contain hydrogenase because of the lack of measurable hydrogen evolution in air. However in work on cultured cells of *Rhizobium* ORS 571 isolated from Sesbiana nodules estimates of the amounts of energy consumed by the cells in fixing nitrogen have been made. The cultured cells were induced to form nitrogenase and hydrogenase and various parameters were measured[19]. The rate of nitrogen fixation by these cells was very high being 1.5 μmol ethylene formed/mg dry wt./hr. It was calculated that 70% of the electrons donated to nitrogenase were allocated to the reduction of protons. Thus a low RE can be maintained in the presence of hydrogenase. The correlation of high activity and low efficiency thus extends to cultured rhizobial cells and it is thus unlikely that the explanation of low RE lies in any nodule property.

With the demise of the hydrogen inhibition hypothesis it is necessary to erect another hypothesis to stimulate work which will lead to an understanding of the low RE of legume root nodules. With such an understanding it may be possible to devise ways of improving the efficiency of nitrogen fixation in legume nodules with consequent benefits to legume crop yields.

The correlation of low RE with a high activity of nitrogenase suggests either an effect of lack of substrate to maintain the high rate or the accumulation of a product which cannot be assimilated or diffuse away as fast as it is produced. The substrate nitrogen has been eliminated as a possibility and so also has the product hydrogen. Ammonia is the other product. Ammonia has no effect on nitrogenase *in vitro*. However, it has been shown, in the case of *Azotobacter vinelandii*, that ammonia affects the transmembrane potential, $\Delta\psi$, and has an inhibitory effect upon nitrogen fixation at lower concentrations than are required to affect the ATP:ADP ratio.[11,12,13] It has been concluded that the reduction of nitrogenase in *Azotobacter* is dependent upon $\Delta\psi$. In similar experiments on the affect of ammonia on *Rhizobium leguminosarum* bacteroids nitrogenase inhibition was not found as the ammonia was pumped out of the cells. However when methylammonium was introduced inside the cells under anaerobic conditions, to prevent it being pumped out, and the cells returned to aerobic conditions it was found that this ammonium analogue had the same effect as that on *Azotobacter*. Nitrogenase activity was determined by acetylene reduction.

When bacteroids are fixing nitrogen the ammonium is produced inside the cells and then is pumped out into the peribacteroid space and diffuses into the cytoplasm where it is assimilated. The situation could arise when the cells are fixing nitrogen rapidly that they are unable to pump it out as fast as it is produced and ammonia accumulates in the cells and affects the reduction of nitrogen. The hypothesis is that the effect of ammonia in such a case is not on the inhibition of total electron flow but an inhibition of acetylene and nitrogen reduction through a lowering of $\Delta\psi$ but which is still sufficient to reduce protons. Before doing experiments to prove or disprove this hypothesis it is worthwhile to calculate the possibility that the cells can accumulate sufficient ammonia to overcome the pumping mechanism and to accumulate enough to affect the $\Delta\psi$. The data required for this calculation may be found in the work by Bergersen[4] and are given in Table 2.

Starting from a zero concentration and assuming no efflux of ammonia, in one minute the cells would accumulate ammonia to 0.75 mM. This clearly shows the need for the extrusion of ammonia from the cells demonstrated by Laane and his colleagues. It also shows that should the

Table 2. Data for the calculation of the rate of production of ammonia by *Rhizobium* bacteroids

Rate of N_2 fixation	23 μmol C_2H_4/h/g f wt.*
Ratio C_2H_4:N_2	4
Ratio infected:non-infected tissue	2
Host cell volume	12×10^{-8} cm^3 *
Number of envelopes which contain bacteroids per cell	8×10^3 *
Mean number of bacteroids/envelope	4.74*
Volume of bacteroid	1.21×10^{-12} *
Rate of ammonia production by bacteroids	0.75 μmoles/min/cm^3

* Data taken from Bergersen[4].

pumping system not be adequate to deal with the ammonia produced by nitrogenase and ammonia leaking in from the cytoplasm, ammonia could readily accumulate to the concentration values that gave the effects observed by Laane et al.[13]. This theory predicts that, at these lower concentrations of ammonia which inhibit nitrogen and acetylene reduction, proton reduction would be relatively unaffected. This has not been reported as having been looked for. Experiments with Hup⁻ strains of *Azotobacter* would be needed.

Occurrence of hydrogenase

As strain E100 was used on pea var Feltham First in the experiments to determine the apparent K_m of root nodules for hydrogen uptake it is necessary to add this note with regard, to its expression of hydrogenase. On finding hydrogen uptake with nodules Feltham First peas, contrary to previous work with this variety[2], the possibility of a mislabelling of seed was explored and seeds of this variety were obtained from two other sources and hydrogen uptake confirmed. Pea var Meteor had been used in previous experiments with *Rhizobium* strain ONA 311 and shown to have high hydrogen uptake activity. In combination with strain E100 no hydrogenase activity was detectable. The results of these preliminary experiments are given in Table 3. It should be noted that the methods used for determining hydrogen uptake were not as sensitive as those used by other workers[2] and lack of uptake as indicated in Table 3 does not mean that hydrogenase activity was entirely absent but that it was very low.

Table 3. Hydrogen exchange by the nodules of two pea varieties

Variety	H_2 uptake	H_2 evolution
Meteor	0	15.2 ± 3.7
Feltham First	5.0 ± 0.53	0

Results expressed as μmoles/g fresh wt./hr. $\pm$ standard error.
Atmosphere: 0.74 N_2, 0.21 O_2, 0.05 H_2.

It has been found that when var Alaska shoots were grafted upon roots of Feltham First hydrogenase activity was stimulated[3]. Thus it would appear that for the functioning of hydrogenase in strain 128C53 some shoot factor was required. This factor was presumably not required by strain E100 as hydrogenase activity was present with the normal host, Feltham First, shoots. Meteor peas with *Rhizobium* strain ONA311 formed hydrogenase but not with E100. Different *Rhizobium* strains therefore required different factors from their hosts in order to fully express the Hup$^+$ gene. These facts emphasise the complexity of the bacterial host interaction. The relationship of pea variety and hydrogenase activity will be explored further.

References

1   Arp D J and Burris R H 1979 Purification and properties of the particulate hydrogenase from the bacteroids of soybean root nodules. Biochim. Biophys. Acta 570, 221–230.

2   Bedmar E J, Edie S A and Phillips D A 1983 Host plant cultivar effects on hydrogen evolution by *Rhizobium leguminosarum*. Plant Physiol. 72, 1011–1015.

3   Bedmar E J and Phillips D A 1984 A transmissible shoot factor promotes hydrogenase activity in *Rhizobium* symbionts. Plant Physiol. 75, 629–633.

4   Bergersen F J 1982 Root nodules of legumes: Structure and functions. Research Studies Press. Chichester. pp 164.

5   Dixon R O D 1972 Hydrogenase in legume root nodule bacteroids: occurrence and properties. Arch Mikrobiol. 85, 193–201.

6   Dixon R O D, Blunden E A G and Searle J W 1981 Intercellular space and hydrogen diffusion in pea and lupin root nodules. Plant Sci. Lett. 23, 109–116.

7   Dixon R O D and Blunden E A G 1983 The relative efficiency of nitrogen fixation in pea root nodules. Plant and Soil 75, 131–138.

8   Fuchsman W H and Hardy R W F 1973 Nitrogenase-catalysed acrylonitrite reductions. Bioinorgan. Chem. 1, 195–213.

9   Hadfield K L and Bulen W A 1969 Adenosine triphosphate requirement of nitrogenase from *Azotobacter vinelandii*. Biochemistry 8, 5103–5108.

10 Hageman R V and Burris R H 1980 Electron allocation to alternative substrates of *Azotobacter* nitrogenase is controlled by the electron flux through dinitrogenase. Biochim. Biophys. Acta 591, 63–75.

11 Laane C, Haaker H and Veeger C 1978 Involvement of the cytoplasmic membrane in nitrogen fixation by *Rhizobium leguminosarum* bacteroids. Eur. J. Biochem. 87, 147–153.

12 Laane C, Haaker H and Veeger C 1979 On the efficiency of oxidative phosphorylation in membrane vescicles of *Azotobacter vinelandii* and of *Rhizobium leguminosarum* bacteroids. Eur. J. Biochem. 97, 369–377.

13 Laane C, Krone W, Konings W, Haaker H and Veeger C 1980 Short-term effects of ammonium chloride on nitrogen fixation by *Azotobacter vinelandii* and by bacteroids of *Rhizobium leguminosarum*. Eur. J. Biochem. 103, 39–46.

14 Postgate J R 1982 The fundamentals of nitrogen fixation. Cambridge University Press, Cambridge, 252 p.

15 Schubert K R and Evans H J 1976 Hydrogen evolution: A major factor affecting the efficiency of nitrogen fixation in nodulated symbionts. Proc. Nat. Acad. Sci. USA 73, 1207–1211.

16 Schubert K R, Jennings N T and Evans H J 1978 Hydrogen reactions of nodulated leguminous plants. II. Effects on dry matter accumulation and nitrogen fixation. Plant Physiol. 61, 398–401.

17   Sinclair T T and Goudriaan J 1981 Physical and morphological constraints on transport in nodules. Plant Physiol. 67, 143–145.

18   Silverstein R and Bulen W A 1970 Kinetic studies of the nitrogenase-catalysed hydrogen evolution and nitrogen reduction reactions. Biochemistry 9, 3809–3815.

19   Stam H, van Verseveld H W, de Vries W and Stouthamer 1984 Hydrogen oxidation and efficiency of nitrogen fixation in succinate-limited cultures of *Rhizobium* ORS 571. Arch Microbiol. 139, 53–60.

20   Walker C C, Partridge C D..P and Yates M G 1981 The effect of nutrient limitation on hydrogen production by nitrogenase in continuous cultures of *Azotobacter vinelandii*. J. gen. Microbiol. 124, 317–327.

21   Zumft W G and Mortensen L E 1975 The nitrogen fixing complex of bacteria. Biochim. Biophys. Acta 416, 1–52.

Plant and Soil 100, 157–169 (1987).
Ms. 100-14

Metabolism and translocation of fixed nitrogen in the nodulated legume

C.A. ATKINS
Department of Botany, University of Western Australia, Nedlands WA 6009, Australia

Key words Legume Nitrogen fixation Nodule Translocation

Summary Nitrogen (N_2) fixed by Rhizobium bacteroids in the legume nodule is excreted as ammonia to the surrounding host cell where it is efficiently assimilated into the amide group of glutamine. Generally glutamine is a minor exported solute of nitrogen, being further metabolised to asparagine in temperate species and to the ureides, allantoin and allantoic acid in tropical species. These solutes serve as the principal translocated forms of nitrogen in xylem. Compartmentalisation of the pathways of nitrogen metabolism and the role of ammonia in regulation of their activity is examined in nodules of both asparagine-forming (*Lupinus albus* L.) and ureide-forming (*Vigna unguiculata* L. Walp) symbioses.

Introduction

Current research progress in the area of nitrogen fixation by the legume: Rhizobium symbiosis has benefitted enormously from an application of the tools of molecular biology and especially those of recombinant DNA technology[22,31,45]. This has greatly enhanced knowledge of the genetics of the microsymbiont, particularly in relation to initial stages of infection and early nodule development. At later stages, at or beyond the onset of nitrogen fixation, the controlling influences in sustained symbiotic activity rely on the host plant and, while it is timely to ask how the tools of molecular biology can be used to examine more closely the genetic interaction between Rhizobium and its host, it is clear that many of the answers lie in a knowledge of both the limitations to and factors controlling nodule development[2,3]. Central to an understanding of nodule functioning is the integration and regulation of the nutritional complementation of the partners to the symbiosis, and recent research, highlighting the structural and biochemical features of this integration, is presented.

Structural considerations

The nutritional advantage to a plant in having a close association with a diazotroph is exploited most successfully in the legume: Rhizobium symbiosis. The host, through a series of specific morphological and biochemical responses to invasion by the bacterium, produces a unique structure, the nodule, which provides an ideal environment for sustained nitrogenase activity. In many legume species successful nodulation results in the provision of adequate nitrogen to maintain maximum rates of plant growth and, in the case of most agriculturally-adapted grain legumes, enough nitrogen for a high yield of protein-rich seed. Essential components of the host response comprise a controlled microaerobic environment close to the site of nitrogenase activity, a continuous supply of oxidizable plant assimilates and an efficient mechanism for the utilisation of ammonia.

Although legume nodules show a wide diversity of growth form and structure[18], in a general sense all contain a central medulla made up of largely infected cells and an outer cortex containing vascular traces, which carry sugars into the nodule in phloem and the products of nitrogen fixation away to the host in xylem. Structural attributes of the nodule which help sustain nitrogenase activity are evident in the high proportion of infected cells in the medulla, the distribution of air spaces, the juxtaposition of infected and uninfected cells, the distribution of leghaemoglobin, the intense development of vascular elements and, in some cases, extensive development of lenticell structures aiding ventilation. Furthermore some nodules maintain a defined and active meristem allowing continued growth and development of the organ throughout the life cycle of the host.

All these features are thought to contribute to efficient nodule functioning but perhaps the most critical of all is the arrangement of bacteroids within a host cell membrane, the peribacteroid membrane (PBM) (Fig. 1). This arrangement effectively separates the Rhizobium from its host while allowing intensive two-way traffic of materials between the cytosol of the infected cell and that of the many thousands of bacteroids enclosed in the cell. It could be said that in this way the host keeps Rhizobium 'outside' while deriving the benefits of nitrogen fixation. Under conditions where these PBM's do not form, such as in symbioses with particular Rhizobium mutants[24,46], or under conditions of incipient nodule senescence where they are rapidly broken down[9], nitrogen fixation either does not begin or promptly ceases.

The PBM is derived from the plant cell plasmalemma[25,35] which encloses bacteroids as they leave the infection thread[35]. It seems likely therefore that the PBM would retain some of the properties of the latter, engaging in processes of active transport and accelerating the rate of movement of selected low molecular weight solutes. There is limited evidence for this in the form of PBM-specific proteins[37], one of which, in soybean, has been identified as a H^+-ATP'ase[15]. A wide range of metabolites traverse the membrane including nitrogen, oxygen, ammonia and

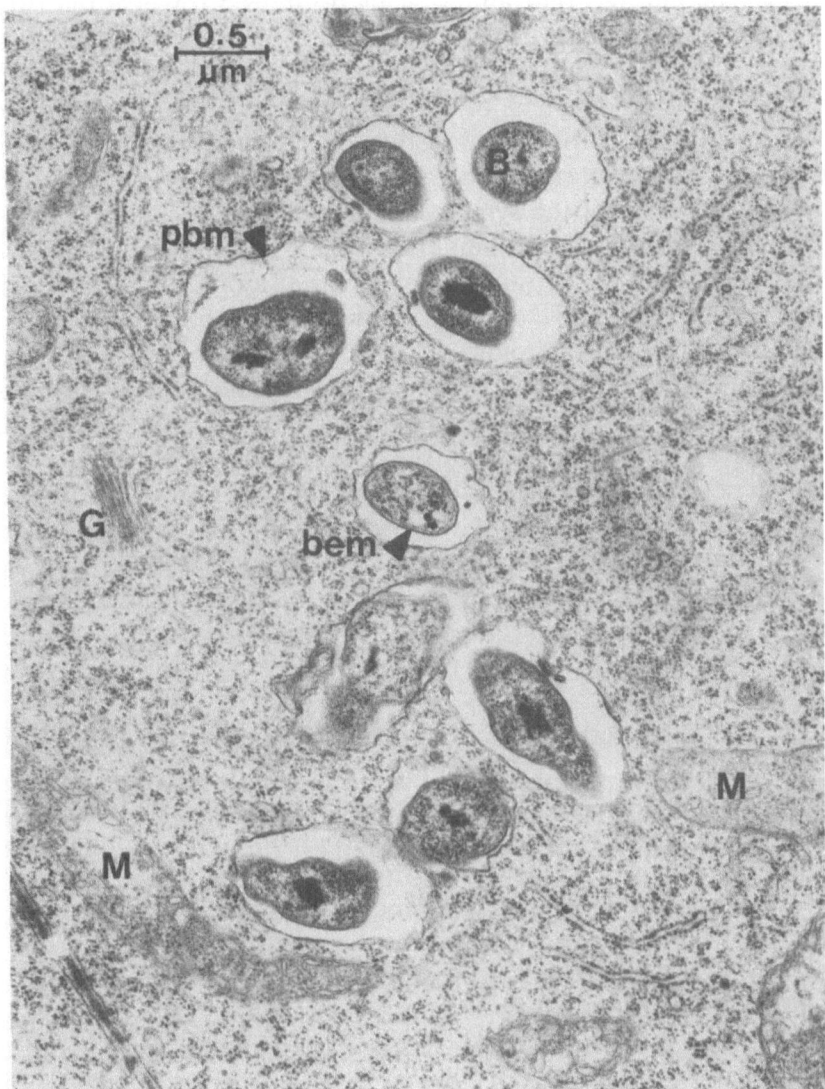

Fig. 1. Electron micrograph of a thin section of young *Canavalia ensiformis* nodule. **M**, mitochondria; **B**, bacteroid; **G**, Golgi body; **pbm**, peribacteroid membrane; **bem**, bacterial envelope membranes. Magnification × 30,000.

oxidizable carbon substrates, (possibly as organic acids[6]). A recent model for nutrient exchange[21] has suggested that in fact amino acids (such as glutamate) transported from the host to the bacteroid serve to provide energy for nitrogen fixation. While this has not been established in terms of energy metabolism, studies using ^{15}N labeled seed of cowpea[9] have demonstrated that cotyledon-N in the young seedling contributes a significant proportion of the N needed for bacteroid growth and from this it may be inferred that amino acids do indeed traverse the PBM. Larger molecules, such as haem, are known to be transported from the bacteroid to the host[19] and there is some evidence which suggests limited transport of the apoprotein of leghaemoglobin (or of the complete haemoprotein) in the opposite direction into the peribacteroid space[12]. Furthermore, if there is any exchange of genetic information between the component genomes of the symbiosis, fragments of nucleic acids could also traverse the complex of membranes enclosing the bacteroids.

Direct regulation of the rate of movement of a range of solutes acids through processes of active transport and exchange, possibly coupled to the activity of membrane-bound ATP'ases, would thus appear to be a critical component in the control of nitrogen fixation. The PBM could serve as the plant's primary means to regulate the nutrition of the Rhizobium bacteroid and as such the course and success of symbiotic development. Selectivity of solute transport could however also be a property of the bacterial envelope membranes (see Fig. 1), especially the inner plasma membrane. While undoubtedly this is the case for free-living Rhizobium there is some evidence that the envelope membranes of isolated bacteroids are fragile and relatively permeable to low molecular weight solutes[44] possibly offering only limited resistance to solute movement.

Most models of nodule functioning envisage ammonia being transported or excreted from the Rhizobium bacteroid to the host where it is assimilated. This is consistent with the properties of isolated bacteroids; levels of the ammonia assimilatory enzymes glutamine synthetase (GS) and glutamate synthase (GOGAT) are generally very low[17], the high-affinity ammonia uptake system found in the free-living organism is absent[23] and $^{15}NH_3$ is exported as a result of $^{15}N_2$ reduction[13]. While in most cases these properties were demonstrated using bacteroids isolated without their enclosing PBM's, the data from in vivo ^{15}N or ^{13}N labeling studies are consistent with ammonia assimilation in the host[28]. One of the consequences of ammonia transport out of the bacteroid is that ammonia repression of nitrogenase synthesis is precluded. Other consequences related to the generation of membrane potentials as a result of ammonium ion translocation and assimilation[23] have yet to be critically evaluated.

Assimilation of ammonia and synthesis of translocated solutes

The mechanism of ammonia assimilation, involving host cell GS (Fig. 2), is apparently common amongst legume nodules despite the fact that the composition of exported nitrogenous solutes in xylem may be very different. Glutamine, though always present in xylem exudates collected from nodulated root systems or from nodules themselves[28], is usually not the major exported product of fixation. Under conditions of high rates of fixation this amide usually accounts for around 10% of translocated N with the major role for export from nodules being taken over by products of glutamine metabolism, asparagine or the ureides, allantoin and allantoic acid (Fig. 2). The further metabolism of glutamine to form these 'transport-solutes' is highly compartmentalised[40,41] involving a number of plant cell organelles and even co-operation of different cell types within the nodule medulla.

In lupin (*Lupinus álbus* L.) nodules, where asparagine is the major product of fixation[28], glutamine-N is utilised by a plastid-located GOGAT to produce glutamate with the amino-N transferred to aspartate either in plastids or in the cytosol[40]. Asparagine synthetase then utilises aspartate and the amide-N of glutamine in a cytosolic reaction to yield asparagine[40]. However in soybean (*Glycine max.* L. Merr.) nodules, which export both ureides and asparagine, the asparagine synthetase, like GOGAT, is apparently a plastid enzyme[16].

Studies of ureide biosynthesis in nodules have centered on cowpea and soybean. In each case allantoin is formed by *de novo* purine nucleotide synthesis followed by purine oxidation[7]. The component enzymes of the *de novo* pathway, together with those of supporting reactions which generate glutamate, serine, glycine and aspartate from glutamine-N, are

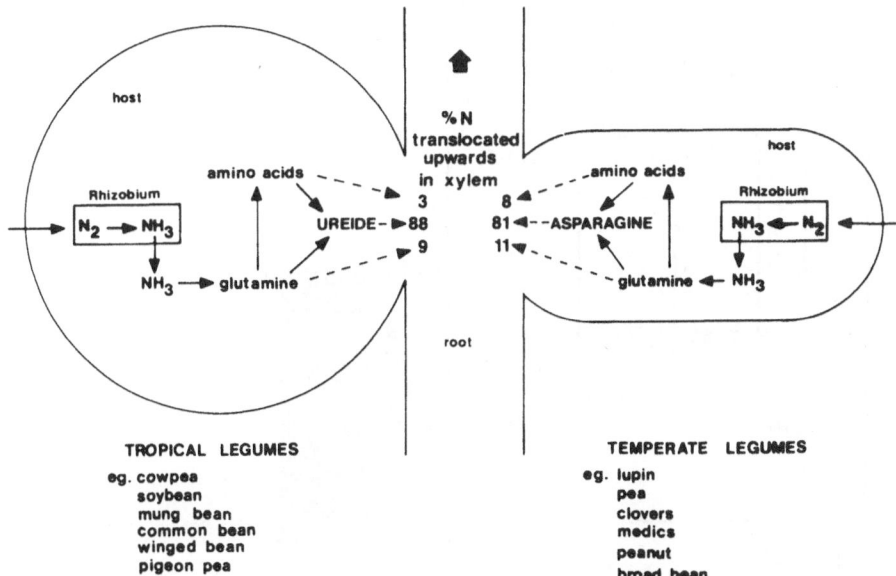

Fig. 2. Assimilation of ammonia into translocated solutes of nitrogen in temperate and tropical legume nodules. Unbroken arrows designate metabolism, broken arrows designate transport.

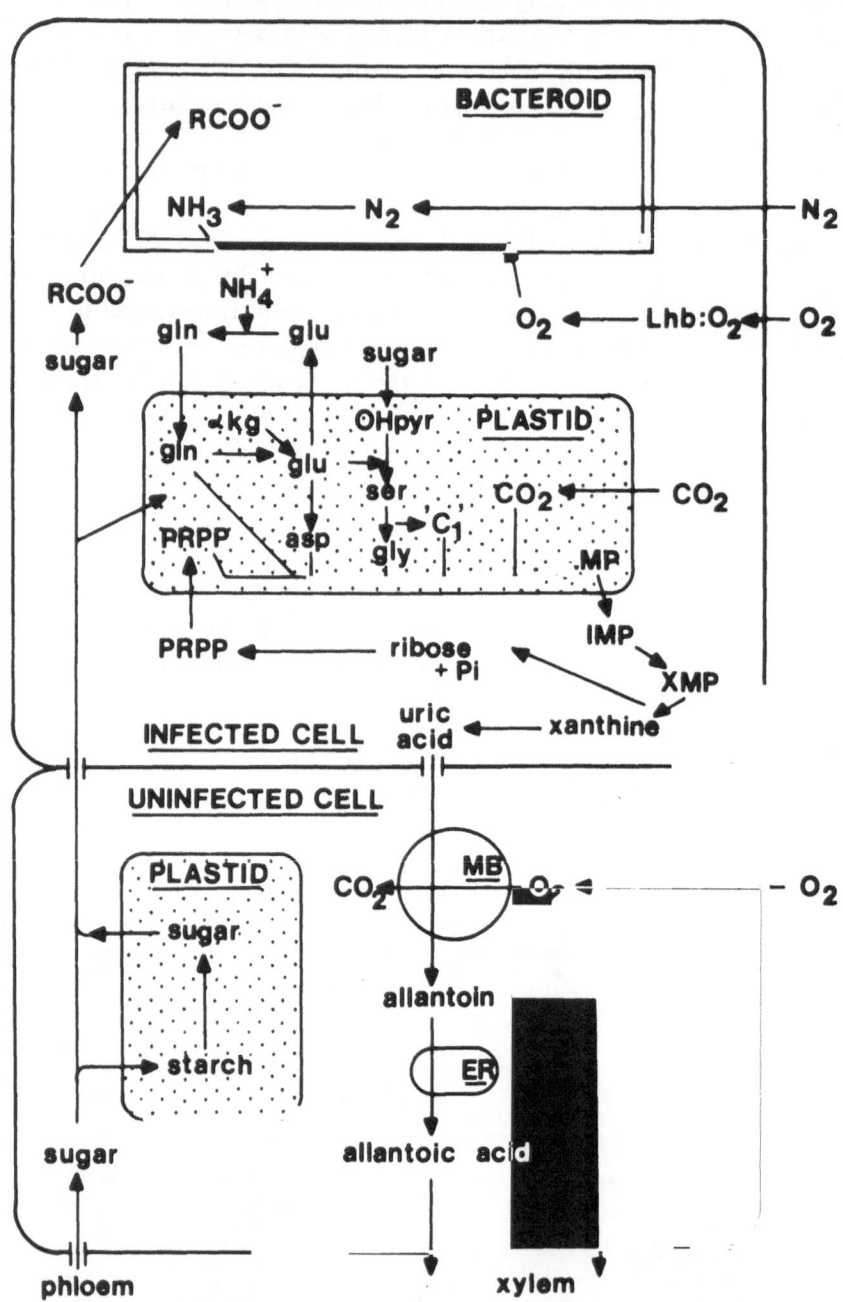

Fig. 3. Inter- and intracellular compartmentalisation of component reactions of ureide synthesis from fixed nitrogen in the cowpea nodule. RCOO, organic acid; **Lhb**, leghaemoglobin; **gln**, glutamine; **glu**, glutamate; **α-kg**, alpha ketoglutarate; **OH pyr**, hydroxypyruvate; **asp**, aspartate, **ser**, serine, **gly**, glycine; **PRPP**, phosphoribosylpyrophosphate; **IMP**, inosine monophosphate; **XMP**, xanthosine monophosphate; **MB**, microbody; **ER**, endoplasmic reticulum.

found in plastids (Fig. 3). The first purine nucleotide product of this sequence, which also depends on PRPP, carbon dioxide and a source activated C_1 units as methenyl- and formyl-THFA, is IMP. IMP oxidation via NAD-specific IMP and xanthine dehydrogenases[7,8,39] to form uric acid is cytosolic (Fig. 3). Further oxidation of uric acid to form allantoin involves a flavoprotein, urate oxidase, which is located in the microbodies of uninfected medulla cells[14,26], thus requiring transfer of the substrate out of the infected cell to adjacent uninfected cells (Fig. 3). The specialisation of these interstitial cells is reflected in their high concentration of large microbodies compared to cells hosting Rhizobium[27]. It seems that the occurrence of urate oxidase in uninfected rather than in infected cells is a consequence of the relatively higher free oxygen levels in the latter. Although there is a high rate of oxygen transfer to respiratory sites in infected cells, their high rate of oxygen consumption together with the occurrence of leghaemoglobin ensures that the free oxygen levels are extremely low ($10-200\,nM$). The Km (oxygen) for urate oxidase from both cowpea and soybean nodules is around $30\,\mu M$[32], indicating that the enzyme would not function effectively in the infected cell but could do so in non-leghaemoglobin-containing cells where free oxygen levels might be expected to approach those in equilibrium with air (around $200\,\mu M$). In cell free extracts of cowpea nodules ureide synthesis is limited by oxygen supply[47] and preliminary data indicate that this may also be the case *in vivo* (Atkins C A and Dakora F unpublished results).

The participation of uninfected cells in other processes of nodule functioning associated with nitrogen fixation and their relative significance and specialisation over a range of symbioses is not known. The proportion of cells which remain uninfected in the nodule medulla varies markedly between symbioses and, while there is some evidence that they have a lower ploidy level than those which are infected with Rhizobium[25], it seems likely that, in some cases at least, their organization is a consequence of the surrounding infected cells[27]. Uninfected cells show high levels of plastid-located starch grains (Fig. 3) suggesting their further specialisation in providing a store of assimilate for controlled release to infected cells at times of limited phloem delivery[33].

Little is known of the route or mechanisms involved in transfer of nitrogenous solutes from their site of synthesis to the vascular elements which carry them out of the nodule to the host plant. Specialised pericycle transfer cells, which could facilitate vein loading of a symplastic flux of nitrogenous solutes, have been found adjacent to the vascular bundles in nodules from a number of asparagine-based symbioses[30]. This is not the case however in those forming ureides (*i.e.* in members of the Phaseoleae[42]) and, although the role of transfer cells has not been established, their absence suggests an alternative mechanism for solute transfer. Newcomb *et al.*[27] have identified in the uninfected cells of soybean nodules a highly characteristic and extensive development of tubular endoplasmic reticulum. This could serve a similar 'export' function to that presumed to result from the greatly enlarged plasma membrane surface afforded by transfer cells. Perhaps significantly the final enzyme in ureide biosynthesis in nodules, allantoinase (Fig. 3), has been isolated in association with the endoplasmic reticulum from soybean nodules[20], placing the enzyme in an ideal location to further process allantoin.

Role of ammonia in nodule development

One of the significant and important responses of the legume host is that of the excretion of ammonia by Rhizobium; the plant providing an enzymic mechanism for its efficient utilisation, forming organic solutes of nitrogen which are suitable for translocation. Even though the component reactions of asparagine biosynthesis or the formation of ureides via purine metabolism are not unique to nodule tissues[1], the level of their activity is greatly amplified in nodules and it has been suggested[34,36] that the increase is a response to ammonia production. In the case of ureide synthesis ammonia production elsewhere in these plants, *e.g.* due to NO_3 assimilation in roots and leaves or to photorespiration, does not occasion induction of the ureide pathway, but rather ammonia is assimilated into amides[1,4]. Thus this particular response to the need for ammonia assimilation is, in a sense, 'nodule specific'.

Studies of 'ineffective' symbioses formed by mutant strains of Rhizobium and their comparison with 'effective' counterparts have been useful in identifying processes involved in the early steps of nodule initiation. However the use of these mutants has shed little light on the role of fixed N, produced by bacteroids, on the course and control of nodule growth and development. In most cases ineffective associations are arrested at or soon after rhizobia are released from the infection thread[25] and prior to the onset of fixation. A technique of culturing legumes with their inoculated root systems in an atmosphere of 80% argon:20% oxygen (v/v) provides a convenient means of forming an 'ineffective' symbiosis which is genetically identical to an 'effective' counterpart grown in normal air[9,11]. Such a comparison has allowed an assessment of the role of ammonia, formed in and excreted by the rhizobia, on the course of symbiotic development.

Prior to the onset of their premature senescence at about 16 days after initiation, nodules grown in argon:oxygen are remarkably similar to those grown in air. The lack of nitrogen does not affect release of rhizobia from infection threads, their enclosure by PBM's, or the rapid appearance of both leghaemoglobin and normal levels of nitrogenase activity[11]. In the case of argon:oxygen-grown nodules nitrogenase catalyses a high rate of hydrogen production and rapid consumption of oxidizable carbohydrate. In ineffective symbioses, in which nitrogenase is not active, there is a marked accumulation of starch in the infected cells[25]; this is not the case in the argon:oxygen-grown nodule. Studies of a wide range of enzymes in these nodules formed in the absence of nitrogen, and thus of ammonia production, indicate that although activities of GS, GOGAT, asparatate aminotransferase, xanthine dehydrogenase and urate oxidase are detectable the normally elevated levels which accompany the onset of fixation are not induced[11]. For the enzymes of *de novo* purine synthesis repression is most marked with pathway levels in argon:oxygen being at most one hundredth those in normal air.

The importance of ammonia production by the symbiosis in pacing the development of its capacity for nitrogen metabolism is seen also in the response of these pathways to an interruption in ammonia production by exposing a mature symbiosis to argon:oxygen for 3 d or so and then returning the root system of the plants to air[5]. Such treatment of both lupin and cowpea caused no significant effect on nodule growth, on the levels of plant cell or bacteroid protein, leghaemoglobin content or nitrogenase activity. In the case of lupin, export of asparagine in xylem ceased within a day[29] and in cowpea ureide export was likewise rapidly arrested[5]. Surprisingly in both symbioses the levels of plant GS were unaffected by the absence of ammonia production, whereas enzymes of glutamine utilisation (GOGAT and asparagine synthetase in lupin and GOGAT and *de novo* purine synthesis in cowpea) as well as those of purine oxidation were markedly reduced by the treatment. In the case of asparagine synthetase, activity was reduced to zero. Following transfer of plants back to air export of nitrogenous solutes from both symbioses resumed promptly. In each case, however, the major product was glutamine, reflecting the relative stability of glutamine synthetase. Subsequently glutamine declined as a proportion of exported nitrogen as enzymes of asparagine and ureide synthesis increased in activity. The effect of argon:oxygen treatment was however not restricted to enzymes of nitrogen metabolism as activities of PEP carboxylase were also modulated.

These data indicate a far-reaching effect of ammonia production on the maintenance and regulation of the symbiotic state. They also imply that many nodule enzymes have a relatively high rate of turnover, possibly with half lives around 24 hr. At present the nature of the interaction or of the participating effector molecules is not known. While a number of different mechanisms and sites for regulation (translational, transcriptional, positive and negative enzyme modulation or protein turnover) could be involved, it is clear that the pacing of asparagine or ureide formation to ammonia production is a central feature of a successfully functioning symbiosis. Interestingly when nitrogen fixation is suppressed by nitrate addition (a treatment which would not necessarily cause a sharp reduction in ammonia production in nodules) the level of enzymes of ammonia utilisation and ureide biosynthesis remains high[38].

In addition to a role in regulation of pathways of nitrogen metabolism in the nodule, ammonia produced by nitrogenase also serves as a source of nitrogen for protein synthesis both in the bacteroid and host cells. Nodule formation has a considerable demand for nitrogen at a time when, in the young legume seedling, supply of nitrogen is severely limited. Estimates from studies with *Phaseolus vulgaris* L.[43] indicate that as much as one-third of the nitrogen in the germinating seed is required for nodule formation. Using seed labeled with ^{15}N in the previous generation the relative contribution of seed reserves and first-formed products of fixation in nutrition of bacteroids and nodule plant cells has been determined for cowpea[9]. At the earliest times of sampling (about 6 d after nodule initiation) more than 50% of the nitrogen in bacteroids and the plant fraction of the nodule was derived from cotyledons. Quite rapidly however newly fixed nitrogen contributed to development with more than 90% of the bacteroid N being derived from current fixation within a few days.

The importance of ammonia production to nodule development can also be seen in plants grown in argon:oxygen. Although rather similar to those in air up to 16 d after initiation they subsequently showed arrested development and progressive ultrastructural degeneration consistent with premature senescence[9]. The symptoms of progressive nitrogen deficiency in the plant itself (inhibited shoot growth, reduced leaf protein, Rubisco and chlorophyll levels) were readily reversed by supplementing the plants with small amounts of combined nitrogen (as nitrate). This 'extra' N did not prevent degeneration of nodules in argon:oxygen, merely delaying the onset of senescence for a short time. Although protein levels in the nodule were greatly stimulated the increase was restricted to the plant cell fraction with bacteroids apparently being unable to utilise (or have access to) this supplementary source of nitrogen. These observations have direct relevance to the novel proposal by Kahn *et al.*[21] that amino acids, derived in the infected plant cell cytosol from fixed nitrogen, are translocated back to the bacteroid providing both a source of nitrogen and, under conditions of mature bacteroid functioning, a source of oxidisable carbon. Clearly in argon:oxygen the absence of ammonia produced by nitrogenase restricts the transfer of nitrogen from the host to Rhizobium suggesting that indeed an outward flux of ammonia is a component of the normal nutrition of the developing bacteroid. Whether or not there is a reciprocal ammonia:glutamate exchange[21] remains to be determined.

Acknowledgements Research from the author's laboratory was supported by grants from the Australian Research Grants Scheme and the Wheat Industry Research Council of Australia. The electron micrograph of a Canavalia nodule was provided by J Kuo, Electron Microscopy Centre, University of Western Australia.

References

1   Atkins C A 1982 Ureide metabolism and the significance of ureides in legumes. *In* Advances in Agricultural Microbiology. Ed. N S Subba Rao. pp 53–88. Oxford and IBH, New Delhi.

2   Atkins C A 1984 Efficiences and inefficiencies in the legume/Rhizobium symbiosis — A review. Plant and Soil 82, 273–284.

3   Atkins C A 1986 The legume: Rhizobium symbiosis — Limitations to maximising nitrogen fixation. Outlook on Agric. 15, 128–134.

4   Atkins C A, Pate J S, Griffiths G J and White S T 1980 Economy of carbon and nitrogen in nodulated and non-nodulated (NO_3-grown) cowpea (*Vigna unguiculata* (L.) Walp.). Plant Phys. 66, 978–983.

5   Atkins C A, Pate J S and Shelp B J 1984 Effects of short-term N_2 deficiency on N metabolism in legume nodules. Plant Physiol. 76, 705–710.

6   Atkins C A and Rainbird R M 1982 Physiology and biochemistry of biological nitrogen fixation in legumes. *In* Advances in Agricultural Microbiology. Ed. N S Subba Rao. pp 25–51. Oxford and IBH, New Delhi.

7   Atkins C A, Rainbird R M and Pate J S 1980 Evidence for a purine pathway of ureide synthesis in N_2-fixing nodules of cowpea (*Vigna unguiculata* L. Walp.) Z. Pflanzenphysiol. 97, 249–260.

8   Atkins C A, Ritchie A, Rowe P B, McCairns E and Sauer D 1982 *De novo* purine synthesis in nitrogen-fixing nodules of cowpea (*Vigna unguiculata* [L.] Walp) and soybean (*Glycine max* [L.] Merr). Plant Physiol. 70, 55–60.

9   Atkins C A, Shelp B J, Kuo J, Peoples M B and Pate J S 1984 Nitrogen nutrition and the development and senescence of nodules on cowpea seedlings. Planta 162, 316–326.

10   Atkins C A, Shelp B J and Storer P J 1985 Purification and properties of inosine monophosphate oxidoreductase from nodules of cowpea (*Vigna unguiculata* L. Walp). Arch. Biochem. Biophys. 236, 807–814.

11   Atkins C A, Shelp B J, Storer P J and Pate J S 1984 Nitrogen nutrition and the development of biochemical functions associated with nitrogen fixation and ammonia assimilation of nodules on cowpea seedlings. Planta 162, 327–333.

12   Bergersen F J and Appleby C A 1981 Leghaemoglobin within bacteroid-enclosing membrane envelopes from soybean root nodules. Planta 152, 534–543.

13   Bergersen F J and Turner G L 1967 Nitrogen fixation by the bacteroid fraction of breis of soybean root nodules. Biochim. Biophys. Acta. 141, 507–515.

14   Bergmann H, Preddie E and Verma D P S 1983 Nodulin-35: a subunit of specific uricase (uricase II) induced and localised in the uninfected cells of soybean nodules. EMBO J. 2, 2333–2339.

15   Blumwald E, Fortin M G, Rea P A, Verma D P S and Poole R J 1985 Presence of host-plasma membrane type H^+-ATPase in the membrane envelope enclosing the bacteroids in soybean root nodules. Plant Physiol. 78, 665–672.

16   Boland M J, Hanks J F, Reynolds P H S, Blevins D G, Tolbert N E and Schubert K R 1982

Subcellular organization of ureide biogenesis from glycolytic intermediates and ammonium in nitrogen-fixing soybean nodules. Planta 155, 45–51.

17 Brown C M and Dilworth M J 1975 Ammonia assimilation by Rhizobium cultures and bacteroids. J. Gen. Microbiol. 86, 39–48.

18 Corby H D L, Polhill R M and Sprent J I 1983 Taxonomy. *In* Nitrogen Fixation. Vol. 3: Legumes. Ed. W J Broughton. pp 1–35. Clarendon Press, Oxford.

19 Dilworth M J and Appleby C A 1979 Leghaemoglobin and Rhizobium hemoproteins. *In* A Treatise on Dinitrogen Fixation Sect I and II, Eds. R W F Hardy, F Bottomley and R C Burns. pp 691–764. Wiley, NY.

20 Hanks J F, Tolbert N E and Schubert K R 1981 Localization of enzymes of ureide biosynthesis in peroxisomes and microsomes of nodules. Plant Physiol. 68, 65–69.

21 Kahn M L, Kraus J and Somerville J E 1985 A model of nutrient exchange in the Rhizobium-legume symbiosis. *In* Nitrogen Fixation Research Progress Eds. H J Evans, P J Bottomley and W E Newton. pp 193–199.

22 Kondorosi A and Johnston A W B 1981 The genetics of Rhizobium. Int. Rev. Cytol. Supple. 13, 191–224.

23 Laane C, Krone W, Konings W, Haaker H and Veeger C 1980 Short-term effect of ammonium chloride on nitrogen fixation by *Azotobacter vinelandii* and by bacteroids of *Rhizobium leguminosarum*. Eur. J. Biochem. 103, 39–46.

24 Mackenzie C R and Jordan D C 1974 Ultrastructure of root nodules formed by ineffective strains of *Rhizobium meliloti*. Can. J. Microbiol. 20, 755–758.

25 Newcomb W 1981 Nodule morphogenesis and differentiation. Int. Rev. Cytol. Supple. 13, 247–298.

26 Newcomb E H and Tandon S R 1981 Uninfected cells of soybean root nodules: ultrastructure suggests key role in ureide production. Science 212, 1394–1396.

27 Newcomb E H, Tandon S R and Kowal R R 1985 Ultrastructural specialization for ureide production in uninfected cells of soybean root nodules. Protoplasma 125, 1–12.

28 Pate J S and Atkins C A 1983 Nitrogen uptake, transport and utilization in nitrogen fixation. *In* Ecology and Nitrogen Fixation, Vol. 3. Ed. W J Broughton. pp 245–298. Clarendon Press, Oxford.

29 Pate J S, Atkins C A, Layzell D B and Shelp B J 1984 Effects of N_2 deficiency on transport and partitioning of C and N in a nodulated legume. Plant Physiol. 76, 59–64.

30 Pate J S, Gunning B E S and Briarty L G 1969 Ultrastructure and functioning of the transport system of the leguminous root nodule. Planta 85, 11–34.

31 Postgate J R 1982 The fundamentals of nitrogen fixation. Cambridge University Press, UK.

32 Rainbird R M and Atkins C A 1981 Purification and some properties of urate oxidase from nitrogen-fixing nodules of cowpea (*Vigna unguiculata* L. Walp.). Biochim. Biophys. Acta. 659, 132–140.

33 Rainbird R M, Atkins C A and Pate J S 1983 Effect of temperature on nitrogenase functioning in cowpea nodules. Plant Physiol. 73, 392–394.

34 Robertson J G, Farnden K J F, Warburton M P and Banks J A M 1975. Induction of glutamine synthetase during nodule development in lupin. Aust. J. Plant Physiol. 2, 265–272.

35 Robertson J G, Lyttleton P, Bullivant S and Grayston G F 1978 Membranes in lupin root nodules. I. The role of golgi bodies in the biogenesis of infection threads and peribacteroid membranes. J. Cell Sci. 30, 129–149.

36 Robertson J G, Warburton M P and Farnden K J F 1975 Induction of glutamate synthase during nodule development of lupin. FEBS Lett. 55, 33–37.

37 Robertson J G, Warburton M P, Lyttleton P, Fordyce A M and Bullivant S 1978 Membranes in lupin root nodules. II. Preparation and properties of peribacteroid membranes and bacteroid envelope inner membranes from developing lupin nodules. J. Cell Sci. 30, 151–174.

38 Schuller K A, Day D A, Gibson A H and Gresshoff P M 1986 Enzymes of ammmonia assimilation and ureide biosynthesis in soybean nodules: effects of nitrate. Plant Physiol. 80, 646–650.

39 Shelp B J and Atkins C A 1983 Role of inosine monophosphate oxidoreductase in the formation of ureides in nitrogen-fixing nodules of cowpea (*Vigna unguiculata* L. Walp). Plant Physiol. 72, 1029–1034.

40   Shelp B J and Atkins C A 1984 Subcellular location of enzymes of ammonia assimilation and asparagine synthesis in root nodules of *Lupinus albus* L. Plant Sci. Lett. 36, 225–230.

41   Shelp B J, Atkins C A, Storer P J and Canvin D T 1983 Cellular and subcellular organization of pathways of ammonia assimilation and ureide synthesis in nodules of cowpea (*Vigna unguiculata* L. Walp). Arch. Biochem. Biophys. 224, 429–441.

42   Sprent J I 1981 Functional evolution in some papilionoid root nodules. *In* Advances in Legume Systematics. Eds. R M Polhill and P H Raven. Royal Botanic Gardens, Kew, UK. pp 671–676.

43   Sprent J I and J A Raven 1985 Evolution of nitrogen-fixing symbioses. Proc. Roy. Soc. Edin. 85B, 215–237.

44   Sutton W D and Patterson A D 1979 The detergent sensitivity of Rhizobium bacteroids and bacteria. Plant Sci. Lett. 16, 377–385.

45   Verma D P S and Long S 1983 The molecular biology of Rhizobium-legume symbiosis. Int. Rev. Cytol. Supple. 14, 211–245.

46   Werner D, Morschel E, Stripf R and Winchenback B 1980 Development of nodules of *Glycine max* infected with an ineffective strain of *Rhizobium japonicum*. Planta 147, 320–329.

47   Woo K C, Atkins C A and Pate J S 1981 Ureide synthesis in a cell-free system from cowpea (*Vigna unguiculata* [L.] Walp.) Studies with oxygen, pH and purine metabolites. Plant Physiol. 67, 1156–1160.

Plant and Soil 100, 171–181 (1987).
Ms. 100-15

Co-evolution of the legume-Rhizobium association

T.A. LIE, D. GÖKTAN*, M. ENGIN**, J. PIJNENBORG and E. ANLARSAL**
Department Microbiology, Agricultural University, Wageningen, The Netherlands

Key words Co-evolution Gene pool Nitrogen fixation *Pisum sativum* L. *Rhizobium leguminosarum*

Summary A number of examples is given demonstrating the co-existence of pea genotypes and their specific Rhizobium strains isolated within the same region. *R. leguminosarum* strains compatible with the cultivated pea have a narrow symbiotic range and they are widely distributed in European soils. This is presumably due to the narrow genetic base of the cultivated pea and its wide-spread cultivation in European soils. Rhizobium strains capable of nodulating a primitive pea line from Afghanistan were only found in soils of the Middle East and Central Asia. A more restricted distribution of specific Rhizobium strains was found for fulvum peas from Israel. Rhizobium strains effective with the fulvum pea were found in Israeli soils. A good example of co-evolution due to geographical isolation was found in south Turkey. Here a pea line was found which can form an effective symbiosis with local Rhizobium strains but not with strains from other parts of Turkey.

Introduction

Genetic variation within cultivated pea lines (*Pisum sativum* L.), with regard to symbiotic response to Rhizobium, is very low[6]. In contrast, large variation was detected in wild and primitive pea plants, and several lines were found which have a defect either in nodule formation or nitrogen fixation[2,3,4,5,10,12]. Our studies[3,5] revealed that this is due to the use of Rhizobium strains isolated from cultivated plants, which are normally available in the bacterial collections. These Rhizobium strains, coined as European strains[7], are the true symbiotic partner for the cultivated pea, but they are often incompatible with wild and primitive pea lines which is expressed as a failure in nodulation or nitrogen fixation[3,4]. Wild and primitive pea plants occur naturally in the Middle East (M.E.) and Central Asia, and we showed that compatible Rhizobium strains (Middle East strains) can be isolated from soils from this region[3,4,8]. Rhizobial strains have been isolated from soils in Turkey and other M.E. countries, capable of nodulating pea cv. Afghanistan, previously reported to be non-nodulated[3,4]. Similarly, for fulvum peas, which fail to fix nitrogen with European Rhizobium strains, highly effective Rhizobium strains were isolated in Israel, where these peas are

Permanent addresses:
 * Engineering Faculty, Ege University, Izmir, Turkey.
** Agricultural Faculty, Cukurova University, Adana, Turkey.

Table 1. Symbiotic specialization of the cultivated pea cv. Rondo from the Netherlands (NL), a primitive pea cv. Afghanistan and a fulvum pea. Fu 3 from Israel, in symbiosis with *Rhizobium leguminosarum* strains PF$_2$ (Netherlands), Tom (Turkey) and F$_{13}$ (Israel) respectively

Pisum sativum	*Rhizobium leguminosarum* strains		
	PF$_2$ (NL)	Tom (Turk)	F$_{13}$ (Israel)
cv. Rondo (NL)	E	E	E/I
cv. Afghanistan	–	E	E/I
cv. Fu 3 (Israel)	I	I	E

–, no nodules; I, ineffective; E/I, partially effective and E, effective.

still present in natural habitats[4]. These results, summarized in Table 1, suggest that there is a strong adaptation between local plants and indigenous Rhizobium strains, presumably as a result of co-evolution, culminating in an effective symbiosis.

In this paper a review of our earlier work together with some recent data is given to support the co-evolution theory.

Geographical distribution of *Rhizobium leguminosarum* strains of the cultivated pea cv. Rondo

Like many of our crop plants, there is still much uncertainty as to the ancient origin of the cultivated pea *Pisum sativum* L. This leguminous plant belongs to the earliest of the cultivated plants in the Fertile Crescent in the Middle East, and according to archeological evidence it was already under cultivation in Neolithic settlements as early as 6000 BC[9]. In the Middle Ages this plant became a prominent crop in Europe where the temperate climate is favourable for its growth. It was even stated that at that time the pea was as common as cereals in Europe[9]. All the commercial varieties grown today are derived from the pea varieties developed in Europe[9].

Due to selection and the repeated use of a limited number of successful plant genotypes in plant breeding stations, our modern cultivated pea, like many of our crop plants, has a narrow genetic base. This is probably the reason why so little symbiotic variation was detected in pea lines present in world pea collections, which consist predominantly of cultivated lines[2,6,10,12]. For example, except for the two primitive pea lines from Afghanistan and Iran, all the other 40 pea lines tested formed effective nodules with a Dutch Rhizobium strain, selected especially for the Dutch pea cv. Rondo. Moreover, when calculated per gram of nodule tissue, very little variation was found in the amount of nitrogen fixed among these pea lines[6].

A striking resemblence in symbiotic properties was shown for two pea lines from the Netherlands and Indonesia, respectively. The latter was

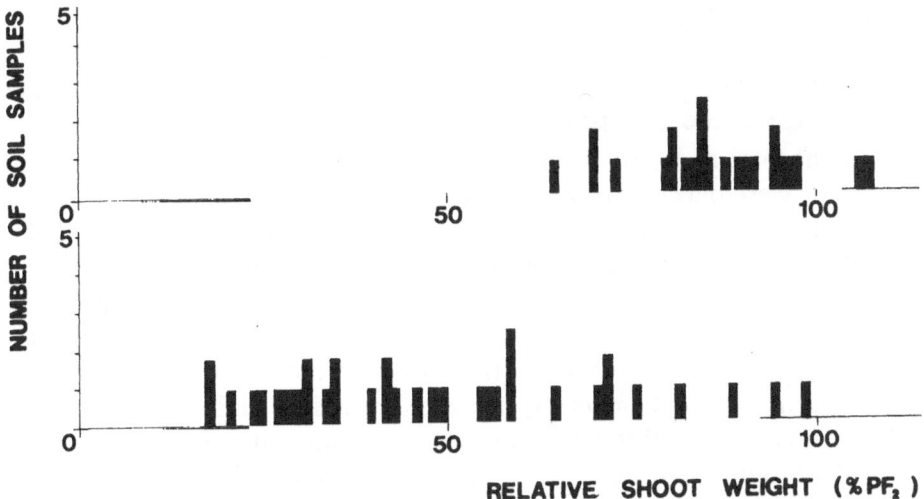

Fig. 1. Frequency distribution of soil samples from Europe (top) and Turkey (bottom) containing *Rhizobium leguminosarum* populations of different symbiotic effectiveness as expressed on pea cv. Rondo. Plants inoculated with the Dutch Rhizobium strain PF_2 serve as control (100%).

presumably introduced from the Netherlands long ago and was propagated for many generations in soils lacking *R. leguminosarum* strains. Nevertheless, when tested in Wageningen, the Indonesian pea line formed nodules with all Dutch *R. leguminosarum* strains. Moreover, the symbiotic response and the differential effect of temperature on the symbiosis, when assayed with five Rhizobium strains from geographically different regions, were essentially the same in both pea lines[8].

The wide-spread cultivation of peas in Europe[9] is presumably the reason why *R. leguminosarum*, the symbiont for peas and some other legumes, is so general in European soils. Most agricultural soils in the Netherlands contain high populations of *R. leguminosarum* strains effective for peas, and inoculation is only required in newly reclaimed soils of the polders and peaty soils in the eastern provinces.

The distribution of *R. leguminosarum* in soils from Europe and some Middle East countries was studied using the Dutch pea cv. Rondo as a test plant[7] (Fig. 1). Rhizobium strains present in soils from the Netherlands, Sweden, Italy and some other European countries contain effective Rhizobium strains for pea cv. Rondo. It is of interest to note that the best Rhizobium strains for the Dutch pea were not only found in Dutch soils, but equally present in soils from Sweden and Italy[7]. The range of variation among the European rhizobial populations, as expressed on the pea cv. Rondo, is rather narrow. They all belong to a group ranging from *ca.* 65 to 110%, relative shoot weight, when compared to the best Dutch Rhizobium strain (Fig. 1)[7]. In contrast, there is a much wider variation among the Turkish Rhizobium strains: there are many completely ineffective, partly effective and also some effective strains (Fig. 1)[7]. The latter, effective Rhizobium strains, are within the same range as the European strains.

The narrow range of the European Rhizobium strains may indicate that they were derived from a limited number of Rhizobium genotypes. We assume that during the migration of leguminous plants from the Middle East to Europe, a casual transport of Rhizobium cells took place, probably as a soil contaminant of the pods or seeds. Under conditions of low soil nitrogen in the new environment these Rhizobium cells may act as a selective agent favouring those plants which are the most efficient in nitrogen fixation. Conversely, these plants may also favour the growth of the most compatible Rhizobium strains. Since the number of plant and Rhizobium genotypes in the new environment is relatively small, selection of the best associations could have been achieved in a relatively short time.

Geographical distribution of specific *Rhizobium leguminosarum* strains in the Middle East

Wild and primitive forms of peas can be found in a region with the Middle East as its Centre, including areas of Central Asia, the Mediterranean and Ethiopia[11] (plate 1a). These areas, known as Centres of Diversity , Centres of Origin or Gene centres, are very rich in genetic variants of plants[11]. Our studies[3,4,7] demonstrate that these Gene centres are also rich in natural variants of Rhizobium.

The presence of specific Rhizobium strains in the soil can be detected by using special plants which require these specific Rhizobium strains for the symbiosis. The following plants were used: a primitive pea from Afghanistan, a fulvum pea from Israel and a recently found pea line from south Turkey, presumably belonging to the humile group.

Pea cv. Afghanistan

This pea line (Plate 1b) is representative of a group of primitive peas from Afghanistan, which are unable to form root nodules with European *R. leguminosarum* strains[2,3,5,10,12]. With the exception of a single plot in Denmark[1] all the other soils studied in Europe do not contain Rhizobium cells capable of forming root nodules on this pea. The exceptional plot in Denmark had been grown with peas 7 years ago, and the number of Afghanistan-nodulating Rhizobium cells are very low. In 1 g soil only two bacterial cells were found capable of nodulating the Afghan pea whereas at the same time *ca.* 50 000 Rhizobium cells were present capable of nodulating the cultivated pea[1].

Plate 1a. Seeds of some wild and primitive pea lines. For comparison the cultivated pea cv. Rondo (centre). Top: black-seeded type of pea ecotype abyssinicum. Middle left: pea cv. Afghanistan. Middle right: pea cv. Iran. Bottom left: pea ecotype fulvum and bottom right: a small seeded type of pea ecotype humile.
Plate 1b. Flower of pea cv. Afghanistan.
Plate 1c. Top view of an amphicarpic form of pea ecotype fulvum with normal aerial pods (green) and yellow-brown flowers. The soil is partly removed to show the white subterranean pod and the two white subterranean stems which form a hook with a tiny flower at the end.
Plate 1d. Pea ecotype fulvum in symbiosis with *Rhizobium leguminosarum* strains from different geographical regions. From left to right: Uninoculated (control), three European Rhizobium strains PF$_2$, 310a and 313, strain Tom from Turkey and strain F 13 from Israel, respectively.

The results shown in Table 2 show that Rhizobium strains capable of nodulating the Afghan pea is very general and widespread in soils of the Middle East. There is a large variation in the N$_2$-fixing capacity of these Middle East Rhizobium strains. For example, from soils of Yemen 27 Rhizobium strains were isolated, and of these, 7 strains were completely ineffective, 8 strains partly effective and 7 very effective on pea cv. Afghanistan (unpublished results).

Fulvum pea
   The fulvum peas are a distinct group of wild peas with yellow-orange flowers (Plate 1c) and occuring naturally in Israel, Lebanon, Syria and some parts of Turkey. However, only some fulvum lines from Israel were available for the study of the symbiosis[4]. An interesting feature of this group is that some fulvum plants are amphicarpic *i.e.* in addition to the normal aerial reproductive branches. subterranean shoots were also formed with small pale flowers, which eventually turn into white pods

Table 2. Geographical distribution of *Rhizobium leguminosarum* strains capable of nodulating the primitive landrace pea cv. Afghanistan and the modern Dutch pea cv. Rondo

Country	Number of soil samples	Number of soils containing Rhizobium for	
		cv. Afghan.	cv. Rondo
North-West and South Europe	55	none	52
North and South America	3	none	3
Middle East			
Crete	8	8	8
Israel	58	55	55
Iran	6	6	6
Turkey	145	140	140
Syria	10	7	nd.
Yemen	3*	3	3
Asia:			
Hindukush region (N-Pakistan)	1	1	1
Lhasa-Tibet	1	1	1

* these 3 samples were collected quite near to each other.
Data from Lie, 1978, supplemented with unpublished results

with normal seeds (Plate 1c). Another interesting fact is that the fulvum lines, so far studied, can form root nodules with European Rhizobium strains, but fail to fix nitrogen[4]. As expected highly effective Rhizobium strains can be found in soils from Israel where these plants are growing[4] (Plate 1d). Using a fulvum pea line as a test plant it was found that fulvum-specific Rhizobium strains are only present in Israel (Table 3). Unfortunately, so far there are no data yet about the distribution of these specific bacteria in Syria, Lebanon and some parts of Turkey, where these plants also occur naturally.

Table 3. Symbiotic effectiveness of *Rhizobium leguminosarum* strains in soils from different geographical regions on the fulvum pea line Fu 3 and the cultivated pea cv. Rondo

Country	Number of soil samples	*Pisum sativum* L.	
		Fu 3	Rondo
Crete	8	I	E/I
Egypt	2	I	E/I
Ethiopia	1	I	E/I
Hindukush	1	I	E/I
Iran	6	E/I	E/I
Israel	18	E	I and E/I
Mali	1	–	–
The Netherlands	6	I	E
Sweden	1	I	E
Turkey	8	I	E/I

–, no nodules formed; I, ineffective; E/I, partially effective and E, effective.
Data from Lie, 1981, supplemented with unpublished results

Table 4. Variation in symbiotic nitrogen fixation in wild peas from Israel, belonging to different ecotypes, in symbiosis with *Rhizobium leguminosarum* strains from the Netherlands (PF_2), Turkey (Tom) and Israel (F_{13}), respectively

Pisum sativum L. Ecotype	Line	*Rhizobium leguminosarum* strain*		
		PF_2	Tom	F_{13}
Fulvum	Fu 3	I	I	E
	Fu 8	I	I	E
	Fu 62	I	I	E
	Fu 63	I	I	E
	Fu 97	I	I	E
	JI 849	I	I	E
	JI 224	I/E	I	E
	Fu 27	I	E	E
	Fu 29c	I	E	E
Humile	H 18	I**	I**	E
	H 28	I	I	E
	L 1292	I	I	E
	L 936	I	E	E
Elatius	L 1293	I	E	E
	JI 64	I	E	E

* I, ineffective; I/E, very low effective; E, effective
** only a few nodules formed.
Data from Lie *et al.* 1984

A number of other wild pea lines from Israel, belonging to other ecotypes, were also studied in symbiosis with some Rhizobium strains (Table 4). In all cases the best symbiosis was obtained from Israeli strain F13, which was selected for the fulvum pea, and the worst association was obtained with the Dutch Rhizobium strain. With the Turkish strain an intermediate response was obtained. These results provide a nice example of convergence, presumably due to the selective force of indigenous Rhizobium strains in Israeli soils.

Pea cv. Adana

Recently, a wild pea, presumably belonging to the humile type, was collected in the Cukurova plain in south Turkey, near Adana. This plain is virtually isolated from the north by the Taurus mountains, and in the south by the Mediterranean sea. This pea line, coined as pea cv. Adana, proved to be highly specific in its requirement for Rhizobium. With the European Rhizobium strains tested, no nodules or only a few small, white nodules were found. Moreover, pea cv. Adana prove to be also 'nodulation-resistant' to Rhizobium strains from West Turkey and Central Anatolia. Highly effective nodules were only obtained with Rhizobium strains from the Cukurova plain and eastwards towards Antakya. So far, no soil samples were tested from East Turkey, but some

Table 5. Geographical distribution of *Rhizobium leguminosarum* strains capable of nodulating pea cv. Adana from South-Turkey

Country	Number of soil samples	Nodulation
The Netherlands	3	None
Sweden	1	None
Italy	3	None
Greece	3	None
Turkey:		
— West coast (Near Izmir, Mulga Fethiye)	32*	With most soil samples no nodules. In a few cases small, white, ineffective nodules.
— Central Anatolia (near Konya and Ankara)	5*	With most samples no nodules. In one case a few small white nodules.
— South Turkey (Cukurova plain around Adana and eastwards towards Antakya)	18*	With 17 samples large red very effective nodules. In one case no nodules).
Syria (North of Aleppo)	8*	Mainly small ineffective and partly effective nodules
Yemen	1*	partly effective nodules
Tibet (Lhasa)	1*	white ineffective nodules

* all these samples contain *R. leguminosarum* strains capable of nodulating pea cv. Afghanistan

soil samples collected near Aleppo, north Syria, near the Turkish border contain Rhizobium strains capable of nodulating pea cv. Adana, but the nodules are not very active (Table 5 and Fig. 2).

A comparison was made of the symbiotic response of pea cv. Adana with another humile pea, cv. Ankara, collected some years ago near the airport of Ankara. All Turkish strains from west, central and south Turkey will form effective nodules on pea cv. Ankara, but as expected, only the strains from south Turkey are compatible with the Adana pea (Table 6). Surprisingly, as shown in Table 7, two Israeli Rhizobium strains are also capable of forming an effective symbiosis with the Adana pea.

To explain these results we assume that there must have been some contact in previous times between south Turkey and Israel over the sea. On the other hand the Taurus mountains prove to be an effective barrier for the Rhizobium strains from the Cukurova plain and the Anatolian highlands.

Concluding remarks

The presence of highly compatible Rhizobium strains in soils of regions where leguminous plants occur naturally, or have been cultivated

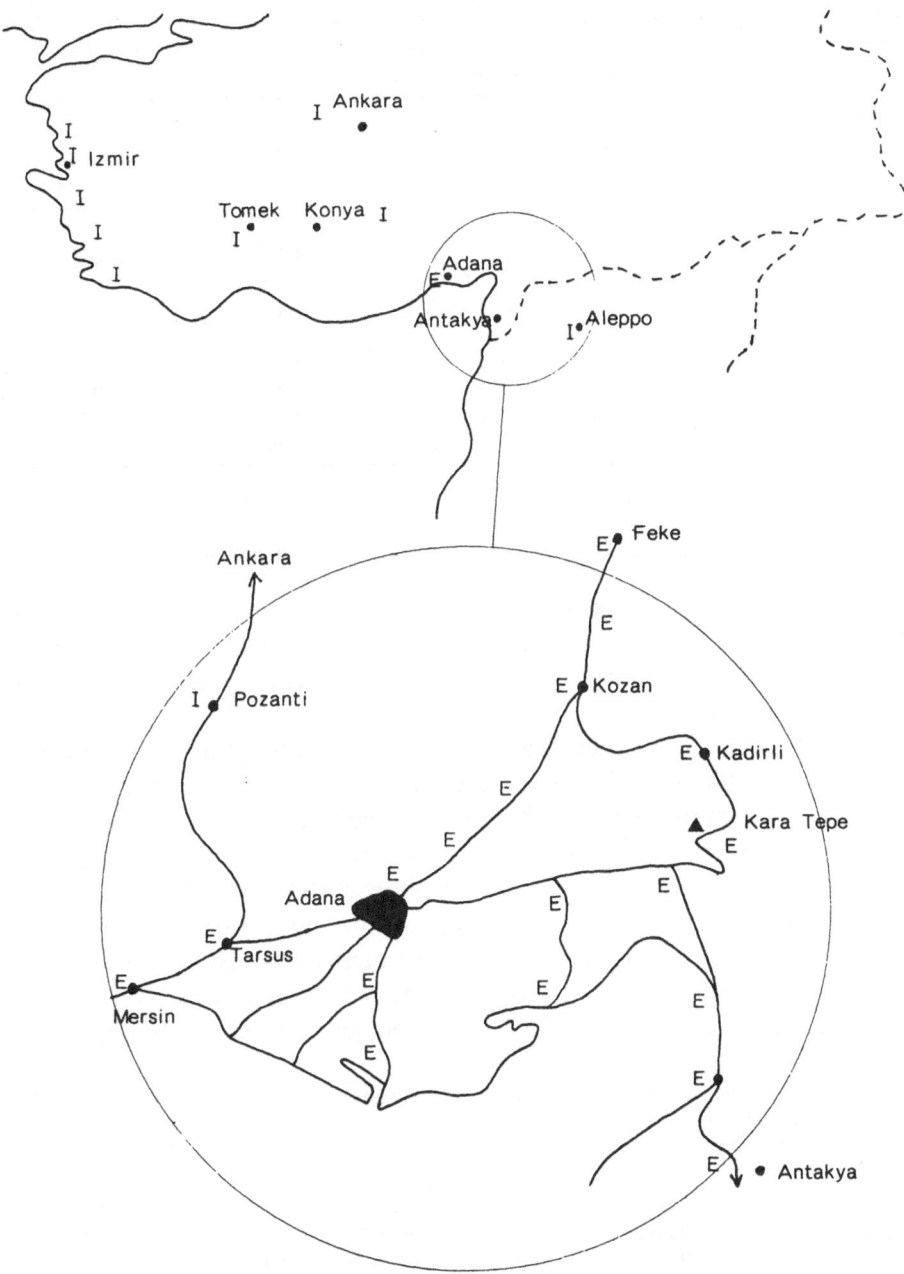

Fig. 2. Geographical distribution of *Rhizobium leguminosarum* strains in Turkey, capable of forming effective (E) or ineffective (I) nodules on pea cv. Adana.

for a long period of time, is best explained by the co-evolution theory. The world-wide distribution of *R. leguminosarum* strains, compatible with the cultivated pea, is directly related to the wide-spread cultivation

Table 6. Symbiotic nitrogen fixation of two *Pisum sativum* lines from Central Anatolia near Ankara (pea cv. Ankara) and from south Turkey near Adana (pea cv. Adana), in association with *Rhizobium* strains from different parts of Turkey and from the Netherlands

Rhizobium strain	Region	Shoot fresh weight of *Pisum sativum* L. (g/plant)	
		cv. Ankara	cv. Adana
None		0.69*	0.58*
PF$_2$	The Netherlands	0.96*	0.61*
Tom	Tomek, Anatolia, Turkey	8.36	0.55*
Ank	Ankara, Anatolia, Turkey	8.02	0.55*
IV KP	Izmir, W. Turkey	3.28	1.47*
XX	Mulga, W. Turkey	9.89	0.45*
XXIII	Marmaris, W. Turkey	6.58	0.71*
Ad^k	Adana, South-Turkey	8.63	6.01
KT 39	Kara Tepe, South Turkey	7.95	6.89

* Nodulation absent or in a few plants 1–3 small white, ineffective nodules.
Plants are yellow and suffer from N-starvation.
In other cases plants are dark green and many big red nodules were present

of the pea, first in Europe and then in different parts of the world. These Rhizobium strains, coined as European strains, show little genetic variation which is presumably a reflection of the narrow genetic base of the cultivated plant.

On the other hand these data show that a high genetic variation of leguminous plants is also associated with a high genetic variation of Rhizobium strains. This is the case in regions known as plant gene centres, where many wild and primitive plants, related to our crop plants, exist in the natural environment. This plant material is extremely important as a genetic source to be used in plant breeding stations. We suggest that Rhizobium strains from these gene centres are equally important and more efforts should be devoted to the study, exploitation and perhaps also conservation of this valuable gene pool of Rhizobium.

Table 7. Nitrogen fixation of pea cv. Adana inoculated with *Rhizobium leguminosarum* strains from different geographical regions

Rhizobium	Region	Nodulation	mg N/plant
PF$_2$	Netherlands	No nodules	1.5
Tom	Turkey	No nodules	1.5
Ad^k	Turkey	Many big red nodules	27.6
Antakya	Turkey	Many big red nodules	22.5
Ng 1a	Israel	A few big red nodules	5.4
F$_{13}$	Israel	Moderate number of big red nodules	21.4

Plants were grown under aseptic conditions in a N-free nutrient solution for 8 weeks. Mean of 6–10 plants per treatment

Acknowledgements Thanks are due to Jan Joost Kessler and Aafje van des Molen for collecting the soil samples in Yemen and in Tibet, respectively. This research is partly financed by OECD.

References

1 Jensen E S, Sørensen L H and Engvild K C 1986 Danish *Rhizobium leguminosarum* strains nodulating 'Afghanistan' pea (*Pisum sativum*). Physiol. Plant. 66, 46–48
2 Lie T A 1971 Symbiotic nitrogen fixation under stress conditions. Plant and Soil, Special Volume, 127–137
3 Lie T A 1978 Symbiotic specialization in pea plants: The requirement of specific Rhizobium strains for peas from Afghanistan. Ann. Appl. Biol. 88, 462–465
4 Lie T A 1981 Gene centres, a source for genetic variants in symbiotic nitrogen fixation: Host induced ineffectivity in *Pisum sativum* ecotype fulvum. Plant and Soil 61, 125–134
5 Lie T A 1984 Host genes in *Pisum sativum* L. conferring resistance to European *Rhizobium leguminosarum* strains. Plant and Soil 82, 415–425
6 Lie T A, Hille Ris Lambers D and Houwers A 1976 Symbiotic specialization in pea plants: Some environmental effects on nodulation and nitrogen fixation. *In* Symbiotic nitrogen fixation in plants. Ed P S Nutman. pp. 319–333, University Press, Cambridge
7 Lie T A and Göktan D 1984 Gene centres, a source for genetic variants in symbiotic nitrogen fixation: The symbiotic response of the cultivated pea to *Rhizobium leguminosarum* strains from Europe and the Middle East. Plant and Soil 82, 359–369
8 Lie T A, Akkermans A D L and Egeraat A W S M van 1984 Natural variation in symbiotic nitrogen fixing *Rhizobium* and *Frankia* spp. Antonie van Leeuwenhoek 50, 489–503
9 Marx G A 1977 Classification, genetics and breeding. *In* The physiology of the garden pea. Eds J F Sutcliffe and J S Pate. pp 21–43, Academic Press, London, New York, San Francisco
10 Ohlendorf H 1983 Selektion auf Resistenz von *Pisum sativum* L. gegen *Rhizobium leguminosarum* Stamm 311d. Z. Pflanzenzücht. 90, 204–221
11 Vavilov N I 1951 The origin, variation, immunity and breeding of cultivated plants (*English translation*), The Ronald Press Co., New York
12 Young J P W, Johnston A W B and Brewin N J 1982 A search for peas (*Pisum sativum* L.) showing strain specificity for symbiotic *Rhizobium leguminosarum*. Heredity 48, 197–201
13 Winarno R and Lie T A 1979 Competition between Rhizobium strains in nodule formation: Interaction between nodulating and non-nodulating strains. Plant and Soil 51, 135–142

Plant and Soil 100, 183–212 (1987).
Ms. 100-17

Endophyte transmission and activity in the Anabaena-Azolla association

J.H. BECKING
Department of Soil Science and Plant Nutrition, Agricultural University, De Dreijen 3, 6703 BC Wageningen, The Netherlands and ITAL, Research Institute of the Ministry of Agriculture and Fisheries, P.O. Box 48, 6700 AA Wageningen, The Netherlands

Key words *Anabaena azollae* Azolla Desiccation Macrosporocarps Microsporocarps Nitrogen fixation Sexual reproduction Survival Azolla

Summary The survival of Azolla was studied in an artificial system which simulated the soil/water interface and the desiccation of soil during a fallow period in lowland rice culture. Tests with non-sporulating and sporulating Azolla fronds showed that Azolla only survives with sporulated fronds. At their reappearance the Azolla fronds already harboured the Anabaena endophyte. A detailed light microscopic and transmission electron microscopic study of macro- and microsporocarp formation and development revealed that the endophyte is transmitted by the macrosporocarps and not by the microsporocarps. The Anabaena cells within the macrosporocarps are found just below the indusium cap. These cells are not nitrogen-fixing akinetes. The free-living Anabaena cells at the stem apex and below the overarching developing leaves do not bear heterocysts and accordingly are non nitrogen-fixing. During the development of the leaf the Anabaena enters the leaf cavity, but later the pore of this cavity closes and the imprisoned cyanobacteria are lysed before the leaf decays. As the Azolla leaves age a nitrogen- fixing capability is successively built up concomitantly with the production of heterocysts. Heterocyst frequencies of 40–50% can be found in *Anabaena azollae*. Usually a gradient of nitrogen-fixing capacity occurs along the Azolla rhizome with two distinct peaks at leaf number 7/8 and at leaf number 13/14 from the apex.

Introduction

Azolla is a small heterosporous aquatic fern which floats freely on the water surface. It contains within its leaf cavities a symbiotic cyanobacterium: *Anabaena azollae* Strasburger. Due to its nitrogen fixing capacity the cyanobacterium is able to fulfill the nitrogen requirements of the association as Azolla can thrive on nitrogen-free nutrient solution.

Azolla is of great interest to tropical agronomy, in particular with regard to lowland rice cultivation. The fern is able to double its biomass in 3–5 days and therefore has the capacity to contribute substantially to the nitrogen status of the soil and through this to the rice crop. The importance of Azolla for lowland rice cultivation has been evaluated in numerous investigations[2,3,10,12,17,19].

Nitrogen inputs of 110–330 kg N/ha per annum have been obtained under optimal conditions. However in agricultural practice it is more likely that lower values are obtained, since there is not unlimited growth of Azolla under field conditions. Further, there is a period in which the soil is fallow and then Azolla has to start growth from its propagules. The latter situation was the starting point of the present work. The aim was to see how Azolla is re-introduced to the soil after dying off and how the symbiotic association is maintained. In order to study this, the desiccation of the wetland rice soil was simulated and the survival and reappearance of Azolla was investigated. A morphological and anatomical study was conducted to monitor and evaluate the survival and reappearance of Azolla and its assocation after a fallow period.

Materials and methods

Plant and soil

Two batches of *Azolla filiculoides* Lamarck plants — one with sporocarp formation, the other without sporocarps — were allowed to desiccate at the surface of a paddy soil. The superficial Azolla biomass was also partly mixed into the soil. The paddy soil was a mud of loamy soil belonging to a previous rice-growing experiment. At the beginning of the experiment the soil was placed in plastic trays (aquaria) and covered with a layer of water. After several days the soil was drained and was allowed to dry to the moisture level of the atmosphere (air-dry) in order to simulate the desiccation of the soil during the fallow period after the rice harvest in agricultural practice.

The soil containing the Azolla plants was allowed to dry for about two months at 25–27°C. Subsequently, it was remoistened and reinundated to conform to the practice in lowland rice cultivation. The reappearance of Azolla plants at the surface of the water layer was followed. Moreover, the presence of Azolla structures in the water layer was investigated. The observed Azolla plants and the Azolla structures in the water layer were examined morphologically and anatomatically, with light microscopy and with transmission electron microscopy.

In addition, in mature Azolla plants the formation and development of sexual reproduction organs such as macro- and microsporocarps was followed morphologically and anatomatically. During the development of the sexual cycle of Azolla, the fate of the Anabaena micro-symbiont was also traced in the various stages.

Structural studies

Light microscopy. Tissue parts, growing-points of stems (apices), leaves or entire sporocarps in various developmental stages were fixed, either in glutaraldehyde (2.5%) plus paraformaldehyde (2.0%) in potassium phosphate buffer, $0.1 M$ (pH = 7.5), or in Karpenchenko (chromic-acetic-formic acid) or in Flemming (chromic-acetic-osmic acid) fixation solutions. After washing overnight in buffer, the tissue segments were dehydrated in ethanol and subsequently embedded in paraffin-wax of melting-point 52°C or in an Epon-Araldite mixture. Sections c. 10–12 μm thick were cut with a Leitz microtome or an LKB ultratome. In the paraffin sections the paraffin-wax was removed by xylol prior to staining. The sections were stained with Heidenhain's haematoxylin (0.1% in water), Safranin and Fast-Green combinations (0.1%, in alcohol) or Toluidine Blue (1.0%) in borax (1.0%). Some preparations were examined unstained. Also wet mounts of living material and fresh tissue squashes were examined.

Electron microscopy. Small tissue segments were fixed in a glutaraldehyde plus paraformaldehyde mixture (see above) in cacodylate buffer (0.1 M) or phosphate buffer (0.1 M), pH = 7.5. Previous to embedding the segments were post-fixed with osmium tetroxide (1.0%). Dehydration of the segments proceeded with ethanol and propyleneoxide followed by embedding in Epon, Epon-Araldite mixture or Spurr resin. After polymerization of the resin at 50–60°C, sections were cut with an LKB ultratome microtome, about 50 nm thick. After sectioning the preparations were stained with uranyl acetate (2.0%) and Reynolds' basic lead citrate (c. 2.0%), or with a potassium permanganate solution (0.6%).

Nitrogenase activity

Nitrogen fixation was determined with the acetylene reduction assay (ARA). Whole sporocarps or single leaves were incubated individually or a number together in 1 ml air-tight plastic hypodermic syringes in a gas volume of 0.4–0.5 ml of air supplemented with C_2H_2 (10 vol. %) at 27–28°C 1–3 h. The leaf and sporocarp samples were incubated in the light (9,000 lux).

Ethylene and acetylene concentrations were measured gas-chromatographically with a Varian aerograph Model 1740 fitted with a hydrogen flame ionization detector and a Porapak T (80/100 mesh, Waters Associated Inc., Framingham, Mass., U.S.A.) column, and operated at a temperature of 100°C.

Results

Azolla re-appearance experiments

After re-inundation of the desiccated paddy soil in none of the samples using non-sporulated Azolla fronds, Azolla growth re-appeared. Apparently all Azolla plants were killed during the drought period.

In the soils treated with sporulated fronds, Azolla seedlings appeared after rewetting and re-inundation of the soil. The time required for reappearance after rewetting was c. 20–30 days. After 3.5 weeks, the water surface of the tanks was completely covered with Azolla seedlings. The first Azolla seedlings appeared 19–20 days after remoistening of the soil. Since the soil was probably seeded with macrosporocarpic structures already with adhering microsporocarpic massulae, the process of gametophyte formation, fertilization of the ovum and germination of Azolla, takes at least 19–20 days.

Hence, survival of Azolla after a drought period is associated with the production of sexual organs. Sporocarp formation is obviously a prerequisite for the continuation of Azolla growth. Examination of the Azolla seedlings showed that they already contained the Anabaena endophyte (see later). This observation induced us to examine in more detail the sexual reproduction of Azolla and the transmission of Anabaena endophyte during this cycle.

Reproduction cycle of Azolla sporophytes

Under certain external conditions of light intensity, temperature and day-length and some growth conditions, sporocarp formation is initiated in Azolla plants. The exact factors governing this transition are not yet well understood.

In our artificial laboratory (greenhouse) conditions sporocarp production of Azolla seems to be induced by high light intensity, relatively high temperature and a shortening of day-length. Moreover, some biological factors are necessary such as the formation of dense mats of mature plants causing overcrowding and production of multilayers. Under these external conditions and ecological stress, all our *Azolla filiculoides*

186 BECKING

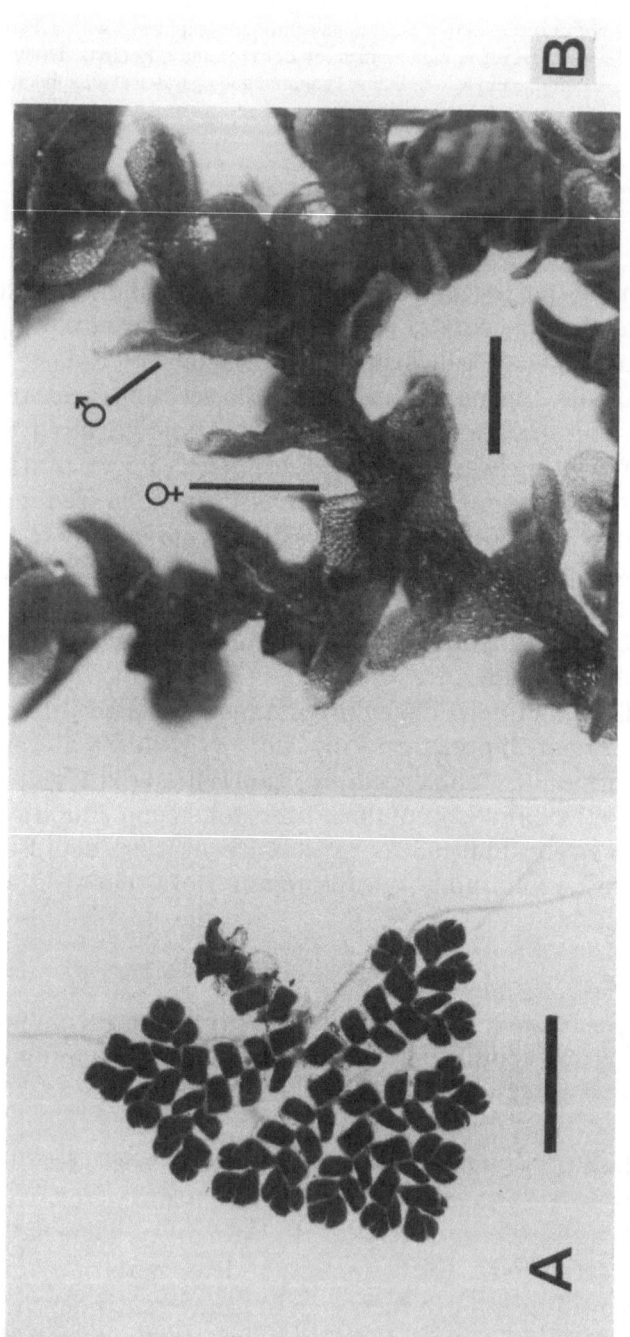

Fig. 1. A. General morphology of *Azolla pinnata* R. Brown showing the main rhizome, the alternate branching, and the leaves. Bar = 0.5 cm. B. Macrosporocarps (♀) and microsporocarps (♂) of *Azolla filiculoides* Lamarck as seen from the lower side of the rhizome. Bar = 1.0 mm.

clones, which are of temperate origin, showed every year in the months of July and August massive sporulation. However, none of our clones of the tropical (Asiatic) *Azolla pinnata* were induced to produce sporocarps under these conditions. Thus, the conditions needed for every *Azolla* species or clone to achieve sporulation might be different.

Azolla plants have a rather regular morphological structure with a main rhizome bearing alternate branches, which, in turn, bear lateral branches (Fig. 1A). Sporocarps are initiated at the point of attachment of branches and replace the lower lobe of the first leaf of a branch. The dorsal lobe forms a flap-like structure, the involucre, enveloping the sporocarp. Sporocarps are borne in pairs, either both of the same sex, or one male and the other female (Fig. 1B). Female sporocarps are called macro- (or mega-) sporocarps, the male organs microsporocarps. Sporocarps are short-stalked and have at their base a number of filamentous hairs. The sporocarps receive a vascular bundle, which penetrates a short way into the stalk.

Macrosporocarp development

Macrosporocarps are oval-shaped with pointed ends, they resemble a bottle or an urn (Fig. 2). In mature stage they are about 0.75–1.0 mm in length and 0.5 mm in breadth. They are yellowish green in colour, becoming more yellowish at the tip having a dark brown upper part of the indusium. The latter part becomes quite hard due to lignification and tannin deposits (Plate 1A).

Macrosporocarp development was followed from the very early stages, morphologically and anatomically. A young macrosporocarp containing a functional macrosporangium is depicted in Fig. 3A. At this stage the macrosporocarp is only 0.30–0.35 mm in length and 0.13–0.14 mm in width. The 2-cell thick sporocarpic wall or indusium (see a) is well developed and its upper portion is already closed and slightly lignified as made evident by the dark colour of the macrosporocarp tip. In an earlier stage the 2-cell layered indusium surrounding the macrosporangium is opened at the tip by a small pore. Through this opening cyanobacteria can slip into the cavity formed by the indusium enveloping the macrosporangium. In a young macrosporocarp, the imprisoned Anabaena cells surround the columella stalk inside the macrosporocarp from the upper part of the bottom of the cavity. It is not difficult to conceive the way of penetration of the cyanobacteria through the pore, because the formation of a sporocarp and its indusium originates from a modified leaf primordium and at these sites cyanobacteria are always present (see later: leaf development). In Fig. 3B representing a later stage, the functional macrospore is not visible since the columella

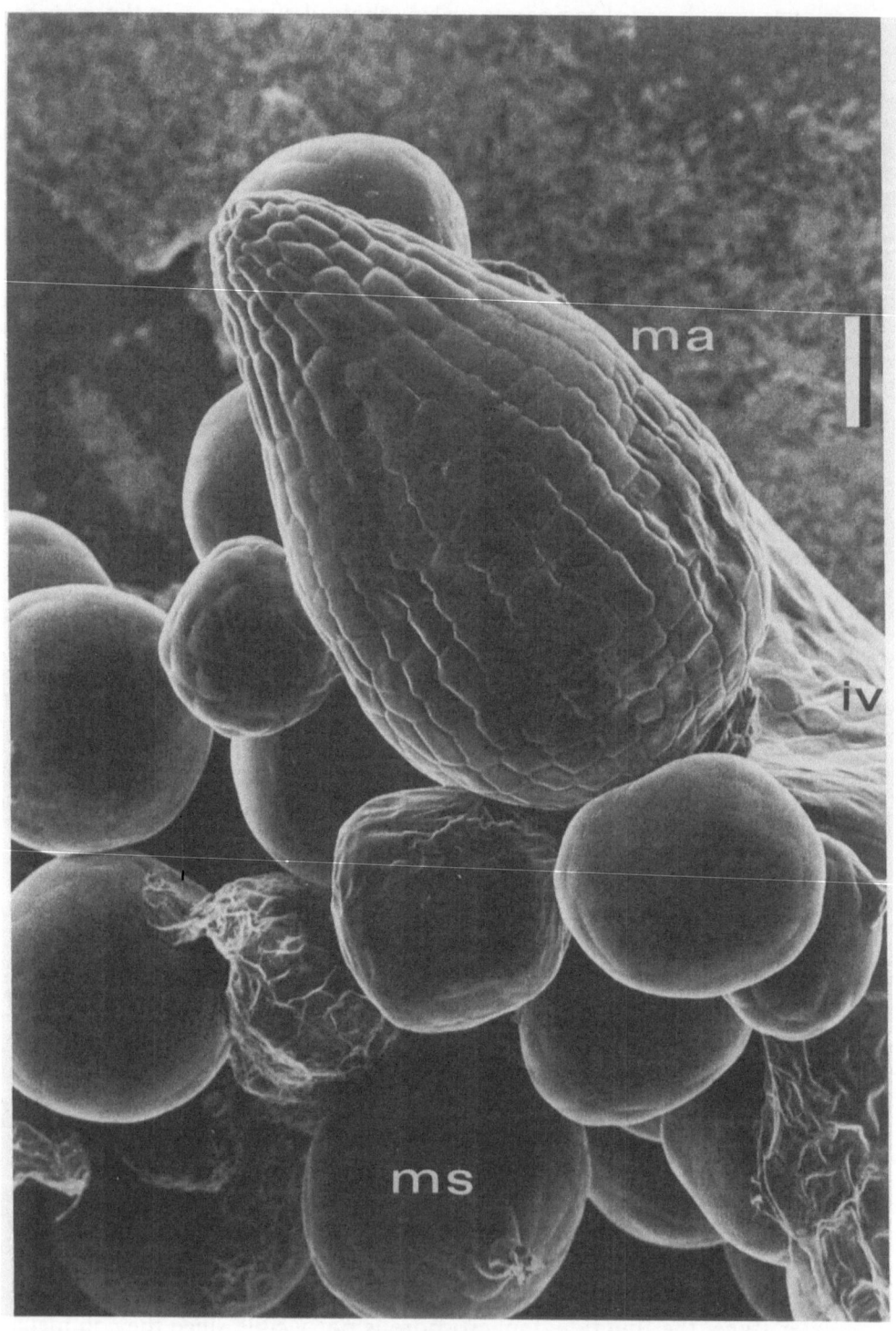

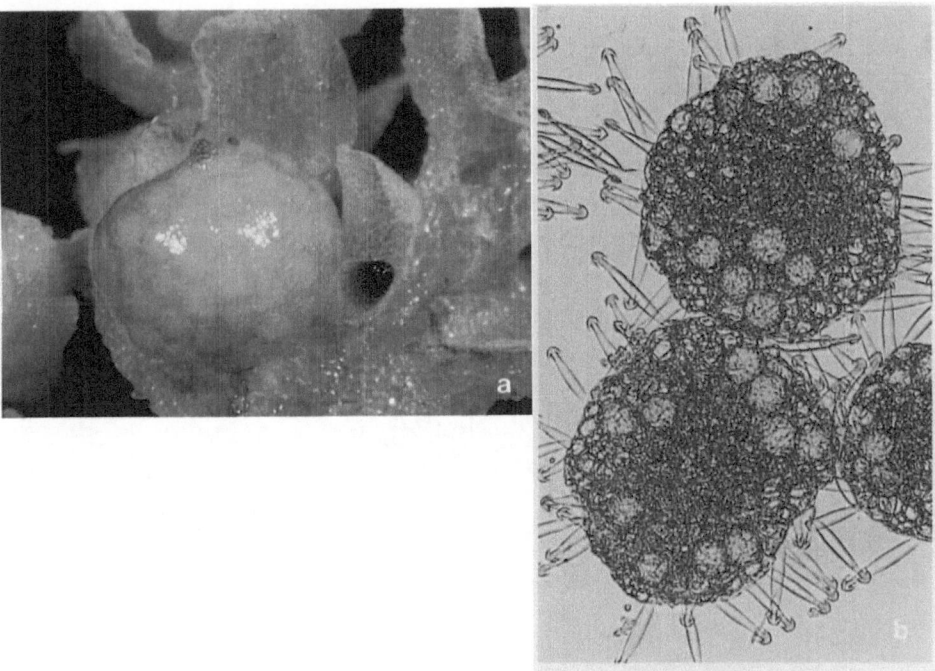

Plate 1.A. A pair of sporocarps, male and female, *in situ* in the axil of the lower-lobe of an *Azolla filiculoides* frond. The dorsal leaf-lobe is visible as a flap-like involucre. As clearly illustrated the urn-shaped macrosporocarp (female) is much smaller than the spherical microsporocarp (male). The dark indusium tip of the macrosporocarp due to lignification and tannin deposits can be clearly seen. Magnification × 20.

B. Free massulae of *Azolla filiculoides* showing their alveolar structure in which the microspores (light spheres) are embedded and the anchor-like projections, so-called glochidia, extending from their surface. The latter structures play an important role in the attachment of the massulae to the exposed rough outer-wall of the macrospore (epispore) of the macrosporocarp. In this way microspores are able to come in close contact with the single macrospore of a macrosporocarpic structure. Magnification × 270.

stalk bearing it, is outside the plane of sectioning. As is evident from morphological examination, the cyanobacteria penetrating the cavity below the indusium are vegetative cells bearing no heterocysts.

In the ripe macrosporocarp (0.75–1.0 by 0.42 mm) the enlarged macrospore and the developing alveolar bodies have pressed the cyanobacteria to the tip of the macrosporocarp, where they finally occupy a space just below the indusium. Fig. 4 (A and B) shows a nearly mature macrosporocarp of about 0.50 by 0.13 mm in size. Serial sections showing various stages of macrosporocarp development indicate that the

Fig. 2. Scanning electron micrograph of a rhizome of *Azolla filiculoides* Lamarck. The urn-shaped macrosporocarp (ma) is visible and is surrounded by microsporangia (ms), which has been released as a result of the collapse of the microsporocarpic outer wall. The flap-like involucre (iv), which initially partly encloses the macrosporocarp, is also visible.
Bar = 100 μm.

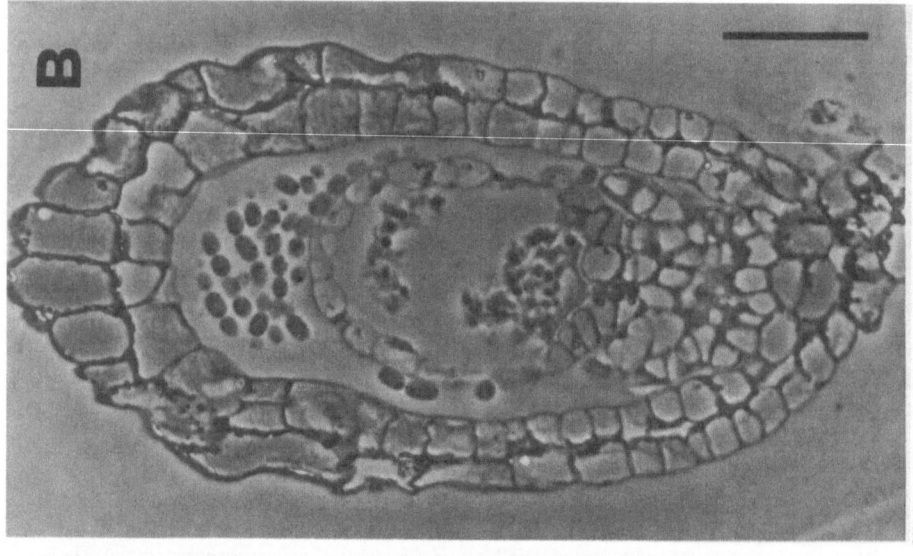

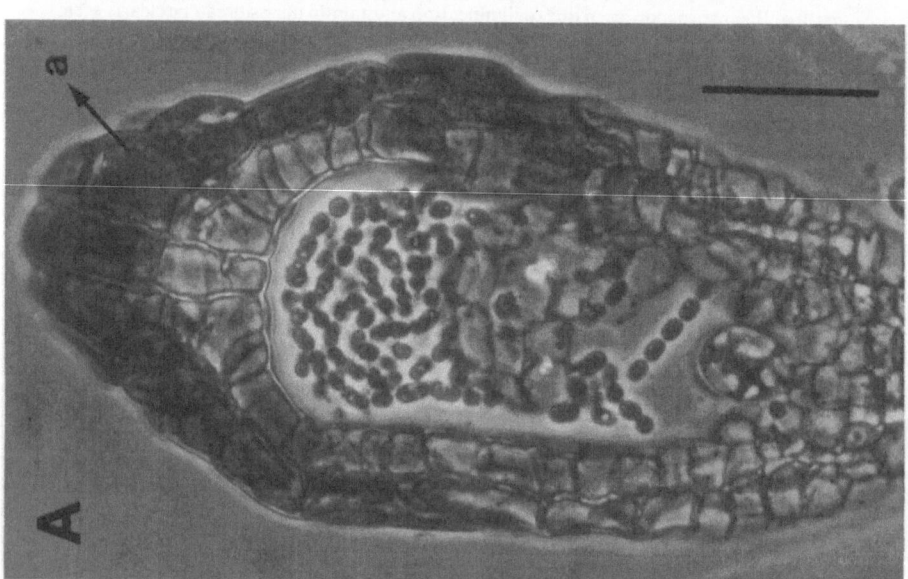

Fig. 3. **A.** Young developing macrosporocarp. The *Anabaena azollae* filaments fill the entire macrosporocarpic cavity.
B. Young macrosporocarp at a later stage of development. The *Anabaena* cells are forced to the upper part of the cavity by the developing macrospore and the alveolar bodies forming the floating bodies. The 2-cell thick sporocarpic wall or indusium can be clearly seen (see a). Bar = 40 µm.

cyanobacteria actually multiply within the sporocarp cavity after the indusium is closed at the tip. In young macrosporocarps cyanobacteria cells are always present in rows or filaments (see Fig. 1A). In a mature macrosporocarp normally only single cyanobacterial cells or very short chains comprising 2–3 cells, have been observed. This situation may be due to the expanding alveolar bodies or floating structures in the upper portion of the macrosporocarp causing a rupture of the cyanobacterial chains. However, it is more likely that in later stages of macrosporocarp development a reduction of cell division occurs and the cells tend to change into a resting stage. In young macrosporocarps the cyanobacteria cells appear to be enlarged as compared to vegetative cells of *Anabaena azollae* observed at the stem apex of Azolla. In this stage, both at the apex and in the young macrosporocarp the Anabaena cells manifest themselves as chains or filaments. The cyanobacteria cells at the stem apex are 3.3–4.8 by 4.2–5.5 μm being thus rather round, whereas the cyanobacteria cells within young macrosporocarps are more elongated measuring in average 7.2 (5.9–8.5) by 5.0 (4.6–5.4) μm.

There is again a slight increase in the size of the enclosed cyanobacteria cells to 7.8 (5.9–9.7) by 5.4 (4.9–5.9) μm, when the macrosporocarps become more mature. The Anabaena cells in a mature macrosporocarp are probably in a resting stage, such cells are usually called 'spores' or 'akinetes'. Generally akinetes are characterized by a thick cell wall, which is often brown or yellow in colour and which has sometimes an elaborated and sculptured texture with appendages. In contrast akinetes of *Anabaena azollae* are rather difficult to recognize. They are commonly found in the leaf cavities of older Azolla leaves next to vegetative cells and heterocytes and are only distinguishable by their larger size, being sometimes spherical or several times larger in length than in width. Usually their protoplasm showed reduced amounts of photosynthetic pigments, but there is an accumulation of large quantities of cyanophycin granules, which are proteinaceous reserves. Light microscopic observations have revealed such large spherical or oval cells filled with reserve granules in mature macrosporocarps. Such cyanobacterial cells, showing granulous cell contents, are often clearly visible in macrosporocarps, but they are also among cyanobacterial cells in the leaf cavity of older leaves. Thus these cells are very likely to be akinetes, although their cell-wall thickness is about the same as of ordinary vegetative cells. The latter feature seems to be a particular characteristic of akinetes of *Anabaena azollae* being the reason why many investigators failed to find them. Transmission electron micrographs of the *Anabaena azollae* akinetes occurring in the upper portion of a mature macrosporocarp close under the indusium confirm the enlarged size of these

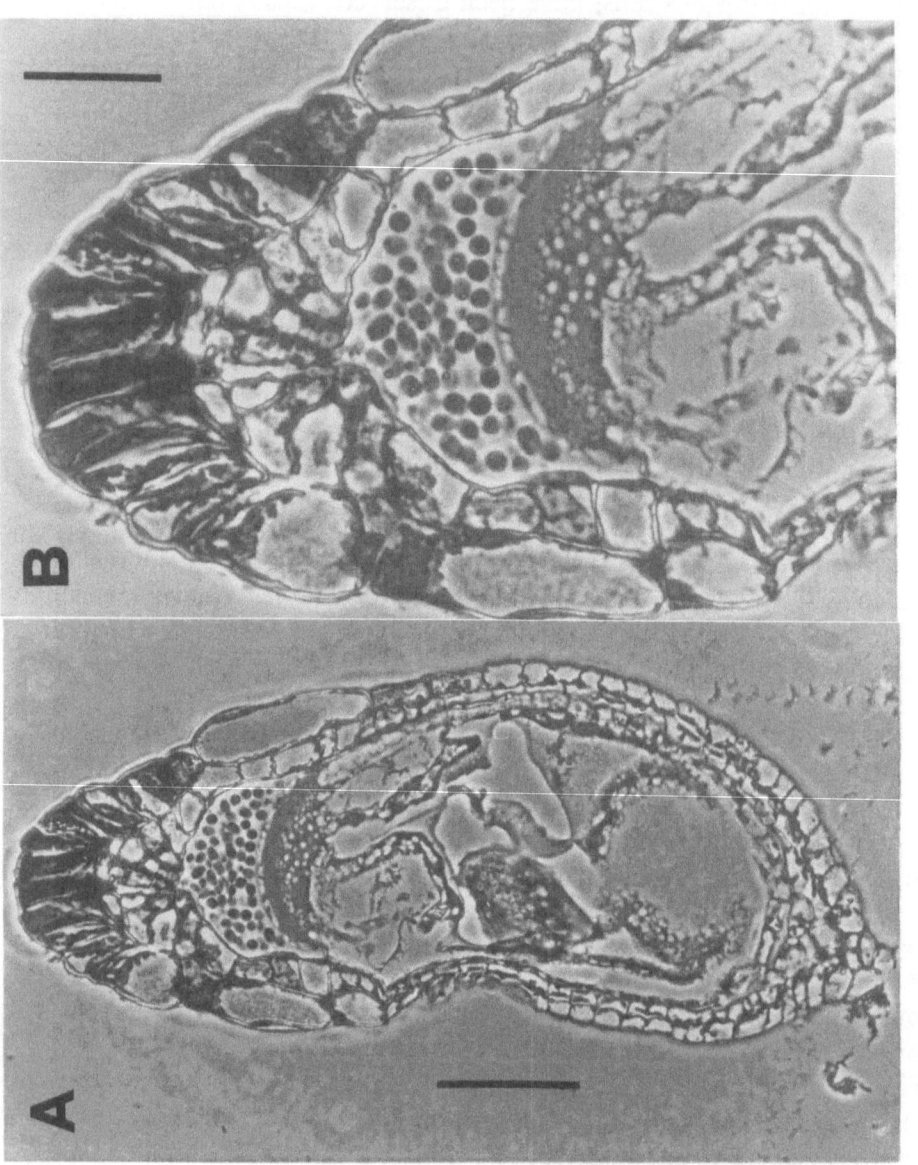

Fig. 4. A. Nearly mature macrosporocarp. The *Anabaena azollae* cells are pushed to the tip of the macrosporocarp and are now situated just below the indusium cap. Bar = 80 μm.

B. Close-up of the apical region of a nearly mature macrosporocarp showing the *Anabaena azollae* akinetes below the indusium cap. Bar = 40 μm.

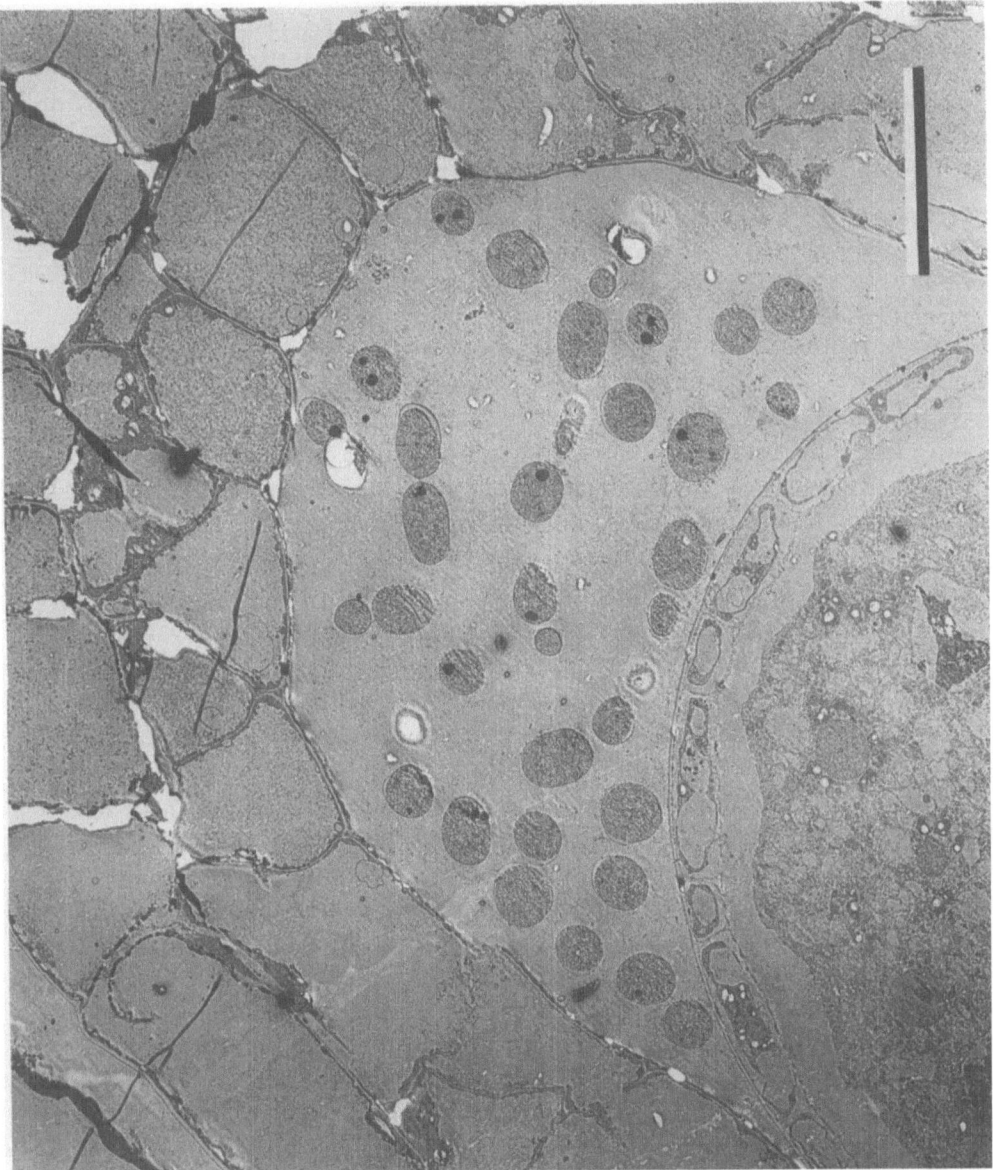

Fig. 5. Transmission electron micrograph of the tip of a mature macrosporocarp of *Azolla filiculoides* Lamarck. The *Anabaena azollae* akinetes below the two cell-layered indusium cap are visible. The akinetes have a relatively large size, but possess thin cell walls. The dark spots within the *Anabaena* cells are reserve materials, probably polyphosphate granules. Bar = 10 μm.

cells and their relatively thin cell walls. Moreover, electron dense areas or spots can be seen which, according to the used staining technique (uranyl acetate and lead citrate), are probably polyphosphate granules (Fig. 5).

Heterocyst-like cell structures were never observed within developing macrosporocarps nor in ripe macrosporocarps. Experimentally it was shown that the cyanobacteria cells inside macrosporocarps have no nitrogen-fixing capacity as evident from acetylene reduction tests (see Table 1). This confirms physiologically the absense of heterocysts in macrosporocarps, because heterocysts are supposed to be the site of the nitrogenase activity.

Microsporocarp development

A mature microsporocarp is more or less globular in shape and measures approximately 2 mm in length and 1.5 mm in breadth. Therefore it is much larger than a mature macrosporocarp.

The early stages of development of microsporocarps and macrosporocarps are very similar. In fact microsporocarps develop from a stage of macrosporocarp's formation. In microsporocarp formation, microsporangial initials arise from the stalk of a macrosporangium, when there is complete degeneration of all macrospores present. If one of the four macrospores survives, a macrosporangium will develop. Once the development of microsporangial initials has started, additional initials appear in basipetal succession. At maturity as many as 120 or more microsporangia may develop inside a single microsporocarp.

Fig. 6A gives a longitudinal section through a young microsporocarp showing the numerous stalked microsporangia developing in basipetal direction on a short placenta. The developing microsporangia may range in number from 50 to 120 or more occupying the whole space within the relatively large, mature microsporocarp. Within the cytoplasm (generally called periplasmodium) of a microsporangium 32 or 64 microspores develop. These spores are aggregated into 3–4 alveolar massulae within each microsporangium. These alveolar bodies are homologous to the float corpuscles of the macrosporocarps. Such a development of spores and alveolar structures (massulae) within a microsporangium is shown in Fig. 6B. Along with the development of massulae, strands of the periplasmodium harden to form tubular or anchor-like structures called glochidia. These glochidia, formed at the inner surface of the massulae, straighten when the microsporangial wall disintigrates. Glochidia have an important ecological function. When they are freed, they become entangled with the long, branched processes of the macrosporic wall or perispore.

Because of the initially similar developmental stages of macro- and microsporocarps, the enclosure of cyanobacterial cells within the indusium near the microsporocarpic tip may be expected. However, we never found any cyanobacterium cells within the indusium wall of the

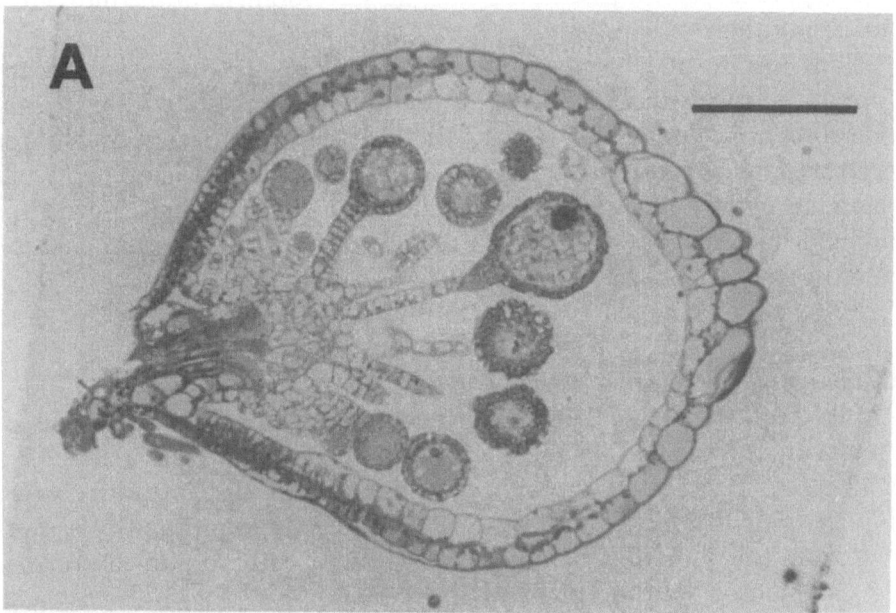

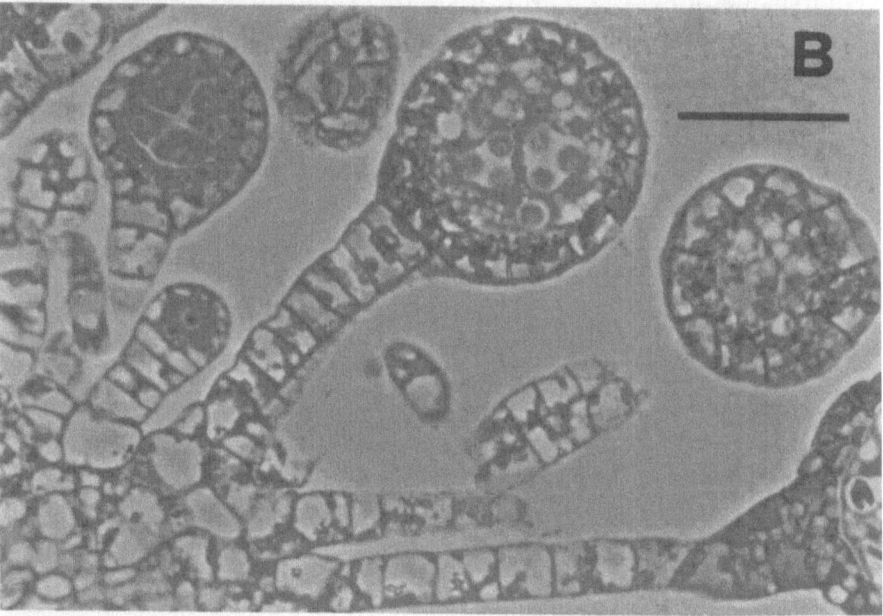

Fig. 6. **A**. Longitudinal section through a young microsporocarp of *Azolla filiculoides* Lamarck. The numerous stalked microsporangia which have developed in basipetal direction on a short placenta are visible. Bar = 160 μm.

   B. Microsporangia of *Azolla filiculoides*. Within the microsporangium the developing massulae containing microspores (dark spots) can be seen. Bar = 40 μm.

microsporocarps, nor associated with the development of the individual microsporangia.

Thus inspite of a similar ontogeny of micro- and macrosporocarp formation, the cyanobacteria are evidently not enclosed within the microsporocarpic indusium. It may be that the cyanobacterial cells are destroyed at a very early stage of microsporocarp development, but remnants of cyanobacterial cells were never observed in microsporocarps by light microscopy (see Figs. 6A and B) nor by transmission electron microscopy.

Gametophyte and zygote development

Like true ferns (Pteridophyta), the first gametophytic structures of Azolla are formed by the macrospore and microspore. These female and male structures produce an archegonium or antheridia, respectively (see Fig. 7). Fertilization of the egg-cell within an archegonium by a spermatozoid (or antherozoid) gives the zygote or embryo and, by further development, a seedling plant.

A microspore remains embedded with its massula during the development to gametophyte. A mature macrospore does not become freed from the indusium and therefore macrogametophytic development also remains completely concealed in the macrosporocarpic complex. This consists of the freed wall of the macrospore (epispore) at its base, and the

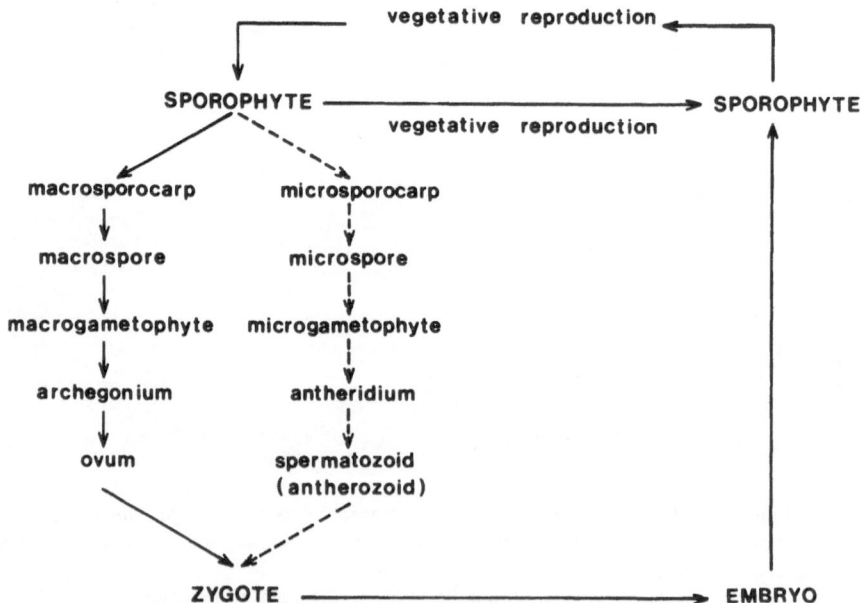

Fig. 7. Life cycle of *Azolla* species. Solid arrows represent the continuation of the association with *Anabaena azollae*. Broken arrows indicate no continuation of the association.

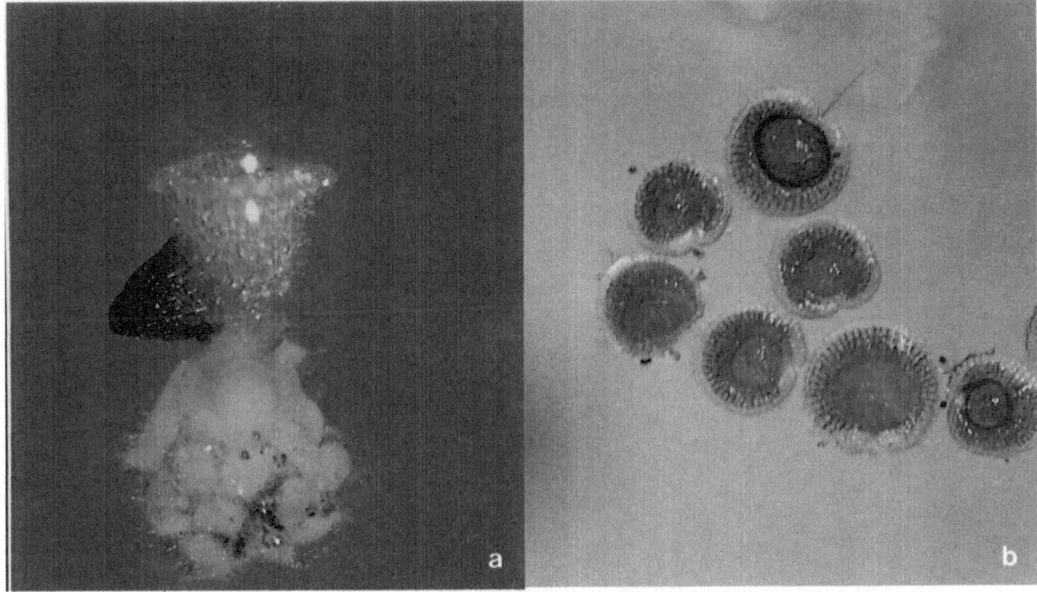

Plate 2.A. An Azolla seedling emerging from a fertilized macrospore of the macrosporocarpic complex, which still shows the many adhering massulae containing the microspores. In the life cycle of a fern, germination is preceded by macro- and microgametophyte formation (see Fig. 7), however, these structures remain concealed in the macrosporocarpic complex. These developments take place in close proximity of each other, as the massulae containing the microspores become entangled with the sculptured processes of the macrospore outer-wall (epispore). In the photograph by the germination stripped-off lignified indusium tip of the macrosporocarpic complex is clearly visible. Magnification × 52.

B. Recently germinated Azolla seedlings floating freely on the water surface. In some of the seedling plants associated Anabaena filaments (see topmost plant) are visible. Magnification × 22.

cap-shaped indusium, dark-coloured by lignification at the tip. The anchor-like ends of the glochidia (Plate 1B) enable massulae containing the microspores to become entangled in the filamentous appendages of the epispore wall and in this way they become firmly attached to the macrosporocarpic complex (Plate 2A). The development of micro- and macrogametophytes in close proximity facilitates the fertilization of the egg produced by the archegonium.

The development of the embryo occurs completely within the macrosporocarpic complex. Only when the seedling plant forms its cotyledon leaf, the first external sign of macrospore germination becomes visible as the developing cotyledon leaf pushes away the dark-coloured lignified indusium cap (Plate 2A). Soon the cup-shaped cotyledon leaf becomes detached from the macrosporocarpic complex leaving behind an empty macrosporocarpic structure and a free-floating Azolla seedling plant on the water surface (Plate 2B). At this stage there is not yet root formation. In the later development of the seedling plant the apical meristem in the

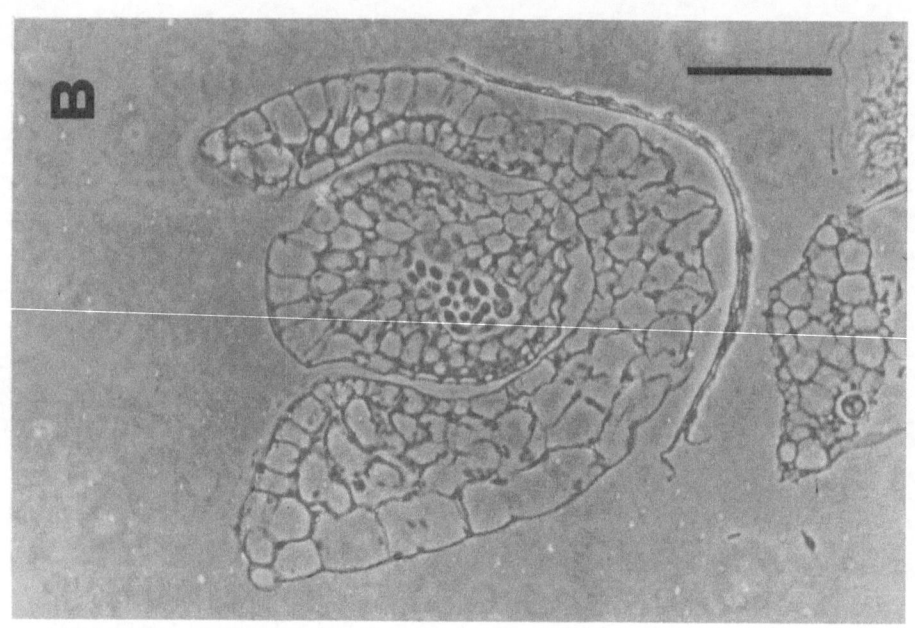

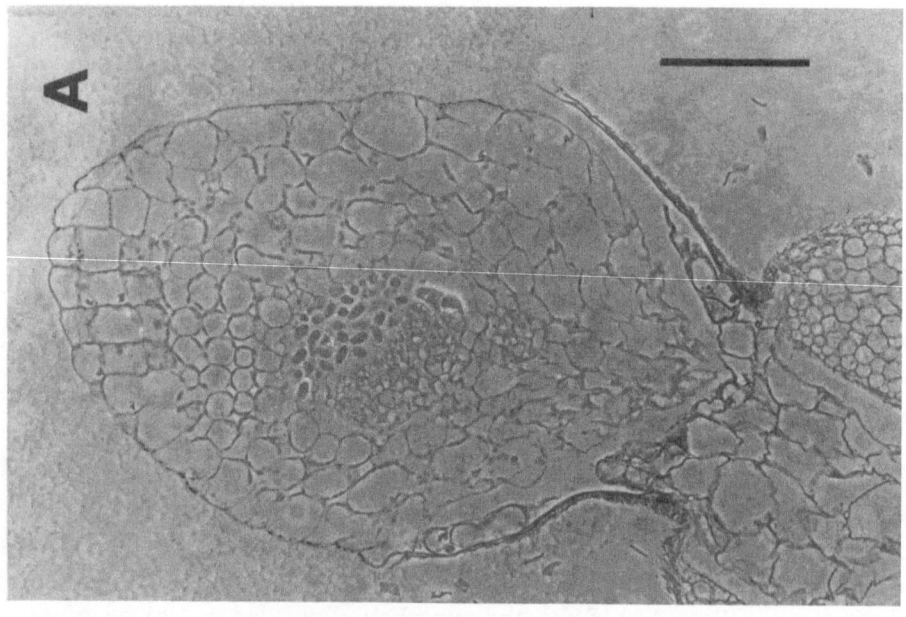

centre of the cotyledon leaf produces the first bilobed leaves containing in their leaf cavities the cyanobacterial endophyte.

Infection of the leaf by the cyanobacterium symbiont occurs when the cotyledon leaf passes the indusium cap containing the akinetes of the symbiont (Fig. 8A, B) (see also Figs. 3B, 4A, B and 5). In this way the cyanobacteria also become associated with the apical meristem including the meristem initials emerging in the centre of the cotyledon leaf which produce the subsequent leaf primordia.

The biological events described above ensure the continuity of the association after sexual reproduction of the Azolla host plant. After the elimination of the Azolla vegetation by a periodic crash of the Azolla mat or by drainage and drying-up of the soil, new Azolla plants will appear which already harbour the symbiont by this mode of sexual reproduction. The first cyanobacterial cells associated with the apical meristem and incorporated into the primordial leaf cavities are derived from the cyanobacterial akinetes present below the indusium cap. These cells were observed (see Fig. 4A, B and Fig. 5) to be about twice as large, *i.e.* 10.4 (4.8–12.1) by 5.5 (4.8–6.1) μm as compared to cyanobacterial cells found in the shoot apex of a mature Azolla plant (Fig. 5A).

Growth of seedling plant

In Azolla seedlings (Plate 2B) the development from the cotyledon stage to the production of the bilobed leaves and the subsequent stem branching can be accurately followed in time, because the morphological events starting from the cotyledon stage were recorded by camera lucida drawings (Fig. 9). Since the moment of cotyledon appearance is known, the growth rate can be calculated. It was found that about 29 mature leaves (average value) were produced in 21 days. This would imply the production of 1.4 bilobed leaves per day under the existing growth conditions in a light cabinet (28°C, 9,000 lux, 12 h light and 12 h dark). It may be that the growth rate of Azolla under natural tropical conditions will be even faster.

The time was also determined between the sowing of the macro-sporocarpic structures (mostly with adhering microsporocarpic massu-

Fig. 8. A. Longitudinal section of an Azolla seedling emerging from the macrosporocarpic complex. The *Anabaena azollae* cells initially present below the indusium cap of the macrosporocarp become associated with the apical meristem initials of the plant. Near the bottom of the photograph the floating bodies of the macrosporocarp can be found. Bar = 80 μm.

   B. Longitudinal section of a later stage of Azolla germination showing the cotyledon leaf and the apical meristem of the plant in its centre producing the first bilobed leaves. The endophyte is already associated with the apical meristem. Near the bottom an adhering massula containing one microspore, can be seen. Bar = 80 μm.

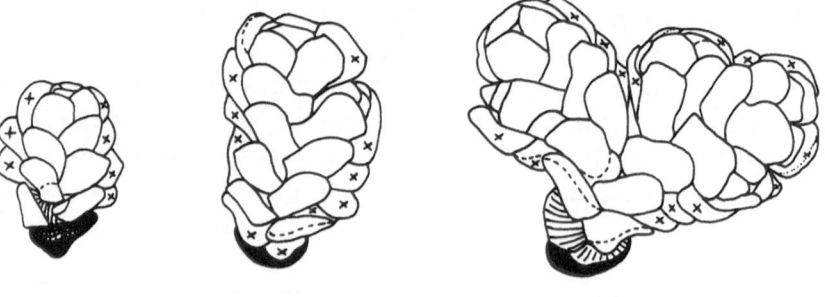

Fig. 9. Development of an Azolla seedling from the cotyledon stage and the first bilobed leaves to the subsequent branching of the main axis as recorded by camera lucida drawings. Bar = 0.5 cm.

lae) and the appearance of the first Azolla seedlings. The development was followed in trays with a water-soil system under normal greenhouse conditions. The first Azolla seedlings in cotyledon-stage appeared 19–20 days after sowing. Thirty days later, the total surface area of the trays (1350 cm^2) was completely covered with Azolla plants. Since 1 cm^2 of Azolla mat represents 2.32 mg dry weight with 3.2% N, the expanding Azolla mat fixes in 30 days a total of 100 mg N or 3.3 mg N per day.

Leaf development

Examination of tissue sections of the apical region of mature Azolla plants revealed that the cyanobacterium symbiont is present near the initials of the apical meristem and under the overarching young developing leaves (Fig. 10A). Apparently, the micro-symbiont grows in complete unity and co-ordination with the appearance of new leaf primordia.

In addition an examination of the sections revealed that during the formation of the leaf cavity, the cyanobacterial symbiont passes from the apical system into these relatively large cavities (Fig. 10B and Transmission electron micrograph Fig. 11). Consecutive sections following the leaf development to maturity from the apex to the base, showed that a pore gives an outside passage in each leaf, but this opening soon becomes closed by a proliferous outgrowth of the margin of the pore (Fig. 12A). Due to this, the symbiotic cyanobacteria become imprisoned in the leaf cavity. In a young leaf, the leaf cavity is completely filled with cyanobacterial filaments (Fig. 12B), but when the leaf subsequently enlarges to maturity, the cyanobacterial filaments form only a lining of the leaf-cavity wall. (Fig. 13). Light microscopy and transmission electron microscopy disclosed that the cyanobacteria cells are actually within a slime layer covering the wall of the leaf cavity (see arrow, Fig. 13). The Azolla host cells bordering the cavity produce bicellular or multicellular glan-

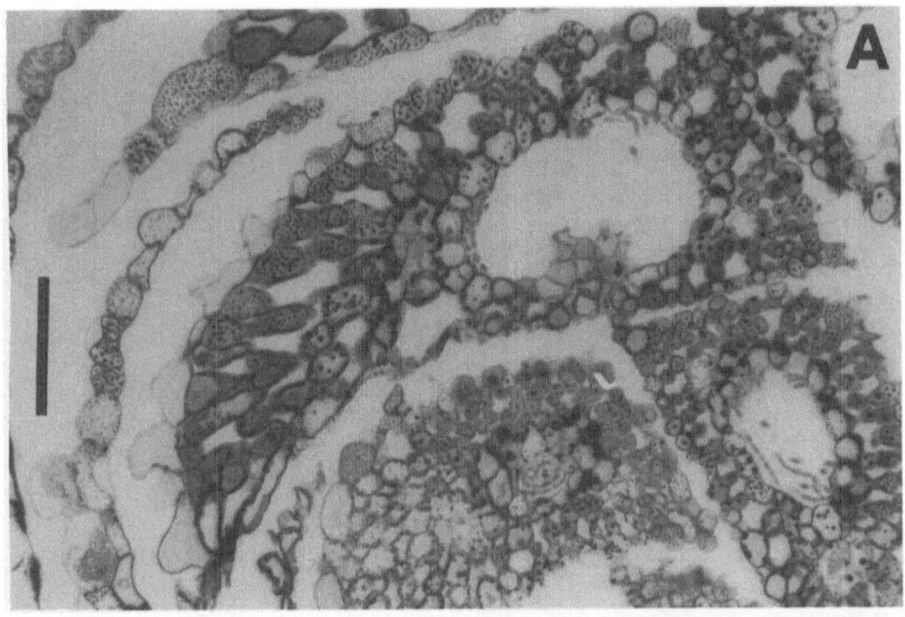

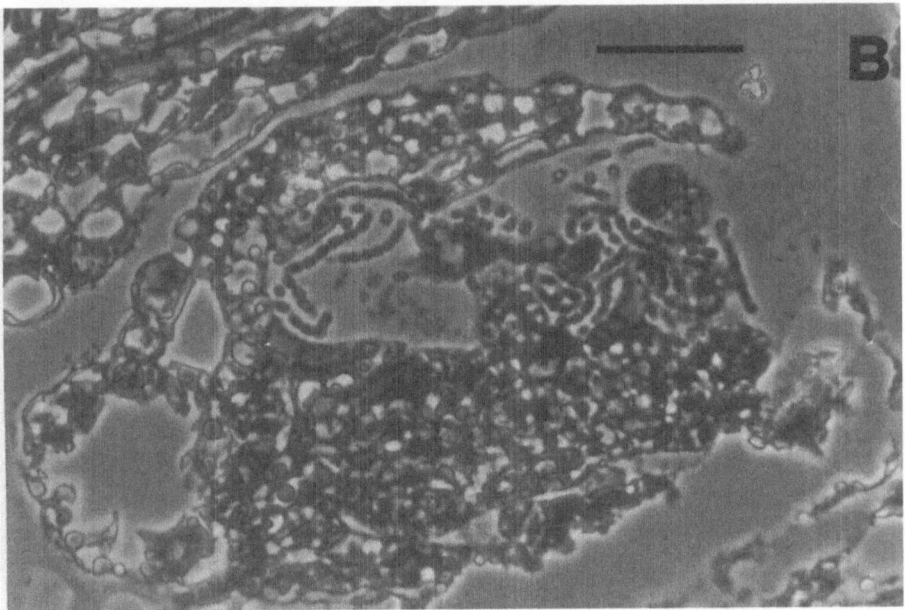

Fig. 10. **A**. *Anabaena azollae* filaments at the leaf apex and under the overarching young developing leaves of *Azolla pinnata* R. Brown. Bar = 40 μm.

      B. The passage of *Anabaena azollae* cells into the leaf cavity during the development of the leaf and its cavity. Bar = 40 μm.

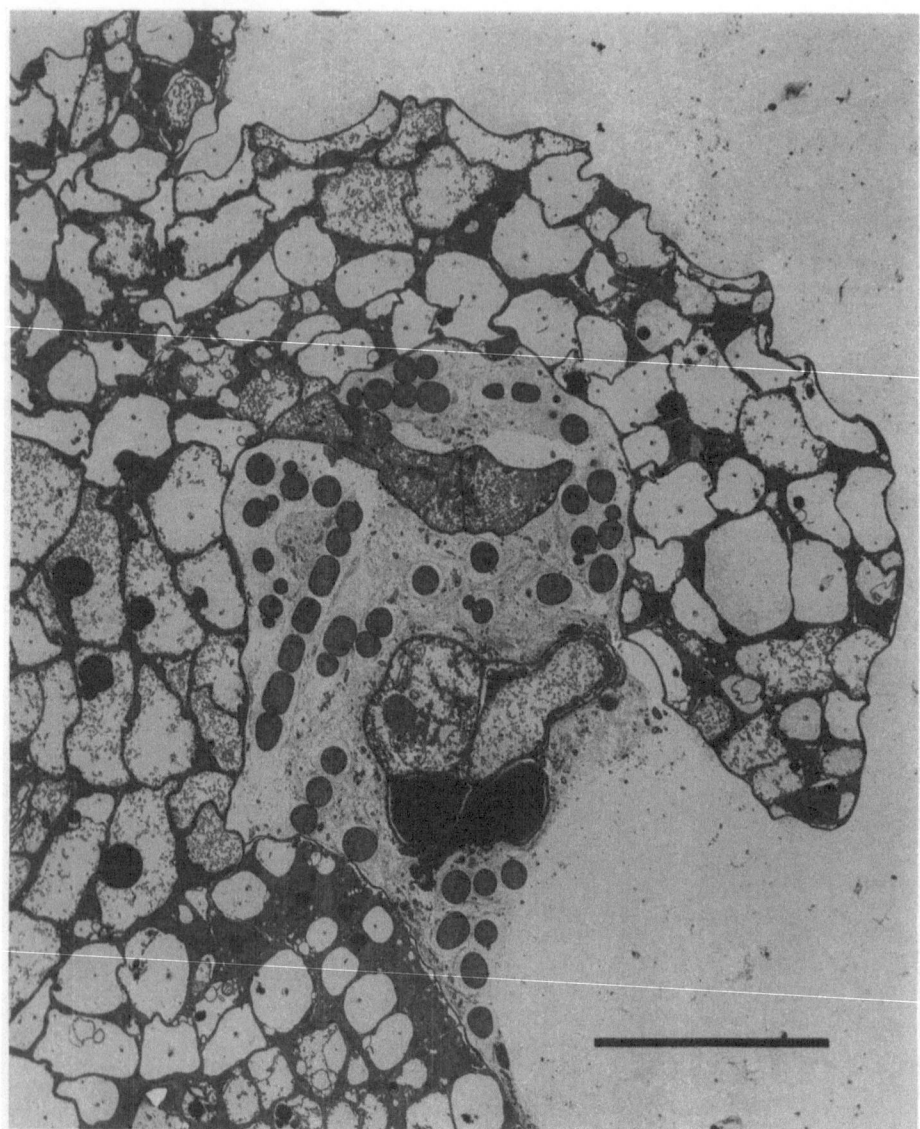

Fig. 11. Transmission electron micrograph.
  Anabaena azollae cells in the developing leaf cavity of an *Azolla pinnata* leaf. A multicellular glandular hair is seen penetrating the leaf cavity and the 'guard' cells of the pore opening at the underside of the upper-lobe are visible. Bar = 0.5 mm.

dular hairs protruding into the slime layer and the leaf cavity (see Scanning electron micrograph Fig. 14). The internal ultrastructure of these hairs as observed by electron microscopy suggests a transfer function of these organelles facilitating the interchange and cross-feeding between Azolla and its cyanobacterial symbiont.

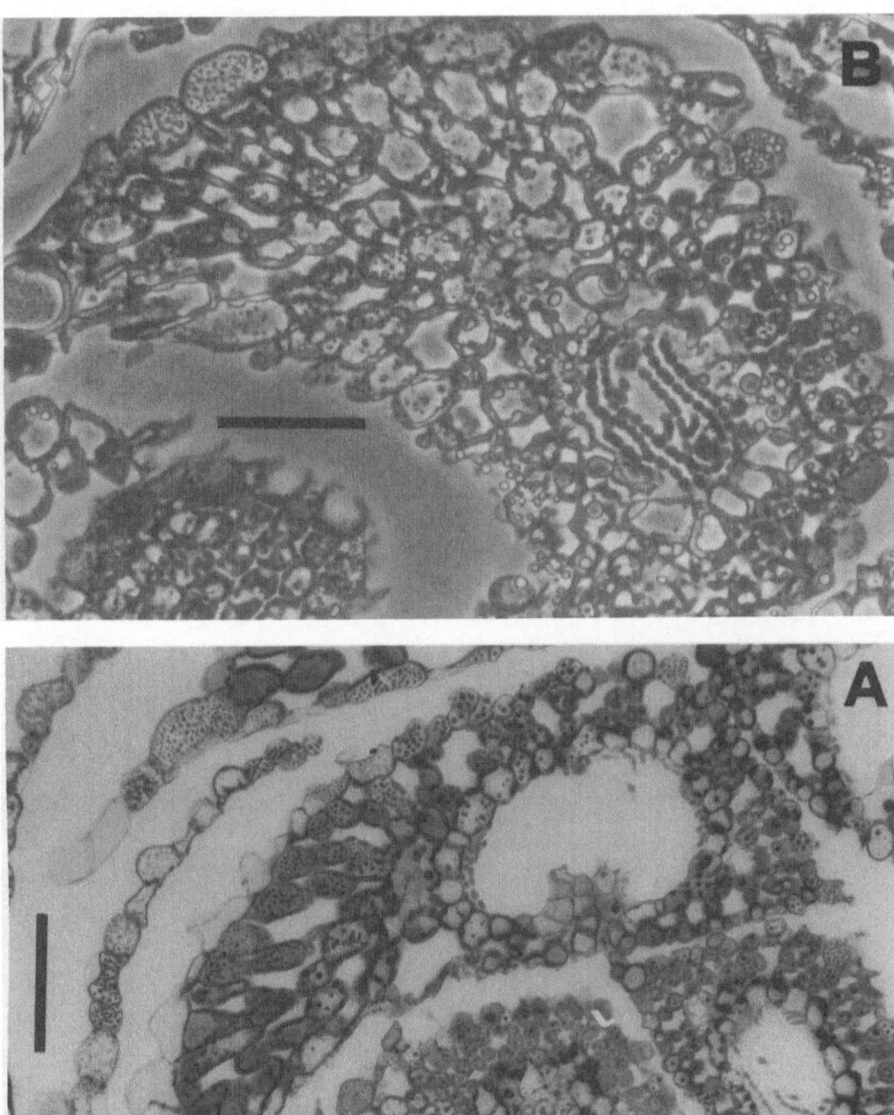

Fig. 12. **A**. Nearly mature leaf of Azolla showing the closure of the pore on the underside of the upper-lobe which imprisons the Anabaena endophytic cells. Bar = 80 5µm.

    B. The *Anabaena azollae* filaments fill completely the developing leaf cavity of a young leaf of Azolla. Bar = 40 µm.

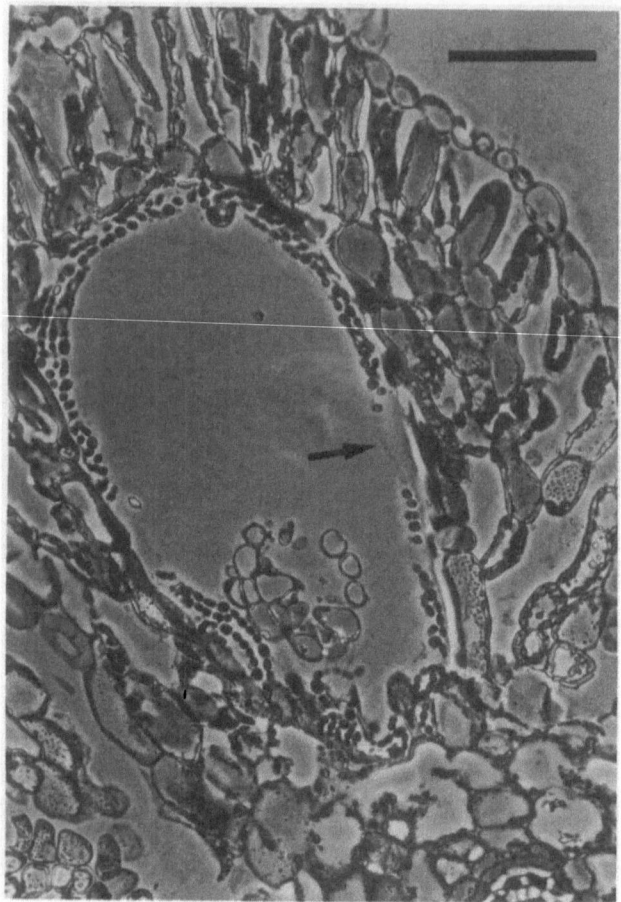

Fig. 13. Mature leaf of Azolla showing a lining of the leaf cavity by the *Anabaena azollae* filaments. The arrow indicates the presence of a mucous layer. Bar = 80.0 μm.

Nitrogen fixation capacity of Azolla megasporocarps, stem apices and leaves

Acetylene reduction tests were performed with megasporocarps of *Azolla filiculoides* containing the cyanobacterial endophyte (see Figs. 3A/B, 4A/B and 5) and apices of mature Azolla plants (Fig. 10A). The tissues were exposed in a 0.4 ml volume of 1 ml air-tight plastic syringes to an atmosphere containing C_2H_2, 10 vol. % (see Materials and methods). The samples were incubated for 3 h at 28°C and 9,000 lux.

The outcome of these experiments is presented in Table 1. As evident from the table in none of these samples could positive nitrogen fixation be ascertained as 3–4 pmol C_2H_4 in 3 h is about the determination limit of the assay. The results obtained are in agreement with the morphologi-

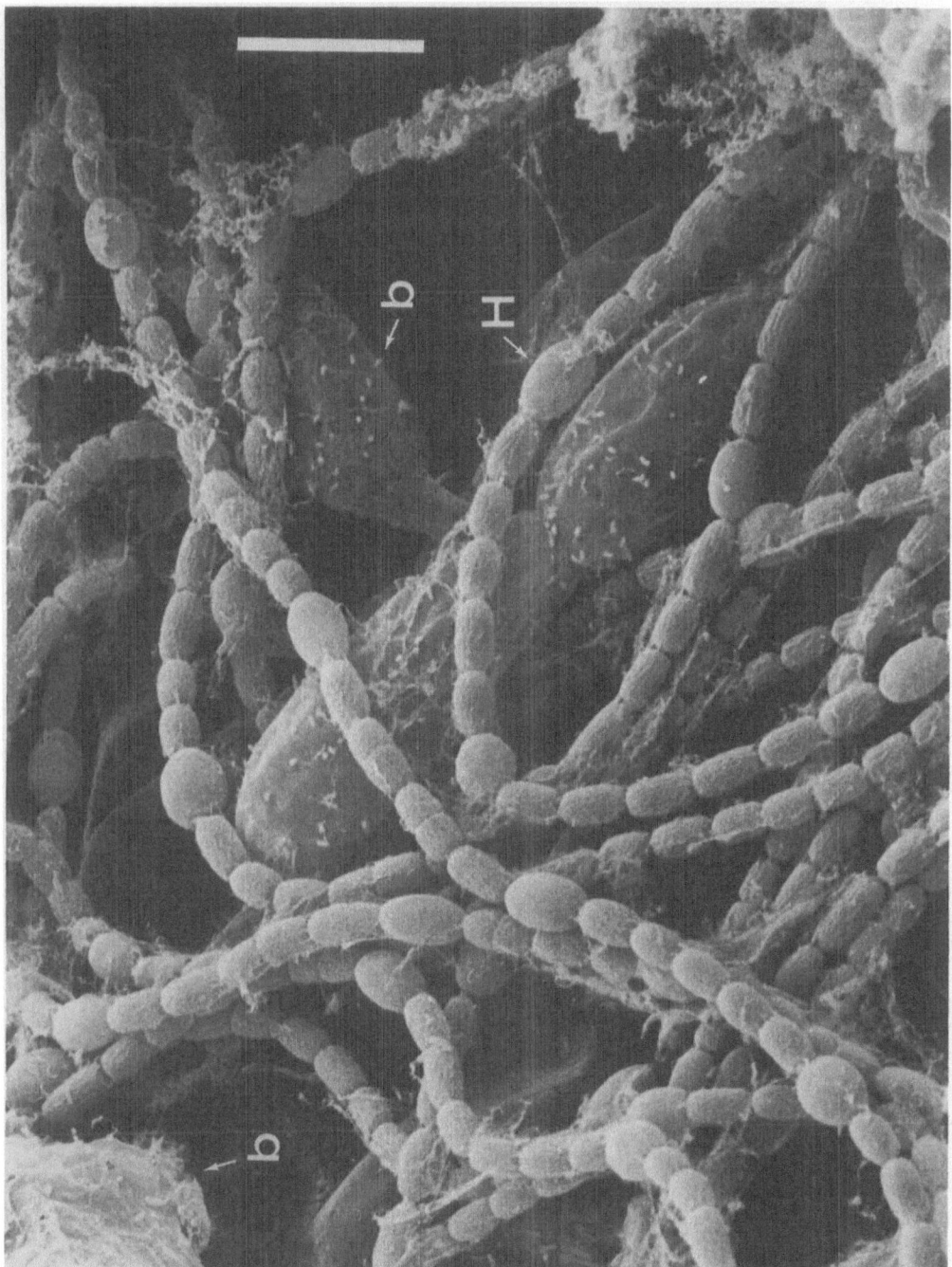

Fig. 14. Scanning electron micrograph.

The *Anabaena azollae* filaments are in close contact with multicellular glandular hairs which protrude into the leaf cavity of the upper-lobe of an Azolla frond. The filaments bear numerous heterocysts (H) and bacterial cells (b) are also visible. Bar = 20.0 μm.

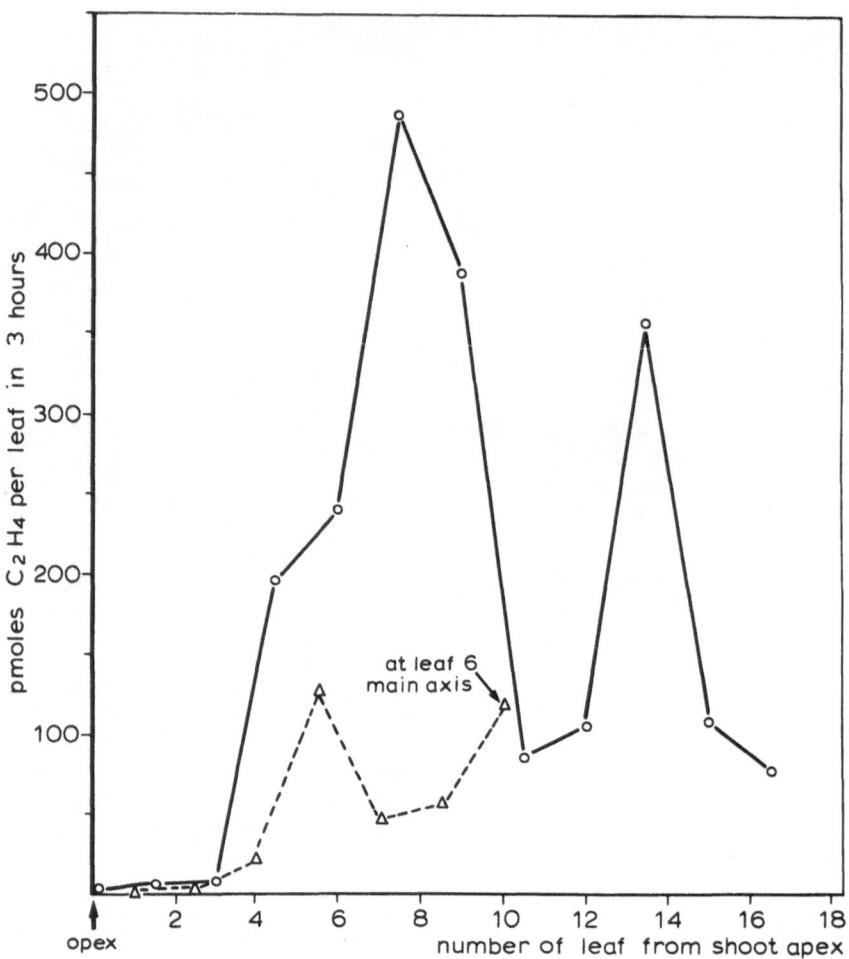

Fig. 15. Graph of the nitrogen-fixation gradient along an *Azolla pinnata* rhizome in the successive leaves from the apex to the base. The investigated rhizome had a main axis with 17 leaves and a lateral branch with 10 leaves.

cal observations, because in sections and squashes of these tissues heterocysts were never encountered.

As reference for positive nitrogen fixation, the nitrogenase activity of single Azolla leaves was investigated. In Azolla, especially in *A. pinnata*, leaf-branching is very regular (Fig. 1A). This makes it possible to study nitrogen fixation and heterocyst frequency per single leaf along the stem. The outcome of such experiments is summarized in Table 1. In addition, in Fig. 15 the nitrogen-fixation activity along a side branch, starting at leaf No. 6 of the main branch, is presented. The experiments indicate that

Table 1. Acetylene reduction patterns in macrosporocarps, apices, and single leaves of Azolla and heterocyst frequency in *Anabaena azollae*

Material	Fresh weight (mg)	pmoles C_2H_4 per sample in 0.4 ml in 3 h	*Anabaena azollae* heterocyst frequency (%)	*Anabaena azollae* total number of cells counted
Macrosporocarps				
12 specimens	0.60	4	0	40
	(12 × 0.05)			
13 specimens	0.66	3	0	40
Apices				
2 pieces	0.64			
	(2 × 0.32)	0	10	50
Leaves				
leaf number from apex				
No. 3		7	20	100
No. 5		194	21	1119
No. 7		485	27	391
No. 9	each	387	31	329
No. 11	0.28	100	37	800
No. 13		356	45	1297
No. 15		107	40	260
No. 17		76	44	774
free living				
Anabaena cylindrica	—	—	heterocyst frequency (%)	total number of cells counted
			6.9	750

along the 17 leaves of the main axis the maximal nitrogen-fixation activity is reached at leaves No.'s 7–8 and a sharp decline appears after this point resulting in a very low nitrogenase activity at leaf No. 10. However, an unexpected rise of nitrogenase activity is observed after leaf No. 10, reaching a màximum value at leaf No. 13 and then again there is a decline. In the side axis of 10 leaves a similar pattern was visible with a peak in nitrogen-fixation activity at leaf No. 6. Since such curves with distinct peaks at certain leaf numbers were found in a large number of experiments, an uneven gradient of nitrogen-fixation activity along the main axis and lateral branches seems to be common in Azolla.

In a large number of samples, comprizing macrosporocarps, apices and single leaves, squashes were made and in light microscopic observations the total number of cyanobacteria cells was counted and the percentage of heterocysts determined. In Table 1 these cell counts and heterocyst frequencies are tabulated. As reference a free-living *Anabaena cylindrica* strain was included in the survey. As evident from the table maximal nirogen fixation was reached in leaf No. 13 from the apex, this leaf is also associated with the largest heterocyst frequency (45%) of the sample. The significance of this figure is rather great as in total 1297 cyanobacteria cells were counted in this sample. In the higher leaf numbers heterocyst frequency remains about the same, although the nitrogenase activity declines. The figures indicate that there is always an increase in heterocyst frequency from the apex to higher leaf numbers. Senescent heterocysts apparently do not disintegrate and are therefore also estimated by the cyanobacteria counts in the successive leaves. A second generation of heterocysts generated from vegetative cells may be responsible for the increase in nitrogen-fixation activity at certain high leaf numbers. In free-living *Anabaena cylindrica* heterocyst frequency is about 7 (av. 6.9, see Table 1), whereas heterocyst frequencies up to 50–53% are occasionally found in *Anabaena azollae*. This may explain the high nitrogen fixation capacity of Azolla fronds.

Azolla ecology and agronomic practice

Mature Azolla plants exposed to particular environmental conditions of temperature, light, *etc.* and suffering high biomass density, are induced to form sporocarps. In temperate regions, high temperature and high light intensity followed by low temperature, may be a trigger for sporocarp formation.

Azolla filiculoides commonly occuring in this temperate region, showed a preference for sporocarp formation in the months July/August. In tropical and subtropical regions low temperature has been reported to induce sporocarp formation in *A. pinnata*. In Taiwan sporulation is in January and February and in eastern Central China local *A. pinnata* strains form spores in the late autumn or early in winter, *i.e.* in the months September/October[16]. In north Vietnam, sporocarps are formed mainly in the months March/April and seem to be associated with high surface density of the Azolla plants[10]. Also in India sporocarps mature in March/April and germinate in September/October[11,15].

Macrosporocarps separated from the mother plant by decay or rupture, sink to the bottom of the pond or the artificial water/soil system. After decay of the lower part of the sporocarpic indusium, the macrospore becomes exposed. Massulae becfome attached to the macrosporic wall and the whole structure now floats up by means of the floating corpuscles below the indusium cap. This may occur before fertilization (Rao[15]) or after fertilization.

Microsporocarps burst easily on pressure. Usually they become detached from the main stem after the microsporangia have attained maturity and sink to the bottom of the tank. The membraneous indusium decays and exposes the microsporangia, which remain attached to the placenta. Finally, the wall of microsporangium also disintigrates and the massulae structures are set free in water. Massulae separate easily from each other and float freely by means of their alveolar structures. The glochidia at the surface of the massulae become entangled with the branched processes of the exposed perispore of the macrospore. Usually many massulae become attached to a single macrosporocarp (see Plate 2A).

In agricultural practice macrosporocarpic and microsporocarpic structures may lay dormant during the period of desiccation of the soil. When the soil is re-inundated the above mentioned events take place in succession.

Discussion

The life cycle of Azolla in nature is still obscure. Azolla appears in the growing season of rice, but disappears during the fallow period. In an experimental set-up simulating the water/soil interface and the desiccation of soil during the fallow period, the re-appearance of Azolla was studied.

Tests with non-sporulating and sporulating Azolla fronds indicated that only sporulating Azolla can survive desiccation treatment. Moreover, when Azolla becomes visible, the symbiotic association of Azolla and *Anabaena azollae* appeared to be completely restored. In view of this, a study of the sexual cycle of Azolla was initiated following the fate and transmission of the endosymbiont. In a detailed study macro- and microsporocarp initiation, development and growth was followed, moreover the subsequent stages producing the embryo and the sporophyte. It was shown that cyanobacterial filaments become enclosed in the macrosporocarp in a very early stage of macrosporocarp formation and that microsporocarps are free from the endophyte. Cells of *Anabaena azollae* in the stage of akinetes are found in mature macrosporocarps in the space below the indusium cap. Previously Strasburger[18] and Campbell[5] produced line-drawing showing that the cyanobacterial cells are incorporated during the formation of the macrosporocarps of Azolla. The histology and cytology of sporocarp initiation and the further development of sporocarps have later been investigated by Pfeiffer[14], Hanning[9], Goebel[7,8], and Duncan[6]. Although cyanobacterial cells are visible in some of the figures presented by the latter authors, they paid no special attention to the occurrence and life cycle of the Anabaena symbiont. More recently, Konar and Kapoor[11] and Ashton and Walmsley[1] made a more detailed study showing the occurrence and the fate of the Anabaena micro-symbiont during sporulation. Konar and Kapoor[11] produced very nice line-drawings of the various stages of sporocarp formation in Azolla and Ashton and Walmsley were the first to produce a micrograph — although at rather low in magnification — of the *Anabaena azollae* cells below the indusium cap. The present study gives the first micrographs of the various developmental stages in the formation of macrosporocarps and shows the process of pushing the endophytic cells initially present in the entire macrosporocarpic cavity to the tip of the macrosporocarp. This occurs by the expanding macrospore and the development of alveolar bodies producing the floating structures. Moreover, the transitions in cell size and the change of the penetrating cyanobacterial filaments to akinetes are clearly documented in this study. A transmission electron micrograph of the akinetes below the indusium tip is presented. This clearly indicates that in contrast to the general morphology of akinetes in cyanobacteria, *Anabaena azollae* akinetes are relatively thin-walled and show no particular appendages or wall sculptures as most akinetes in other cyanobacteria. Akinetes of the Azolla symbiont are only distinguishable from vegetative cells by their larger more oval shape. The presence of proteinaceous and polyphosphate reserve granules is, however, similar to akinetes of other cyanobacteria species.

Detached macro- and microsporocarps sink after detachment to the bottom layer of the water. After disintigration of the lower part of the macrosporocarp and complete decay of the microsporocarpic bodies, including the microsporangial wall releasing the massulae, the macrosporocarpic structures with attached massulae float-up to the water surface. This event takes place before or during fertilization producing the zygote. The Azolla seedlings emerging from the sporocarpic complex immediately float up to the water surface.

Morphological and physiological studies of the *Anabaena azollae* cells at the apices of Azolla plants showed that these cells are relatively small, and lack heterocysts. In agreement with this, these free-living or associative cyanobacterial cells have no nitrogen-fixation capacity. In leaf development the cyanobacterial filaments slip into the large opening of a developing leaf cavity. In the further development of the leaf, the cyanobacterial cells become imprisoned in the leaf cavity, because the leaf pore at the underside of the upper-lobe, closes. Captured cyanobacterial cells never leave the leaf cavity alive, because in the older and senescent leaves these cells become lysed before the leaf disintegrates.

During leaf development and ageing of the Azolla leaf the nitrogenase activity of the association is built up. From the relatively young leaves at the stem apex to those of the more basal parts of the stem, a nitrogen-fixation gradient exists. Highest nitrogen-fixation activity is usually found at leaf No. 7–8 from the apex, where also heterocyst frequency comes close to its maximum, *i.e.* c. 30%. This confirms earlier observations by Becking and Donze[4] showing gradients of nitrogen fixation along the rhizome of Azolla, which activity was correlated with a shift in pigment production of the cells. Because nitrogenase activity at higher leaf numbers decreased without decreasing heterocyst frequency, a proportion of the heterocysts in older leaves probably become inactive due to senescence. These inactive heterocysts are apparently not lysed. At still higher leaf numbers with accordingly an increased heterocyst frequency (up to 45%) a second peak of nitrogen fixation occurs, probably associated with a second generation of active heterocysts.

Acknowledgements   The author is most grateful to drs. J. M. Bongers of the Department of Botany, Agricultural University, Wageningen for allowing me to examine thin sections of Azolla sporocarps prepared by him and for providing the electron micrograph Fig. 5. I am indebted to Technical and Physical Engineering Research Service — TFDL at Wageningen (Miss E. Bouw and Mr F. Thiel) for the electron microscopic work.
The skilful technical assistance of Mr T. W. Strubbe is gratefully acknowledged.

References

1   Ashton P J and Walmsley R D 1976 The aquatic fern Azolla and its Anabaena symbiont. Endeavour 35, 39–43.
2   Becking J H 1975 Nitrogen fixation in some natural ecosystems in Indonesia. In Symbiotic Nitrogen Fixation in Plants. Ed. P S Nutman, Cambridge University Press, pp 539–550.
3   Becking J H 1979 Environmental requirements of Azolla for use in tropical rice production. In Nitrogen and Rice, Publ. International Rice Research Institute, Los Baños, Laguna, Philippines, pp 345–373.
4   Becking J H and Donze M 1981 Pigment distribution and nitrogen fixation in Anabaena azollae. Plant and Soil 61, 203–226.
5   Campbell D H 1893 On the development of Azolla filiculoides Lam. Ann. Bot. 7, 155–187.
6   Duncan R E 1940 The cytology of sporangium development in Azolla filiculoides. Bull. Torrey Bot. Club 67, 391–412.
7   Goebel K 1905 Organographie der Pflanzen. English ed. Translated by I B Balfour, Part 2, Oxford, Oxford Univ. Press 707 pp + 417 figs.
8   Goebel K 1930 Organographie der Pflanzen, insbesondere der Archegoniaten und Samen-pflanzen. Vol. 2, Teil II, Dritte, umgearbeitete Auflage, Jena, Verlag Gustav Fischer, 1103–1362.
9   Hanning E 1911 Über die Bedeutung der Periplasmodien. Flora 102, 210–278.
10 Karamyshev V P 1957 Azolla as a fertilizer for rice (in Russian). Nauka i peredovoi opyt v sel'skom Khozyaistve 7(10), 75–77.
11 Konar R N and Kapoor R K 1974 Embryology of Azolla pinnata. Phytomorphology 24, 228–261.
12 Li Zhuo-Xin 1982 Nitrogen fixation by Azolla in rice fields and its utilization. In Non-symbiotic Nitrogen Fixation and Anorganic Matter in the Tropics, Symposia Papers: 12th International Congress of Soil Science, New Delhi, India, Febr. 8–16, 1982, pp 83–95.
13 Lucas R C and Duckett J G 1980 A cytological study of the male and female sporocarps of the heterosporous fern Azolla filiculoides Lam. New Phytol. 85, 409–418.
14 Pfeiffer W M 1907 Differentiation of sporocarps in Azolla. Bot. Gaz. 44, 445–454.
15 Rao H S 1936 The structure and life-history of Azolla pinnata R. Brown with remarks on the fossil history of the Hydropterideae. Proc. Indian Acad. Sci. 2, 175–200.

16   Shen E Y 1960 *Anabaena azollae* and its host *Azolla pinnatta*. Taiwania 7, 1–7.
17   Singh P K 1977 Multiplication and utilization of fern '*Azolla*' containing nitrogen-fixing algal symbiont as green manure in rice cultivation. Il Riso 26, 125–137.
18   Strasburger E 1873 Über Azolla, Hermann Dabis Verlag, Jena, Germany, 86 pp + Tafel I–VII.
19   Watanabe I, Espinas C R, Berja N S and Alimagno B V 1977 Utilization of the *Azolla-Anabaena* complex as nitrogen fertilizer for rice. IRRI Research paper series, No. 11, 15 pp.

Plant and Soil 100, 213–223 (1987).
© 1987 *Martinus Nijhoff Publishers, Dordrecht.*

Ms. 100-19

^{15}N-Determined effect of inoculation with N$_2$ fixing bacteria on nitrogen assimilation in Western Canadian wheats[1]

R.J. RENNIE
Esso Chemical Canada. Lethbridge, Alberta, Canada T1J 4A9

and J. B. THOMAS
Agriculture Canada Research Station, Lethbridge, Alberta, Canada T1J 4B1

Key words Azospirillum   Bacillus   N$_2$ fixation ^{15}N isotope dilution

Summary A two-year field study was undertaken using ^{15}N isotope techniques to differentiate between stimulation of N uptake and N$_2$ fixation in Western Canadian cultivars of spring wheat (*Triticum aestivum* L. emend Thell) and durum (*T. turgidum* L. emend Bowden) in response to inoculation with N$_2$-fixing bacteria.

Bacterial inoculation either had no effect or lowered the % N derived from the fertilizer and the fertilizer use efficiency. Despite the depression of fertilizer uptake, inoculants did not alter the relative uptake from soil and fertilizer-N pools indicating that bacterial inoculation did not alter rooting patterns. Nitrogen-15 isotope dilution indicated that N$_2$ fixation did occur.

In 1984, % plant N derived from the atmosphere (% Ndfa) due to inoculation with Bacillus C-11-25 averaged 23.9% while that with *Azospirillum brasilense* ATCC 29729 (Cd) averaged 15.5%. In 1985, higher soil N levels reduced these values by approximately one-half. Cultivar × inoculant interactions, while significant, were not consistent across years. However, these interactions did not affect cultivars 'Cadet' and 'Rescue'. In agreement with previous results, 'Cadet' performed well with all inoculants in both years while 'Rescue' performed poorly.

Among 1984 treatments, the N increment in inoculated plants was positively correlated with % Ndfa but no such correlation existed in 1985. N$_2$ fixation averaged over all cultivars and strains was 17.9 and 6.7 kg N fixed ha^{-1} in 1984 and 1985, respectively. Highest rates of N$_2$ fixation were estimated at 52.4 kg N ha^{-1} for 'Cadet' in 1984 and 31.3 kg N ha^{-1} for 'Owens' in 1985, both inoculated with Bacillus C-11-25, an isolate from southern Alberta soils. Inoculation with either of *Azospirillum brasilense* strain Cd (ATCC29729) or 245 did not result in as consistent or as high N$_2$ fixation, suggesting that these wheats had not evolved genetic compatability with this exogenous microorganism. These agronomically significant amounts of N$_2$ fixation occurred under optimally controlled experimental conditions in the field. It is yet to be determined if N$_2$ fixation would occur in response to bacterial inoculation under dryland conditions commonly occurring in Western Canada.

Introduction

Dinitrogen (N$_2$) fixation associated with Canadian wheat has been studied under N-free conditions[17,22,31], in the field[33] and quantified in a Brazilian soil using the principle of ^{15}N isotope dilution[31]. Acetylene-reducing activity, indicative of N$_2$ fixation, has been found associated with wheat roots by several researchers[3,7,8,13,23,24,25,27] and inoculation of wheat with N$_2$-fixing bacteria has resulted in increases in dry matter or N yield[1,2,12,14,15,16,17,21,35,38,39]. Others have found either no yield increase

[1] Contribution from Agriculture Canada Research Station, Lethbridge, Alberta, Canada.

attributable to inoculation with N_2-fixing bacteria[6,18,19] or have indications that any yield increases are due to stimulation of nutrient assimilation and plant growth rather than to N_2 fixation per se[1,12,14,35,39].

The ability of a plant to support and benefit from associated N_2 fixation has been defined as *nis* (*ni*trogen fixation *s*upport)[29]. The phenotypic expression of *nis* has been studied with disomic chromosome substitution lines (DCSL) between cultivars 'Rescue' and 'Cadet'[18,19,22,31,32,33] and 'Chinese Spring' and 'Atlas 66' or 'Cheyenne'[13]. These studies have indicated that several chromosomes are involved in the expression of *nis*. *Nis* has never been used as a basis of selection during breeding of wheats and so the characteristic may have been eroded or lost while breeding under high fertility conditions.

The primary purpose of this study was to use ^{15}N isotope techniques to measure N uptake and N_2 fixation in Western Canadian wheat varieties in response to inoculation with N_2-fixing bacteria. The secondary purpose was to evaluate the effect of the plant genotype on response to inoculation with specific bacteria.

Materials and methods

The effect of inoculation of wheat (*Triticum aestivum* and *T. turgidum*) with three N_2-fixing bacteria (Bacillus C-11-25, *Azospirillum brasilense* Cd or *A. brasilense* 245) was evaluated in the field over two years. Nitrogen-15 techniques were used to quantify plant N derived from the soil, fertilizer and atmosphere. The direct parentage of the wheat varieties evaluated is described in Table 1. Experiments were carried out at the Agriculture Canada Substation, Vauxhall, Alberta, on a Chin

Table 1. Parentage and licensing dates of wheat varieties used in the study

Variety	Date[†]	Parents
T. aestivum (hard red spring)		
Thatcher	1935	Marquis/Iumillo//Marquis/Kanred
Rescue	1946	Apex/S-615
Cadet	–	Merit/Thatcher
Park	1963	Mida/Cadet//Thatcher
Neepawa	1969	Thatcher*7/Frontana
		Thatcher*2/Thatcher*6/Kenya Farmer/3/
		Frontana/Thatcher
Columbus	1980	Neepawa*6/RL 4137
T. aestivum (soft white spring)		
Fielder	1976	Yaktana54a*4//Norin 10/Brevor/3/
		2*Yaqui50/4/Norin 10/Brevor//Baart/Onas
Owens	1984	
T. aestivum (3M)		
HY 320	1985	Fielder Sib/6/2* Yaktana 54A*4//Norin 10/
		Brevor/3/ Twin/5/2* Fielder Sib/3/Twin Sib/
		PI 227916//Twin Sib/4/Gaines/Lembi '53
T. turgidum (durum)		
Hercules	1969	RL 3097/RL 3304/Stewart; RL 3380

[†] Date of introduction or release.

loam (Typic Haploboroll) soil. The 1984 field had been ploughed from virgin grassland in the fall of 1983 and contained 20.4 kg NO_3-N ha^{-1} to the 60-cm depth. The fall and winter of 1984 were very warm and continued mineralization of organic N resulted in 52.5 kg NO_3-N at the same depth in the spring of 1985.

Plots were established as a split-plot design with five replications, wheat cultivars being the main plot and bacterial inoculation the subplot.

Plant growth

Wheat was seeded in 17.5-cm interrows during the first two weeks of May in four-row plots 7.62 m long at a rate of 68 kg seed ha^{-1}. Each plot was bordered by one row of 'Winalta' winter wheat. Phosphorus as triple super phosphate was broadcast and incorporated prior to seeding at a rate of 80 kg P ha^{-1}. The plots were irrigated to field capacity each time the soil moisture tension reached 45 kPa as indicated by mercury manometer tensiometers installed at the 30-cm depth. Microplots 75 cm long in each of rows 2 and 3 were used for bacterial inoculation and ^{15}N subplots. At maturity, a 30-cm section from each of rows 2 and 3 was harvested and the plants were dried at 70°C for 72 h, weighed and ground to pass a 40-mesh sieve. Total N including oxides of N was determined[5].

Bacterial inoculation

Bacillus C-11-25 was originally isolated from the rhizosphere of a DCSL of 'Cadet' wheat[22] grown in a southern Alberta soil that had been in continuous wheat since 1911 without N fertilization. *Azospirillum brasilense* 245 was isolated from the rhizosphere of wheat in Brazil and provided by J. Dobereiner (Empresa Brasileira de Pesquisa Agropecuaria, 23460 Seropedica, Rio de Janeiro, Brazil). *Azospirillum brasilense* ATCC 29729 (strain Cd) was purchased from the American Type Culture Collection. Bacillus C-11-25 was grown in modified Hino and Wilson's medium[11] and *A. brasilense* in Kaliniskaya's medium[4] as previously described[32,33]. Bacterial suspensions were concentrated by centrifugation at 20,000 g for 15 min (10°C), resuspended in sterile phosphate-buffered saline (pH 7.2), rewashed three times and concentrated in sterile H_2O to yield a final count of 3×10^8 bacteria ml^{-1} by Petroff-Hauser counting procedures. The addition of 10^9 cells of *A. brasilense* Sp. 107 had no effect on plant N yield or ^{15}N content with similar wheat lines[18] so these bacterial treatments would not be expected to have any effect when inoculated into these soils.

Two weeks after seedling emergence, 150 ml of bacterial suspension were added to the 75 cm of wheat in the subplot in each of rows 2 and 3. Ten millilitres were injected 10 cm deep into the rooting zone using a 50-ml syringe equipped with a modified 18-gauge needle which had holes on the sides rather than the tip to prevent soil clogging the needle[30].

^{15}N techniques

Ammonium sulphate enriched at 4.7116 (1984) and 4.7656 (1985) atom %^{15}N excess was added at a rate of 10 kg N ha^{-1}. Ten millilitres of ^{15}N solution were injected 10 cm into the soil adjacent to the plant root in the 75-cm subplot of rows 2 and 3 seven days after seedling emergence. The solution was delivered using a 50-ml syringe equipped with a modified 18-gauge needle which had holes in the sides rather than the tip to prevent soil clogging the needle[30]. The ^{15}N was allowed to equilibrate for one week prior to bacterial inoculation (see above).

Nitrogen-15 content of the plant material at maturity harvest was determined by LiOBr conversion[26,36] of NH_4^+ to N_2 followed by analysis of ^{15}N:^{14}N ratios on a VG Micromass SIRA 12 isotope ratio mass spectrometer. All isotope terminology and calculations were according to Rennie *et al.*[34]. Atom %^{15}N excess was calculated with reference to the natural ^{15}N abundance of the atmosphere (0.3663 atom %^{15}N)[20]. The percentages of plant N derived from the fertilizer (Ndff), atmosphere (Ndfa) and soil (Ndfs) were calculated as follows:

$$\% \text{ Ndff} = \frac{\text{atom }\%^{15}\text{N excess (plant)}}{\text{atom }\%^{15}\text{N excess (fertilizer)}} \times 100 \tag{1}$$

$$\% \text{ Ndfa} = \left(1 - \frac{\text{atom }\%^{15}\text{N excess (inoculated plant)}}{\text{atom }\%^{15}\text{N excess (uninoculated plant)}}\right) \times 100 \tag{2}$$

$$\% \text{ Ndfs } = 100 - (\% \text{ Ndff } + \% \text{ Ndfa}) \tag{3}$$

N_2 fixed was calculated:

$$N_2 \text{ fixed } = \frac{\% \text{ Ndfa } \times \text{ N yield (kg ha}^{-1}) \text{ (inoculated plant)}}{100} \tag{4}$$

The average ^{15}N enrichment of the uninoculated plant was used to calculate % Ndfa for both inoculated and uninoculated treatments. This averaging occasionally resulted in negative values for % Ndfa and N fixed in the control treatments.

The efficiency of utilization of applied fertilizer N, $i.e.$, percent Fertilizer Use Efficiency was calculated:

$$\% \text{ FUE } = \frac{\% \text{ Ndff } \times \text{ N yield (kg ha}^{-1})}{\text{N application rate (kg ha}^{-1})} \tag{5}$$

The 'A' value[9] as a measure of soil N availability was calculated:

$$\text{'A' value (kg N ha}^{-1}) = \frac{\% \text{ Ndfs } \times \text{ N application rate (kg N ha}^{-1})}{\% \text{ Ndff}} \tag{6}$$

Statistical analysis

The data from each experiment were evaluated by analysis of variance. Tukey's procedure was used to determine the honestly significant difference (ω) between sample means[37]. To delineate any relationship between cultivars and N utilization (soil, fertilizer, or atmosphere), correlation analysis was performed using Pearson's correlation coefficients to determine significance.

Results and discussion

The N yield was not significantly affected by bacterial inoculation in either year (Table 2). Over all inoculations, only in 1985 was there any difference in N yield among the wheat cultivars tested. The averaged N yields in 1984 were 20% higher than those in 1985 (109 vs. 91 kg N ha^{-1}).

Bacterial inoculation had a significant effect on the % Ndff. In 1984, inoculation either reduced the % Ndff in most cases or had no significant effect (Table 2). In 1985, the same general response to inoculation was observed. However, inoculation with Bacillus C-11-25 did increase the % Ndff of Thatcher and Fielder while *A. brasilense* 245 increased the % Ndff of Thatcher, Fielder and HY 320, and *A. brasilense* Cd increased the % Ndff of Columbus and Fielder. Fielder was the only cultivar to have a positive inoculation response (as % Ndff) to all three N$_2$-fixing bacteria. Percent Ndff averaged 6.4% in 1984 compared to 9.3% in 1985.

In agreement with % Ndff, % FUE was higher (83.3 vs. 67.0%) in 1985 than 1984. Inoculation had no significant effect of % FUE in 1985 but did in 1984. However, in no instance did bacterial inoculation significantly increase % FUE in 1984; rather its effect was to lower % FUE.

Lower % Ndff and % FUE observed in inoculated treatments suggest that if stimulation of rooting did occur, it did not increase the plant's uptake of mineral N from either the soil or fertilizer N pools. In fact, 'A' values (Table 2), which are an estimate of soil N availability[9] indicated that bacterial inoculation had no effect on the availability of soil N. This suggests that bacterial inoculation had no effect of nutrient volume explored by the roots — a corollary is that inoculation did not alter rooting patterns in the soil. Thus, any accretion of plant N yield or ^{15}N dilution due to inoculation may reasonably be assumed to have come from the atmospheric ^{14}N pool *via* N₂ fixation.

Percent Ndfa was calculated using the uninoculated treatment as the non-fixing control plant. Because background N₂ fixation can not be excluded, % Ndfa only measures N₂ fixation occurring in response to the inoculated diazotrophs. However, azospirilla have not been isolated in Western Canadian soils[10,28] and endogenous bacillus levels seldom exceed 10^3 per gram[28] so we presume that background levels of N₂ fixation were minimal. Non-zero values of Ndfa and amount of N fixed by uninoculated treatments are an artifact of the method of calculation (see Materials and methods).

Based on % Ndfa and N fixed (Table 2) it appears that the ability to incorporate atmospheric N₂ into the plant, following inoculation of the rhizosphere with diazotrophic bacteria, is a relatively common trait among Canadian wheat cultivars in the field. Despite high soil N, amounts fixed were large enough to be agronomically significant, both in terms of the average response to inoculation (17.9 kg ha^{-1} in 1984 and 6.7 kg ha^{-1} in 1985) and in terms of the peak response (52.4 kg ha^{-1} in 1984 and 31.3 kg ha^{-1} in 1985).

Inoculation treatments had no direct effect on N yield (Table 2). However, there was a positive correlation of % Ndfa and N₂ fixed with the difference in N yield between inoculated and uninoculated plants in 1984. Therefore, high levels of N fixation can improve the overall N status of the plant (Table 3). In conjunction with the 'A' values, this finding eliminates the possibility that positive estimates of N₂ fixation are an artifact resulting merely from the alteration by the inoculant of plant access to labelled fertilizer and unlabelled soil N pools. In 1985, correlation between % Ndfa and the N increment while still positive was insignificant, possibly because higher availability of soil N and lower average levels of fixation obscured the relationship.

Significant variation in the N₂ fixed by particular cultivar × inoculum combinations within years (Table 2) suggests that the genotype of both the wheat plant and the bacterium might be important to the overall result. In 1984, the % Ndfa due to inoculation with Bacillus C-11-25 averaged 23.9% while *A. brasilense* Cd was 15.5%. Over all inoculation treatments in 1984, 'Thatcher', 'Cadet', 'Neepawa', 'Columbus' and 'Park' averaged 30% of their plant N from associated N₂ fixation. In

Table 2. Nitrogen yield and ^{15}N-derived source of plant nutrients of wheat cultivars inoculated with N_2-fixing bacteria

Cultivar	N_2-fixing strain	N yield		Ndff		FUE		Ndfa		'A' value		N_2 fixed	
		1984	1985	1984	1985	1984	1985	1984	1985	1984	1985	1984	1985
		kg ha^{-1}		%						kg N ha^{-1}		kg ha^{-1}	
Rescue	Unin.	99	77	6.9	10.4	66.2	79.2	0.0	0.0	135	86	3.4	0.9
	C-11-25	96	95	6.6	9.6	59.6	90.3	4.6	8.4	135	86	10.1	8.8
	245	ND	86	ND	10.9	ND	93.8	ND	-4.8	ND	86	ND	-3.9
	CD	115	91	6.1	10.1	68.8	92.0	11.6	3.5	135	86	16.0	3.4
Cadet	Unin.	89	97	7.7	10.8	68.4	104.1	0.0	0.0	120	83	0.9	0.7
	C-11-25	111	105	4.1	8.3	45.0	86.5	47.4	22.9	118	83	52.4	24.6
	245	ND	108	ND	8.7	ND	93.0	ND	19.3	ND	83	ND	20.9
	CD	97	100	5.5	9.4	50.4	94.5	28.7	12.2	120	83	31.5	12.4
Thatcher	Unin.	110	93	9.5	8.7	69.0	80.8	0.0	0.0	95	105	-0.3	0.0
	C-11-25	109	93	5.1	9.5	54.8	87.9	46.4	-8.6	95	110	51.7	7.3
	245	ND	100	ND	9.8	ND	98.3	ND	-12.3	ND	110	ND	-12.3
	CD	95	87	8.3	8.2	78.7	70.7	12.7	5.5	95	111	12.5	5.8
Park	Unin.	100	77	7.9	11.1	78.4	83.5	0.0	0.0	117	80	0.5	1.6
	C-11-25	103	85	5.8	9.0	56.6	77.4	26.6	18.8	117	80	30.9	15.5
	245	ND	94	ND	9.1	ND	85.6	ND	17.8	ND	80	ND	16.7
	CD	122	92	4.7	9.5	57.1	87.2	40.2	-3.4	117	99	49.8	-2.5
Neepawa	Unin.	83	102	8.9	9.2	73.5	93.1	0.0	0.0	102	99	0.1	0.7
	C-11-25	105	95	5.7	7.5	56.9	70.5	35.9	18.8	102	98	40.3	18.7
	245	ND	90	ND	8.7	ND	78.2	ND	6.0	ND	98	ND	5.0
	CD	120	95	6.0	9.2	68.0	87.4	32.6	-0.2	102	99	43.7	-0.1
Columbus	Unin.	99	90	7.8	8.6	77.1	76.6	0.0	0.0	118	108	0.6	0.1
	C-11-25	110	85	5.0	8.3	58.7	70.7	35.6	2.1	119	108	44.2	1.6
	245	ND	83	ND	9.2	ND	75.4	ND	-8.0	ND	108	ND	6.2
	CD	107	89	5.8	10.4	61.0	82.7	25.7	-12.5	118	86	29.0	8.9

Fielder	Unin.	132	89	6.4	8.1	74.6	91.5	0.0	0.0	146	113	1.8	1.2
	C-11-25	101	84	6.0	9.2	60.7	77.6	5.7	11.9	147	113	6.9	10.1
	245	ND	81	ND	10.1	ND	81.6	ND	3.2	ND	113	ND	2.7
	CD	108	87	6.5	9.7	68.2	83.0	−1.4	7.1	146	113	1.7	7.3
Owens	Unin.	128	87	6.5	11.2	77.8	97.0	0.0	0.0	144	79	0.8	0.2
	C-11-25	117	103	6.0	7.7	68.2	79.5	7.2	30.8	145	132	12.1	31.3
	245	ND	81	ND	7.1	ND	58.9	ND	36.0	ND	154	ND	27.9
	CD	94	87	8.0	8.0	70.5	68.4	−23.1	28.3	144	123	−14.4	26.2
Hercules	Unin.	101	94	7.0	10.0	70.3	93.6	0.0	0.0	133	90	0.6	0.3
	C-11-25	123	85	6.0	8.3	73.7	70.2	13.7	17.3	134	90	17.2	14.8
	245	ND	103	ND	7.7	ND	76.7	ND	23.4	ND	89	ND	26.7
	CD	131	92	5.3	8.4	69.1	77.7	24.4	16.0	133	90	32.4	14.5
HY320	Unin.	130	91	6.2	9.8	75.4	89.7	0.0	0.0	151	93	7.0	−0.5
	C-11-25	112	81	5.2	9.7	55.1	77.2	16.1	1.3	151	92	22.5	−8.0
	245	ND	82	ND	10.9	ND	83.4	ND	−5.8	ND	88	ND	−14.4
	CD	115	96	5.9	9.3	64.8	87.6	3.8	4.8	153	92	9.3	4.6
ω(P < 0.05)													
Cultivar		NS	17	1.3	1.4	12.7	16.4	4.4	6.2	NS	20	23.6	12.7
Inoculation		NS	NS	0.5	0.7	5.1	NS	2.8	3.9	NS	NS	9.5	6.5
Cult. × Inoc.		NS	NS	2.5	3.0	24.8	36.0	6.1	8.5	NS	NS	45.9	29.4

Table 3. Relationship between wheat cultivars and indices of N assimilation as determined by Pearson correlation coefficients

Factors	Ndfa		N increment[†]		N_2 fixed		FUE	
	1984	1985	1984	1985	1984	1985	1984	1985
Ndfa	1.00**†	1.00**	0.717**	0.176^{ns}	0.989**	0.965**	−0.606**	−0.478**
N increment	0.717**	0.176^{ns}	1.00**	1.00**	0.749**	0.264^{ns}	−0.176	−0.548**
N_2 fixed	0.989**	0.965**	0.749**	0.264^{ns}	1.00**	1.00**	−0.596**	−0.400**
FUE	−0.616**	−0.479**	−0.176^{ns}	0.548**	−0.596**	−0.400**	1.00**	1.00**

† N increment = N yield (fs) − N yield (nfs).

† Significant at 0.05 (*), 0.01 (**) level of probability or non-significant (ns).

1985, the highest % Ndfa was observed with 'Cadet', 'Owens' and 'Hercules'. Bacillus C-11-25 again had the highest % Ndfa (16.5%) followed by *A. brasilense* 245 (9.9%) and Cd (8.1%). 'Cadet' has a documented history of supporting N₂ fixation by Bacillus C-11-25[31,32,33] and appears to be one of the best cultivars for *nis*. The Bacillus C-11-25 was isolated from the rhizosphere of a DCSL derived from 'Cadet'[22]. This may explain the apparent superior *nis* of the Bacillus with Western Canadian varieties and with 'Cadet' in particular. *Azospirillum* spp. are not commonly isolated from Western Canadian soils[10,28] and, in general, inoculation of Canadian wheat varieties has resulted in either no effect or a detrimental effect on plant yield[31,33]. However, these effects were inconsistent across years (correlation of N₂ fixed for treatments repeated between 1984 and 1985 was zero).

In terms of N₂ fixed, the responses of 'Cadet', 'Rescue', 'Fielder', and 'Hercules' to particular inoculants were all reasonably consistent between years. In addition, and in agreement with previous experience, 'Cadet' generally gave the most positive results while 'Rescue' was among the least responsive[32,33].

However, in other cases, effects of inoculating particular varieties with particular bacteria were inconsistent across years, in some cases radically so (Table 2). Inconsistencies also exist between the present findings and those from the pot experiments reported by Rennie *et al.*[31] for the Cd strain of *A. brasilense* but not for Bacillus C-11-25 (Table 4). In general, 'Cadet' was the active and most reliable supporter of fixation among the wheats (Table 4) while C-11-25 was the most active and the most reliable diazotroph (Table 2).

The difference between 'Cadet' and 'Rescue' has been reported to be under relatively simple genetic control[31,32,33]. The importance of genetic

Table 4. Comparison of % Ndfa with previous data

Wheat cultivar	N₂-fixing bacterium	Rennie *et al*[31]	Vauxhall	
			1984	1985
Rescue	Azo Cd[†]	0.0	11.6	3.5
	C-11-25	NG[‡]	4.6	8.4
Cadet	Azo Cd	15.0	28.7	12.2
	C-11-25	18.2	47.4	22.9
Park	Azo Cd	0.0	40.2	−3.4
	C-11-25	16.1	26.6	18.8
Fielder	Azo Cd	6.7	−1.4	7.1
	C-11-25	12.9	5.7	11.9
Hercules	Azo Cd	0.0	24.4	16.0
	C-11-25	21.0	13.7	11.3

[†] *Azospirillum brasilense* Cd, Bacillus C-11-25.
[‡] No germination.

differences among wheat cultivars has also been reported by other observers[3,13]. However, other unknown factors are also clearly disturbing the relationship between cultivar and bacterium. Until this interaction of cultivar × bacterium combinations with the environment is understood or at least diminished in scope, it will be difficult to attribute variations in the relationship to the genetics of either partner.

The agronomically significant amounts of N_2 fixation, observed in these experiments occurred under optimally controlled experimental conditions in the field. It is yet to be determined if significant amounts of N_2 fixation would occur in response to bacterial inoculation under dryland conditions, which are common in Western Canadian agriculture.

References

1   Avivi Y and Feldman M 1982 The response of wheat to bacteria of the genus Azospirillum. Israel J. Bot. 31, 237–245.
2   Baldani V L D, Baldani J I and Dobereiner J 1983 Effects of Azospirillum inoculation on root infection and nitrogen incorporation in wheat. Can. J. Microbiol. 29, 924–929.
3   Baldani V L D and Dobereiner J 1980 Host-plant specificity in the infection of cereals with *Azospirillum* spp. Soil Biol. Biochem. 12, 433–439.
4   Biggins D R and Postgate J R 1969 Nitrogen fixation by cultures and cell-free extracts of *Mycobacterim flavum* 301. J. Gen. Microbiol. 56, 181–193.
5   Bremner J M 1965 Total nitrogen. *In* Methods of Soil Analysis. Vol. 2. Ed. C A Black pp 1149–1178. American Society of Agronomy, Madison, WI.
6   Cahmakci M L, Evans J J and Seidler R J 1981 Characteristics of nitrogen-fixing *Klebsiella oxytoca* isolated from wheat roots. Plant and Soil 61, 53–63.
7   Döbereiner J 1977 Plant genotype effects on nitrogen fixation in grasses. *In* Genetic Diversity in Plants. Eds. A Muhammed, R Aksel, and R C von Borstel. pp 325–334. Plenum Press, NY.
8   Elliott L F, Gilmour C M, Cochran V L, Coley C and Bennett D 1979 Influence of tillage and residues on wheat root microflora and root colonization by nitrogen-fixing bacteria. *In* The Soil-Root Interface. Edited by J L Harley and R Scott Russell. pp. 243–258. Academic Press, London.
9   Fried M and Dean L A 1952 A concept concerning the measurement of available soil nutrient. Soil Science 73, 263–271.
10 Germida J J 1985 Population dynamics of *Azospirillum brasilense* and its bacteriophage in soil. *In* Nitrogen Fixation with Non-Legumes. Eds. F A Skinner and P Uomala. pp 117–118. Martinus Nijhoff Pubs. Dordrecht, Holland.
11 Hino S and Wilson P W 1957 Nitrogen fixation by a facultative bacillus. J. Bacteriol. 75, 403–407.
12 Inbal E and Feldman M 1982 The response of a hormonal mutant of common wheat to bacteria of the genus Azospirillum. Israel J. Bot. 31, 257–283.
13 Johnson V A and Mattern P R 1977 Genetic improvement of productivity and nutritional quality of wheat. Rep. Res. Findings U.S. Agency Int. Dev. Contract No. AID/ta-C-1093.
14 Kapulnik Y, Kigel J, Okon Y, Nur I and Henis Y 1981 Effect of Azospirillum inoculation on some growth parameters and N-content of wheat, sorghum and panicum. Plant and Soil 61, 65–70.
15 Kapulnik Y, Sarig S, Nur I and Okon Y 1983 Effect of Azospirillum inoculation on yield of field-grown wheat. Can. J. Microbiol. 29, 895–899.
16 Kundu B S and Gaur A C 1980 Establishment of nitrogen-fixing and phosphate-solubilizing bacteria in the rhizosphere and their effect on yield and nutrient uptake of wheat crop. Plant and Soil 57, 223–230.

17 Larson R I and Neal J L Jr. 1978 Selective colonization of the rhizosphere of wheat by nitrogen-fixing bacteria. *In* Environmental Role of Nitrogen-Fixing Blue-Green Algae and Asymbiotic Bacteria. Ed. U Granhall. Vol. 26. pp 331–342. Ecol. Bull. (Stockholm).

18 Lethbridge G and Davidson M S 1983 Root-associated nitrogen-fixing bacteria and their role in the nitrogen nutrition of wheat estimated by ^{15}N isotope dilution. Soil Biol. Biochem. 15, 365–374.

19 Lethbridge G, Davidson M S and Sparling G P 1982 Critical evaluation of the acetylene reduction test for estimating the activity of nitrogen-fixing bacteria associated with the roots of wheat and barley. Soil Biol. Biochem. 14, 27–35.

20 Mariotti A 1983 Atmospheric nitrogen is a reliable standard for natural ^{15}N abundance measurements. Nature 303, 685–687.

21 Mishustin E M 1970 The importance of non-symbiotic nitrogen-fixing microorganisms in agriculture. Plant and Soil 32, 545–554.

22 Neal J L Jr and Larson R I 1976 Acetylene reduction by bacteria isolated from the rhizosphere of wheat. Soil Biol. Biochem 8, 151–155.

23 Nelson A D, Barber L E, Tjepkema J, Russell S A, Powelson R, Evans H J and Seidler R J 1976 Nitrogen fixation associated with grasses in Oregon. Can. J. Microbiol. 22, 523–530.

24 Neyra C A and Dobereiner J 1977 Nitrogen fixation in grasses. Adv. Argon. 29, 1–38.

25 Pederson W L, Chakrabarty K, Klucas R V and Vidaver A K 1978 Nitrogen fixation (acetylene reduction) associated with roots of winter wheat and sorghum in Nebraska. Appl. Environ. Microbiol. 35, 129–135.

26 Porter L K and O'Deen W A 1977 Apparatus for preparing nitrogen from ammonium chloride for nitrogen-15 determinations. Anal. Chem. 45, 514–516.

27 Rai S N and Gaur A C 1982 Nitrogen fixation by *Azospirillum* spp. and effect of *Azospirillum lipoferum* on the yield and N-uptake of wheat crop. Plant and Soil 69, 233–238.

28 Rennie R J 1980 Dinitrogen-fixing bacteria: computer-assisted indentification of soil isolates. Can. J. Microbiol. 26, 1275–1283.

29 Rennie R J 1981 Potential use of induced mutations to improve symbioses of crop plants with N₂-fixing bacteria. *In* Induced Mutations — a Tool in Plant Breeding. International Atomic Energy Agency, Vienna. pp. 293–321.

30 Rennie R J 1986 Comparison of methods of enriching a soil with ^{15}N to estimate N₂ fixation by isotope dilution. Agron. J. 78, 158–163.

31 Rennie R J, de Freitas J R, Ruschel A P and Vose P B 1983 ^{15}N isotope dilution to quantify dinitrogen (N₂) fixation associated with Canadian and Brazilian wheat. Can. J. Bot. 61, 1667–1671.

32 Rennie R J and Larson R I 1979 Dinitrogen fixation associated with disomic chromosome substitution lines of spring wheat. Can. J. Bot. 57, 2771–2775.

33 Rennie R J and Larson R I 1981 Dinitrogen fixation associated with disomic chromosome substitution lines of spring wheat in the phytotron and in the field. *In* Associative N₂-Fixation. Vol. I. Eds. P B Vose and A O Ruschel. pp. 145–154. CRC Press, Inc., Boca Raton, FL.

34 Rennie R J, Rennie D A and Fried M 1978 Concepts of ^{15}N usage in dinitrogen fixation studies. *In* Isotope in Biological Dinitrogen Fixation. pp 107–133. International Atomic Energy Agency, Vienna.

35 Reynders L and Vlassak K 1982 Use of *Azospirillum brasilense* as biofertilizer in intensive wheat cropping. Plant and Soil 66, 217–223.

36 Ross P J and Martin A E 1970 A rapid procedure for preparing gas samples for N-15 determination. Analyst 95, 817–822.

37 Steel R G C and Torrie J H 1960 Principles and Procedures of Statistics. McGraw-Hill, Toronto. 481 p.

38 Subba Rao N S 1980 Crop response to microbial inoculation. *In* Recent Advances in Biological Nitrogen Fixation. Ed. N S Subba Rao. Edward Arnold. pp 406–420. London.

39 Zambre M A, Konde B K and Sonar K R 1984 Effect of *Azotobacter chroococcum* and *Azospirillum brasilense* inoculation under graded levels of nitrogen on growth and yield of wheat. Plant and Soil 79, 61–67.

Plant and Soil 100, 225–236 (1987).   Ms. 100-30

Total and CO-reactive heme content of actinorhizal nodules and the roots of some non-nodulated plants

JOHN D. TJEPKEMA
Department of Botany and Plant Pathology, University of Maine, Orono, ME 04469, USA

and DARWIN J. ASA
Department of Biology, University of Michigan-Flint, Flint, MI 48502, USA

Key words Actinorhizae Alnus Casuarina Frankia Hemoglobin Myrica Nitrogen fixation Root nodules

Summary The concentration of total and CO-reactive heme was measured in actinorhizal nodules from six different genera. This gave the upper limit to hemoglobin concentration in these nodules. Quantitative extraction of CO-reactive heme was achieved under anaerobic conditions in a buffer equilibrated with CO and containing Triton X-100. The concentration of CO-reactive heme in nodules of Casuarina and Myrica was approximately half of that found in legume nodules, whereas in Comptonia, Alnus and Ceanothus the concentrations of heme were about 10 times lower than in legume nodules. There was no detectable CO-reactive heme in Datisca nodules, but low concentrations were detected in roots of all non-nodulating plants examined, including *Zea mays*. Difference spectra of CO treated minus dithionite-reduced extracts displayed similar wavelengths of maximal and minimal light absorption for all extracts, and were consistent with those of a hemoglobin. The concentration of CO-reactive heme was not correlated to the degree to which CO inhibited nitrogenase activity nor was it affected by reducing the oxygen concentration in the rooting zone. However, there was a positive correlation between heme concentration and suberization or lignification of the walls of infected host cells. These observations demonstrate that, unlike legume nodules, high concentrations of heme or hemoglobin are not needed for active nitrogen fixation in most actinorhizal nodules. Nonetheless, a significant amount of CO-reactive heme is found in the nodules of Alnus, Comptonia, and Ceanothus, and in the roots of *Zea mays*. The identity and function of this heme is unknown.

Introduction

Until recently, hemoglobins were not thought to occur in higher plants, except in the root nodules of legumes. However, hemoglobins have now been purified from the root nodules of two different non-legumes, and there is spectroscopic evidence for hemoglobin in the nodules of additional nonlegumes[3,12,26,29]. Moreover, using cloned leghemoglobin cDNA as a probe, there is evidence for cross-hybridizing sequences in genomic DNA from a wide variety of plants, including plants that do not form root nodules[14,20,21,23,24]. All of the above suggests that the gene for hemoglobin is widespread in the plant kingdom. Hemoglobins have also been found in bacteria[1,30].

In the present work we have measured the concentration of total heme and CO-reactive heme in selected actinorhizal root nodules and in roots of non-nodulating plants as a first step in investigating the function of hemoglobin in nonlegumes. Nodule properties that might be related to hemoglobin function were also studied. With a single exception, we found that soluble, CO-reactive heme was a substantial fraction of total heme in all nodules and roots investigated. However, high concentrations of heme were found in only two of the six genera of actinorhizal plants studied.

Materials and methods

Plant growth

All nodulated plants were grown on vermiculite and were watered twice weekly with a one-fourth strength -N Hoagland's solution and with distilled water on other days. Between the time of seed germination and nodule development, the -N solution was supplemented with 1 mM urea. The potted plants were kept in a growth chamber at a constant temperature of 23°C, relative humidity of 70%, light intensity of 300 to 400 μmol m^{-2} s^{-2} (400–700 nm), and photoperiod of 17 h. Casuarina was inoculated with Frankia strain HFPCcI3[32], Alnus with strain HFPArI3[5], Myrica with strain LLR161101, Datisca and Ceanothus with crushed nodules of Ceanothus americanus, and Comptonia was inoculated with crushed nodules of Comptonia that were collected in the town of Orono. The crushed nodules of Ceanothus were from plants whose original inoculum was soil and nodules from a Ceanothus site in Massachusetts. Plants were used in experiments at two to three months after inoculation. Roots of Zea mays were harvested eight days after planting the seeds in a flat of vermiculite.

Acetylene reduction assays

Nitrogenase activity was measured using the acetylene reduction assay[8,13]. Intact potted plants (7.5 cm diameter pots) were placed in 1000-ml tall form beakers that were sealed with a plastic lid and modeling clay. Acetylene was added to the jars to give a mixture of 10% acetylene and 90% air. The jars were incubated at 23°C in the dark, and gas samples were taken at 15, 30, and 45 min after acetylene addition. The concentration of acetylene and ethylene was measured using a gas chromatograph equipped with a flame ionization detector. The rate of acetylene reduction was constant as a function of time.

The effect of CO on nitrogenase activity was measured using the acetylene reduction assay as described above. After a 60-min acetylene reduction assay in the absence of CO, CO was injected into the incubation jars, and the assay was continued for an additional 60 min. In control jars to which no CO was added, the acetylene reduction rate was usually constant during the 2-h assay. Inhibition by CO was calculated from the acetylene reduction rate for individual plants before and after the addition of CO. If there was any change in acetylene reduction rate by the control jars during the assay, this was taken into account in the calculations.

Extraction of total and CO-reactive heme

CO-reactive heme was extracted from root nodules using a modification of the method of Appleby[3]. The extraction buffer (0.1 M potassium phosphate, 1 mM EDTA, 1% Triton X-100, pH 7.4) was equilibrated with CO before use and 5 ml was added to between 50 and 1000 mg of nodules placed in a 15 ml glass centrifuge tube (Corex No. 8441). Then 10 mg of sodium dithionite was added and the tube was sealed onto a Biospec homogenizer (No. 1281-0, 10,000 rpm; 1/2 inch stator) using a rubber stopper. The headspace was flushed with argon or nitrogen gas for 60 s, then the homogenizer was run at room temperature for 40 s. The tube was then removed from the homogenizer, stoppered, flushed with argon or nitrogen, and centrifuged at 4500 g for 30 min at 20°C. The supernatant was analyzed for heme content.

Roots were extracted in the same way, except that 0.2 g of insoluble polyvinylpolypyrrolidone

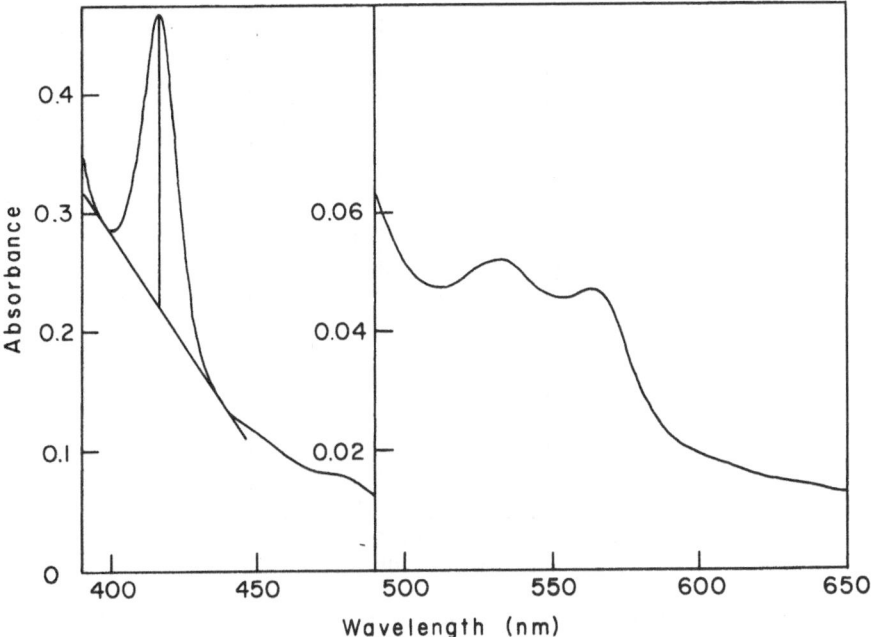

Fig. 1. Absorption spectrum of a CO-equilibrated extract from the nodules of *Myrica gale*. The vertical line at 416 nm is the value of A from which the concentration of CO-reactive heme was calculated.

(Sigma No. P-6755) was added to each tube after homogenization. This reduced background absorbance in the 420 nm region that interfered with quantification of the CO-reactive heme.

Total heme was extracted in the same way as CO-reactive heme, except that 2 ml of pyridine was added to the 5 ml of homogenate immediately after the homogenization was completed. This method was derived from that of Davenport[10].

Spectroscopic methods

Spectra were recorded using a Bausch and Lomb Spectronic 2000 spectrophotometer of 2 nm slit width, set at a 100 nm min^{-1} scan speed. Wavelength and absorbance accuracy were checked with neutral density and holmium oxide filters (Thomas Scientific) and agreed with the factory settings. Samples of about 1.0 ml were measured in semi-micro cuvettes.

For difference spectra, CO was omitted from the extraction buffer, and extracts were added to both sample and reference cells. The sample cell was then gently bubbled with CO.

The concentration of CO-reactive heme was calculated by drawing a "baseline" between the minima at 400 and 435–445 nm on either side of the Soret absorption peak as indicated in Fig. 1. The ΔA was then measured from the baseline to the absorption peak at 416 to 420 nm. To calculate the concentration of CO-reactive heme we assumed a $\Delta E(mM)$ value of 180. This value (C.A. Appleby, personal communication) was determined from pure Parasponia carboxyhemoglobin and is representative of similar hemoglobins from a variety of sources.

Total heme concentration was calculated in a manner similar to that used for CO-reactive heme. The "baseline" was drawn from the minima at about 538 and 573 nm on either side of the alpha peak of pyridine proto hemochrome (see Fig. 2). To calculate total heme concentration, we derived a ΔE (mM) value of 28.7 from an enlargement of Fig. 1 of Paul et al.[19].

Growth of plants at 5 kPa O_2

After initiation of root nodules and growth on vermiculite, seedlings (0.5 to 1.6 g fresh weight)

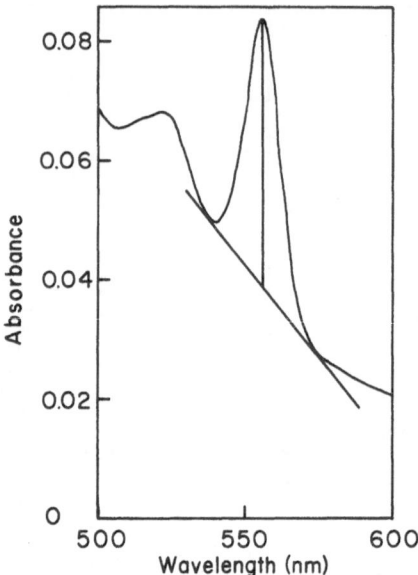

Fig. 2. Absorption spectrum of a pyridine-treated extract from nodules of *Myrica gale*. The vertical line at 556 nm is the value of A from which the concentration of total heme was calculated.

were transferred to water cultures (1/4 strength -N Hoagland's solution). Plastic containers (20 cm diameter, 20 cm high) with snap-on lids were used, with all six plants of a given treatment being placed in the same container. The plants were sealed into split rubber stoppers by wrapping the stems with thin strips of closed cell foam. The stoppers were inserted in holes in the lid, and the cultures were continuously bubbled with either air or a mixture of 95% N_2, 5% O_2, and 0.04% CO_2. A fritted disk was used to disperse the gas and the gas exiting from the cultures bubbled with 5% O_2 was collected for analysis by gas chromatography[31]. Both entering and exiting gas was 5.0% O_2, while entering gas was 0.041% CO_2, and exiting gas was 0.074% CO_2. The gas flow rate was 30 ml min^{-1} or more. The exterior of the containers was covered with aluminum foil to prevent the growth of algae. The water level was kept high, to minimize gas leakage at the seal between the rubber stoppers and plant stems. Seedlings of *Myrica gale* were grown in the water cultures for 14 days, while *Alnus rubra* seedlings were grown for 20 days in the cultures.

Tests for suberin and lignin

Autofluorescence was observed by the method of Berg[4], except that 350–500 nm excitation and 550 nm barrier filters were used. Phloroglucinol staining was as described by Berg, while chromic acid digestion was only 1 day instead of the 4 days used by Berg.

Results and discussion

Extraction of CO-reactive heme

In our initial experiments[26], soluble polyvinylpyrrolidone (PVP) was used instead of Triton X-100 in the extraction buffer. We discontinued use of PVP because it resulted in extracts that absorbed light more strongly at the shorter wavelengths and thus interfered with the assay for CO-reactive heme. However, both methods yielded about the same

Table 1. Total heme, CO-reactive heme, and acetylene reduction rates in actinorhizal nodules, nodules of Lupinus, and roots of *Zea mays*. (Means $\pm$ SE, n = 3 to 6)

Species	Total heme	CO-reactive heme	Acetylene reduction rate
	nmol g^{-1} (fresh wt)		μmol h^{-1} g^{-1} (fresh weight)
Lupinus albus (legume)	197 $\pm$ 11	174 $\pm$ 0.4	–
Casuarina cunninghamiana	95.8 $\pm$ 23	80.1 $\pm$ 4.4	66.8 $\pm$ 2.2
Myrica gale	127 $\pm$ 11	103 $\pm$ 7	43.5 $\pm$ 4.3
Comptonia peregrina	31.5 $\pm$ 2.2	17.4 $\pm$ 0.9	16.4 $\pm$ 0.4
Alnus rubra	23.7 $\pm$ 1.2	15.4 $\pm$ 0.6	43.1 $\pm$ 1.5
Ceanothus americanus	15.6 $\pm$ 2.7	11.6 $\pm$ 1.5	49.0 $\pm$ 3.2
C. americanus – roots only	4.3 $\pm$ 0.4	1.4 $\pm$ 0.2	–
Datisca glomerata	13.0 $\pm$ 0.6	0.0	48.4 $\pm$ 2.1
Zea mays – roots	4.1 $\pm$ 0.3	1.7 $\pm$ 0.1	–

amount of CO-reactive heme: when 5% soluble PVP (Sigma PVP-10, 10,000 mol. wt) was used the yield of CO-réactive heme from nodules of *Myrica gale* was 86 $\pm$ 8 nmol g^{-1}, while with 1% Triton X-100 the yield was 98 $\pm$ 10 nmol g^{-1} (mean $\pm$ SE, n = 3).

CO-reactive heme could also be extracted when Tween 80 or CHAPS (Sigma No. C 3023) were substituted for the Triton X-100. No CO-reactive heme could be extracted in the absence of soluble PVP or the detergents mentioned. The insoluble form of PVP was completely ineffective when used without soluble PVP or detergent.

These results suggest that the CO-reactive heme in actinorhizal nodules is either in an insoluble form when present in the nodule, or has a strong tendency to become insoluble during nodule homogenization. This confirms the results of Davenport[10], who could find only insoluble hemoglobin in actinorhizal nodules.

Concentration of total and CO-reactive heme

The Soret absorption band for the CO-bound heme in our extracts had a peak at 416 to 420 nm. This is consistent with the properties of a hemoglobin and is evidence against certain of the other hemoproteins that are CO reactive[1,17]. Cytochrome P-450, whose CO complex has a peak at 450 nm is ruled out. Likewise, cytochrome a$_3$ is probably not involved, since it is not readily solubilized and its CO peak is at 430 nm. However, there are other CO-reactive hemoproteins that might be present in our extracts. Horseradish peroxidase has an absorption peak at 423 nm, while cytochrome o has a peak at about 416–418 nm[9]. Thus the absorption maximum at 416 to 420 nm in our extracts is consistent with the presence of a hemoglobin, but does not rule out certain other CO-reactive hemoproteins.

The concentrations of total and CO-reactive heme that we found in nodules of Lupinus and Casuarina (Table 1) were comparable to the results of other investigators[3,10,11,12]. This further confirms that actinorhizal nodules such as Casuarina may have hemoglobin contents that approach those found in legume nodules. However, *Myrica gale* was the only other actinorhizal plant where we found this to be true. In Comptonia, Alnus, and Ceanothus, the concentrations of CO-reactive heme were 5 to 8 times lower than for Casuarina. In Datisca we found no CO-reactive heme at all. Except for Comptonia, these low heme concentrations were accompanied by high nitrogenase (acetylene reduction) activities. Thus there was no indication that the low heme contents were due to a defective symbiosis or other abnormal condition.

In work reported previously[28], we found much lower concentrations of CO-reactive heme, even though the concentrations of total heme were about the same as reported in Table 1. The reason is that these earlier extractions were done in the absence of CO. Following the method of Appleby[3], we modified our procedure and now extract in the presence of CO. We find that the absorption bands of the CO-containing extracts are quite stable when the extract is aerated, whereas the absorption bands of CO-free extracts are rapidly lost.

Because we found evidence for hemoglobins in unrelated genera of actinorhizal nodules, it seemed possible that hemoglobins might be found outside of root nodules. We first examined roots of Ceanothus from which all nodules were removed and found a low but measurable concentration of CO-reactive hemoprotein (Table 1). We found similar results in the roots of all non-nodulating plants examined: *Ulmus americana, Acer saccharinum, Helianthus annuus, Triticum aestivum*, and *Zea mays*. *Zea mays* is the only non-nodulating plant included in Table 1, since total heme was determined only for this species. As can be seen, CO-reactive heme is responsible for a substantial fraction of total heme in *Zea mays*.

Difference spectra

Depending on the material being examined, substances other than CO-reactive hemoproteins absorbed light in the 400 to 650 nm wavelength range, and distorted the absorption bands. Thus difference spectra were used to compare CO-reactive hemoproteins from various plant sources. In the case of *Ceanothus americanus* (Fig. 3), the difference spectra clearly showed the presence of a CO-reactive hemoprotein, whereas in extracts and in spectra of nodule slices[26] the absorption bands are partially obscured by other substances.

The absorption maxima and minima for the difference spectra of plus CO minus dithionite reduced extracts were similar for all nodules and

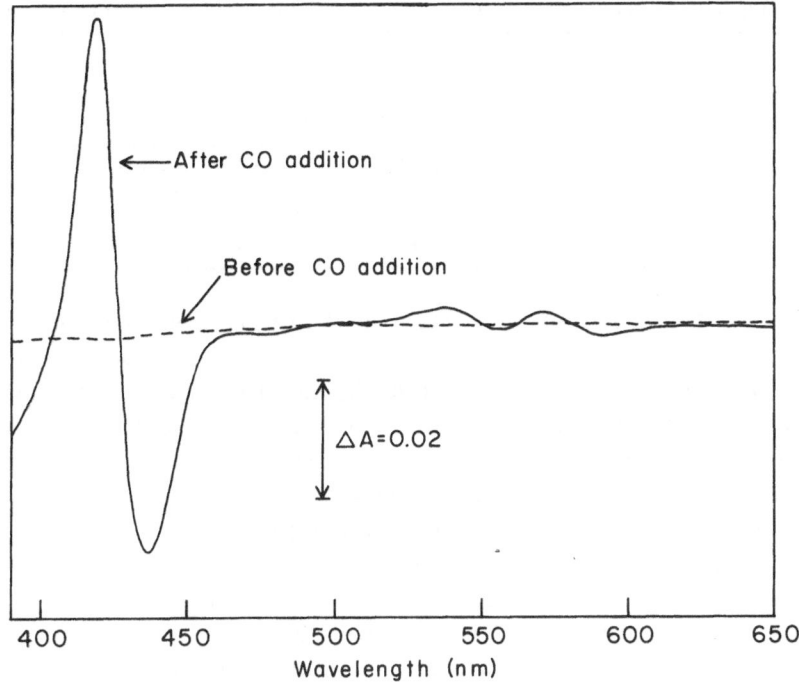

Fig. 3. Difference spectrum of a CO-equilibrated extract from Ceanothus minus a dithionite reduced extract.

roots (Table 2). The wavelengths are consistent with those of a hemog-lobin[1,17], but as discussed above, they could be due at least in part to some other CO-reactive hemoprotein. Whatever the identity of the CO-reactive heme, it has three similarities in all nodules and roots examined: 1. It is soluble in 1% Triton X-100 or 5% soluble PVP. 2. It makes up a substantial fraction of total heme. 3. The plus and minus CO difference spectra are consistent with those of a hemoglobin.

The question of whether the CO-reactive heme is partly or entirely a hemoglobin is best pursued using the purified protein. This has been done for the root nodules of *Casuarina glauca,* and the protein extracted has all the properties of a hemoglobin, including ligand binding, mole-

Table 2. Difference spectra of hemoprotein extracts (CO minus dithionite reduced). The wavelengths of maximum difference are given (nm)

Species	+	−	+	−	+
Casuarina cunninghamiana	420	437	537	557	572
Myrica gale	416	433	532	553	568
Alnus rubra	419	436	536	555	571
Ceanothus americanus	419	437	538	557	571
C. americanus — roots only	418	437	–	–	–
Zea mays – roots	419	438	534	555	570

Table 3. Effect of pO_2 on nodule weight and heme content (mean $\pm$ SE, n = 3 to 6)

	pO_2 ––– kPa	Total heme	CO-reactive heme	Nodule fresh wt
		$nmol\,g^{-1}$ (fresh wt)		mg/plant
Myrica gale	5	99.2 ± 3.5	60.7 ± 4.6	88 ± 7
	20	108 ± 6	70.7 ± 5.2	192 ± 24
Alnus rubra	5	22.0 ± 0.8	8.6 ± 0.4	94 ± 18
	20	23.7 ± 0.9	14.1 ± 0.6	260 ± 42

cular weight, and amino acid sequence[12,15]. Other evidence comes from the absorption spectra of slices of nodules of *Myrica gale, Comptonia peregrina,* and *Alnus rubra*[26]. These display the absorption bands of oxyhemoglobin when taken in an atmosphere of O_2, while the bands disappear when N_2 is substituted for O_2. In spite of this evedence, the identity of the heme found in the present study must be considered an open question until further studies are completed.

Possible functions of hemoglobin in actinorhizal nodules

Much higher concentrations of presumed hemoglobin were found in the nodules of Myrica and Casuarina than in other actinorhizal nodules. Since both of these genera often grow in wet soils, where the oxygen concentration may be reduced[25,26], we hypothesized that hemoglobin might be serving to facilitate oxygen transport at low oxygen concentrations. Such a function would be similar to the function of leghemoglobin in transporting oxygen at the low oxygen concentrations that prevail within the diffusion barrier of legume nodules[2,27]. An increase in hemoglobin concentration caused by reduced oxygen concentration in the root environment would support this hypothesis. We examined Alnus as well as Myrica, since it also grows in relatively wet soils.

Our results provide no support for the measured heme being involved in an adaptation to reduced partial pressures of O_2 (pO_2). As found by others[6,16,25], reduced pO_2 in the root zone caused reduced plant growth (results not shown) and reduced nodule weight (Table 3). In spite of this oxygen stress, there was little effect on the concentration of total and CO-reactive heme.

The nodules of Casuarina are unique among actinorhizal plants in that the endophyte, Frankia, forms no vesicles in the nodule[4,18]. Studies of Frankia strain HPCcI3, isolated from Casuarina nodules, show that vesicles are necessary for nitrogen fixation when the cultures are grown at atmospheric pO_2 (20 kPa), but are absent when the cultures fix nitrogen at 0.1 to $0.3\,kPa\,pO_2$[18]. These results suggest that the pO_2 within Casuarina nodules may be low compared to other actinorhizal nodules. A possible cause is the suberization or lignification of the walls of

Table 4. Evidence for the suberization of the wall of infected cells in selected actinorhizal plants

Species	Test result for lignin-suberin		
	Chromic acid	Phloroglucinol	Autofluorescence
Casuarina cunninghamiana	+ +	+	+ +
Myrica gale	+	+	+
Comptonia peregrina	+	+	+
Alnus rubra	−	−	−
Ceanothus americanus	−	−	−
Datisca glomerata	−	−	−

infected host cells in Casuarina, which could restrict oxygen diffusion into these cells[4]. If so, the hemoglobin in Casuarina nodules might function to facilitate oxygen transport in a zone of low pO_2, as does leghemoglobin in legume nodules.

We decided to investigate the relationship between heme content, suberization, and lignification in the six actinorhizal genera where we measured heme concentration. By comparing Tables 1 and 4 one can see that there is a correlation between the presence of suberization and lignification and elevated heme concentration, except in the case of Comptonia peregrina. Comptonia is closely related to Myrica and we found identical results in the two species when testing for suberin and lignin. However the hemoglobin concentration was much lower in Comptonia, and was similar to that in Alnus, which lacks suberin and lignin. It is possible that these particular nodules of Comptonia were not fully effective, as suggested by their relatively low acetylene reduction rates (Table 1). Overall the data support a relationship between elevated heme content and the presence of suberized and lignified walls, but more information is needed to elucidate the nature of this relationship.

Since hemoglobins have a high affinity for CO, it seemed possible that nodules with little or no hemoglobin might be less sensitive to inhibition by CO than those with high concentrations of hemoglobin. If so, this might be used to study the function of the hemoglobin. As can be seen by comparing Tables 1 and 5, there was no correlation between hemoglobin content and the inhibition of acetylene reduction by CO. The

Table 5. Inhibition of acetylene reduction in actinorhizal and legume nodules by 0.33 kPa of CO (mean ± SE, n = 3 to 6)

Species	Inhibition(%)
Alnus rubra	43 ± 4
Casuarina cunninghamiana	54 ± 1
Datisca glomerata	60 ± 7
Myrica gale	62 ± 4
Ceanothus americanus	95 ± 2
Lupinus albus	90 ± 2
Pisum sativum	82 ± 3

inhibition observed could have been due to the effect of CO on a cytochrome oxidase, hemoglobin, or nitrogenase itself[17,22]. Using ^{15}N, Bond found comparable results for inhibition by CO in nodules of Alnus, Myrica, and Pisum[7].

Conclusions

Our results establish that there are high concentrations of total and CO-reactive heme in the nodules of *Myrica gale* and *Casuarina cunninghamiana*. However, the concentrations of heme in other actinorhizal genera were about 5 to 8 times lower, with no significant differences in acetylene reduction rates by nodules with high and low heme contents. Moreover, Datisca nodules had typical acetylene reduction rates, but no detectable heme that was extractable and CO-reactive. We thus conclude that high heme contents are not essential for nitrogen fixation in actinorhizal nodules.

The CO-reactive heme in *Casuarina cunninghamiana* is probably a hemoglobin, since large amounts of a hemoglobin have been purified from the nodules of *Casuarina glauca*[12]. It is now of great interest to purify and characterize the CO-reactive heme extracted from actinorhizal plants such as *Alnus rubra* and non-nodulating plants such as *Zea mays*. If this proves to be a hemoglobin, it would suggest that hemoglobins have some relatively general function in plants.

Our initial investigations into the function of the CO-reactive heme in actinorhizal nodules have been inconclusive. There is a possible relationship with suberization of the cell wall of the infected host cells, but this needs more study. No correlation was observed between heme content and either root zone pO_2 or the sensitivity of nitrogenase activity to CO inhibition.

The simplest hypothesis is that the CO-reactive heme measured in Table 1 is a hemoglobin and functions in facilitating oxygen transport in tissues where the pO_2 is low. This is the most likely function of the lethemoglobin present in legume nodules[2]. Variations in the concentration of hemoglobin in nodules and roots might be due to variations in the intensity of respiration and the resistance to oxygen diffusion to the sites of respiration.

Acknowledgements We thank Linda Krywy for expert technical assistance, Mary Lechevalier for providing Frankia strain LLR 161101, John Torrey for providing some of the seed and Frankia strains used, Larry Winship for seeds of Datisca, Christa Schwintzer for helpful comments on the manuscript, and Cyril Appleby, Tony Fleming, Jonathan Wittenberg, Beth Mullin, Douglas Johnson and Ian Sussex for helpful discussions and data in advance of publication. This work was supported by US National Science Foundation Grant No. DMB-8315415.

References

1 Appleby C A 1969 Electron transport systems of *Rhizobium japonicum*. II. Rhizobium hemoglobin, cytochromes and oxidases in free-living (cultured) cells. Biochim. Biophys. Acta 172, 88–105.

2 Appleby C A 1984 Leghemoglobin and Rhizobium respiration. Annu.Rev. Plant Physiol. 35,443–478.

3 Appleby C A, Tjepkema J D and Trinick M J 1983 Hemoglobin in a nonleguminous plant, Parasponia: possible genetic origin and function in nitrogen fixation. Science 220, 951–953.

4 Berg R H 1983 Preliminary evidence for the involvement of suberization in infection of Casuarina. Can. J. Bot. 61, 2910–2918.

5 Berry A and Torrey J G 1979 Isolation and characterization *in vivo* and *in vitro* of an actinomycetous endophyte from *Alnus rubra* Bong. *In* Symbiotic Nitrogen Fixation in the Management of Temperate Forests. Eds. J C Gordon, C T Wheeler and D A Perry. pp 69–83. Oregon State University, Corvallis, OR.

6 Bond G 1952 Some features of root growth in nodulated plants of *Myrica gale* L. Ann. Bot. 16, 467–475.

7 Bond G 1960 Inhibition of nitrogen fixation in non-legume root nodules by hydrogen and carbon monoxide. J. Exp Bot. 11, 91–97.

8 Burris R H 1974 Methodology. *In* The Biology of Nitrogen Fixation. Ed. A. Quispel. pp 9–33. North-Holland Publishing Co., Amsterdam.

9 Castor L N and Chance B 1959 Photochemical determinations of the oxidases of bacteria. J. Biol. Chem. 234, 1587–1592.

10 Davenport H E 1960 Haemoglobin in the root nodules of *Casuarina cunninghamiana*. Nature 186, 653–654.

11 Egle K and Munding H 1951 Über den Gehalt an Häminkörpen in den Wurzelknöllchen von Nicht-Leguminosen. Naturwiss. 38, 548–549.

12 Fleming A I, Wittenberg B A, Wittenberg J B and Appleby C A 1985 Hemoglobin in the root nodules of the actinorhizal genus Casuarina. Proc. Aust. Biochem. Soc. 17, 24 (Abstr.).

13 Hardy R W F, Burns R C and Holsten R D 1973 Application of the acetylene reduction assay for measurement of nitrogen fixation. Soil Biol. Biochem. 5, 47–81.

14 Hattori J and Johnson D A 1985 The detection of leghemoglobin-like sequence in legumes and non-legumes. Plant Molec. Biol. 4, 285–292.

15 Kortt, A A, Burns J E, Inglis A S, Appleby C A, Trinick M J and Fleming A I 1985 Hemoglobins from the nitrogen-fixing root nodules of non-legumes and their relationship to the legume hemoglobins. Proc. Aust. Biochem. Soc. 17, 43 (Abstr.).

16 MacConnell J T 1959 The oxygen factor in the development and function of the root nodules of alder. Ann. Bot. 23, 261–268.

17 Mahler H R and Cordes E H 1966 Biological Chemistry. Harper and Row, New York, 872 p.

18 Murry M A, Zhongze Z and Torrey J G 1985 Effect of O_2 on vesicle formation, acetylene reduction and O_2-uptake kinetics in *Frankia* sp. HFPCcI3 isolated from *Casuarina cunninghamiana*. Can. J. Microbiol. 31, 804–809.

19 Paul K G, Theorell H and Åkeson Å 1953 The molar light absorption of pyridine ferroprotoporphyrin (pyridine haemochromogen). Acta Chem. Scand. 7,1284–1287.

20 Roberts M P, Jafar S and Mullin B C 1985 Leghemoglobin-like sequences in the DNA of four actinorhizal plants. Plant Molec. Biol. 5, 333–337.

21 Roberts M P, Jafar S and Mullin B C 1985 Leghemoglobin-like sequences in the DNA of non-leguminous plants. First International Congress of Plant Molecular Biology, Savannah, Georgia, Abstract No. OR-05-02.

22 Rivera-Ortiz J M and Burris R H 1975 Interactions among substrates and inhibitors of nitrogenase. J. Bacteriol. 123, 537–545.

23 Schulman A H 1986 The Detection of DNA sequences Similar to Leghemoglobin Genes in the Genomes of Non-Leguminous Plants. Ph.D. Dissertation, Yale University.

24 Schulman A H and Sussex I M 1985 Detection of DNA sequences from non-legumes with homology to soybean leghemoglobin genes. First International Congress of Plant Molecular Biology, Savannah, Georgia, Abstract No. PO-2-104.

25   Schwintzer C R 1985 Effect of spring flooding on endophyte differentiation, nitrogenase activity, and shoot growth in *Myrica gale*. Plant and Soil 87, 109–124.

26   Tjepkema J D 1983 Hemoglobins in the nitrogen-fixing root nodules of actinorhizal plants. Can. J. Bot. 61, 2924–2929.

27   Tjepkema J D 1984 Oxygen, hemoglobins, and energy usage in actinorhizal nodules. *In* Advances in Nitrogen Fixation Research. Eds. C Veeger and W E Newton. pp 467–473. Martinus Nijhoff, The Haugue.

28   Tjepkema J D and Murry M A 1985 Heme content and diffusion limitation of respiration and nitrogenase activity in nodules of *Casuarina cunninghamiana*. *In* Nitrogen Fixation Research Progress. Eds. H J Evans, P J Bottomley and W E Newton, p 370. Nijhoff, Dordrecht.

29   Tjepkema J D, Schwintzer C R and Benson D R 1986 Physiology of actinorhizal nodules. Annu.Rev. Plant Physiol. 37, 209–232.

30   Wakabayashi S, Matsubara H and Webster D A 1986 Primary Sequence of a dimeric bacterial haemoglobin from Vitreoscilla. Nature 322, 481–483.

31   Winship L J and Tjepkema J D 1983 Simultaneous measurement of acetylene reduction and respiratory gas exchange of attached root nodules. Plant Physiol. 70, 361–365.

32   Zhang Z, Lopez M F and Torrey J G 1984 A comparison of cultural characteristics and infectivity of Frankia isolates from root nodules of *Casuarina* species. Plant and Soil 78, 79–90.

Plant and Soil 100, 237–247 (1987).
Ms. 100-20

Biological control of nematodes: Soil amendments and microbial antagonists

R. RODRIGUEZ-KABANA, G. MORGAN-JONES
Department of Plant Pathology, Auburn University, Alabama Agricultural Experiment Station, Auburn, AL 36849, USA

and I. CHET
Department of Plant Pathology and Microbiology, The Hebrew University of Jerusalem, Faculty of Agriculture, Rehovot, Israel

Key words Agricultural wastes Microbial ecology Nematode ecology Pest management Root-knot nematodes Soil enzymes Waste disposal

Summary Organic matter amendments to soil can be used to manage phytoparasitic nematodes. The most effective amendments are those with narrow C:N ratios and high protein or amine-type N content. For soil with 1.0% (w/w) organic matter amendment there is a direct relation between extent of nematode control and the N content of amendments. A special group of amendments are those containing chitinous materials. Chitin addition to soil results in stimulation of a select microflora capable of degrading the polymer. Several microbial species are known to destroy the eggs of phytonematodes (*Meloidogyne* spp.). Organic matter can be modified by addition of specific compounds or by inoculation with particular microbial species to produce an amendment that will induce suppressiveness.

Introduction

Addition of organic matter to soil for crop improvement is a practice almost as old as agriculture. References to increased yields in response to manuring can be found in classical Greek and Roman literature. In pre-Columbian times, farmers on the American continent practiced manuring to increase yields of corn, potatoes and other crops. Yield benefits from organic amendments were attributed to increased availability of nutrients to the crops by the amendments. Benefits derived from certain types of manures probably resulted also from suppression of plant pathogens, including nematodes. Linford and Oliveira[10] made a systematic study of organic amendments for the management and control of plant-parasitic nematodes. Development of effective fumigant nematicides in the 1940's and 1950's relegated research on organic amendments to secondary importance. Because of this, during the past two decades, research on the use of organic amendments for control of phytonematodes has been mainly conducted in developing nations. Of all papers published on the subject from 1971 to 1981, 72% reported on studies conducted in developing nations; 48% in India[18]. About 39% of these studies dealt with control of root-knot nematodes in vegetables and crops typically grown by small, subsistence-type farmers.

Recent removal of key nematicides (DBCP, EDB) from use in several industrialized countries through regulatory action has spurred research on 'unconventional' nematode management methods. Efforts are being made to develop management strategies that do not rely on nematicides, or are aimed at reducing use of these materials. New knowledge has been acquired on naturally occurring microorganisms capable of destroying eggs or other stages of the life cycle of phytonematodes[5,6,16,17,20,21]. It is within the context of this renewed interest in unconventional technologies for nematode control that there is currently a re-examination of the value or organic amendments. The objective of this paper is to present some concepts based on our experiences with organic amendments. Several reviews on the subject are available[1,2,18,22,28].

Organic amendments and biological control

Addition of organic matter to soil stimulates microbial activity as evidenced by increased populations of actinomycetes, algae, bacteria, fungi, and other organisms, such as microbivorous nematodes. Proliferation of microorganisms results in increased enzymatic activities of the amended soil and accumulation of specific end-product compounds which may be nematicidal[2,23]. The magnitude of microbial stimulation and the qualitative nature of the responding microflora depend on the nature of the organic matter added. Effectiveness of a given amendment for nematode suppression depends on its chemical composition and the species of microorganisms that develop. Some organic materials contain compounds toxic to nematodes, e.g. oil cakes of castor bean (*Ricinus communis*) and neem (*Azadirachta indica*)[22,28]. Also, a large group of other organic materials act against nematodes through microbial activity and release of ammonia. These materials have typically low C:N ratios and have high protein or amine contents. Studies by Mian et al.[13] demonstrated that when organic amendments are added to soil at a rate of 1.0% (w/w) there is a direct relation between the N content, or inverse with the C:N ratio, of the amendments and their effectiveness against phytonematodes (Fig. 1). Although ammonia and organic amendments with low C:N ratios are nematicidal[12,26] the quantities of these materials required to obtain consistent nematode control in the field are large. Urea, for example, is a good nematicide when applied at levels in excess of 300 mg N/Kg soil[22]. Such high rates of urea result in significant accumulations of nitrate and ammoniacal N in soil and may cause phytotoxicity. Phytotoxic effects of amendments with high N content

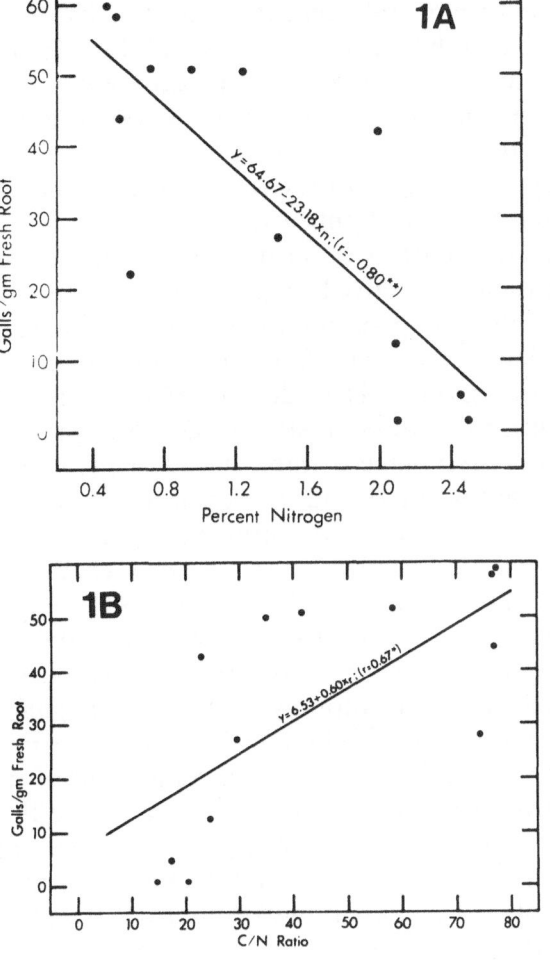

Fig. 1. Relation between the number of root galls caused by *Meloidogyne arenaria* in squash (*Cucurbita pepo*) and the N content (**A**) or the C:N ratio (**B**) of organic amendments added [rate = 1.0% w/w] to soil infested with the nematode.

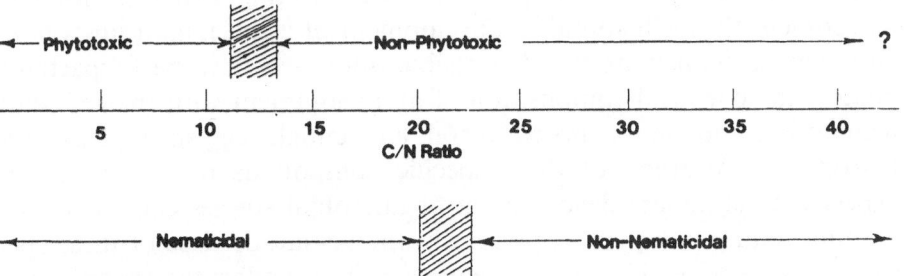

Fig. 2. Schematic representation of the relation between C:N ratio of organic amendments, nematicidal activity and phytotoxicity.

can be reduced by increasing the C:N ratio by adding a supplementary source of carbon readily available to soil microorganisms[9,25]. The additional carbon permits microbial utilization of all the N and slows formation and accumulation of phytotoxic levels of microbial activity end-products. Figure 2 presents a summary of our experiences on the relations between the C:N ratios of amendments, phytotoxicity and nematicidal activity. There is an optimal range of C:N values for amendments in which nematicidal activity occurs without phytotoxicity.

The action of amendments with narrow C:N ratios against nematodes can be conceived as based on a non-selective broad spectrum stimulation of the soil microflora. Most interesting among the nitrogenous amendments that stimulate a specialized soil microflora are those containing chitin or similar mucopolysaccharides. The addition of chitin to soil results in the stimulation of a microflora capable of decomposing the polymer into chitobiose and N-acetyl-glucosamine[15,19,24]. Further microbial activity results in deamination of the sugar and accumulation of ammonia and nitrates[24]. As might be expected, chitin amendment results in very sharp increases in chitinase activity and of other enzymatic activities associated with stimulation of the chitin-decomposing microflora[4,14,24]. Chitin amendments to soil are effective for control of phytonematodes[7,11,14,27,30]. More important, with some exceptions, a relationship has been demonstrated to exist between chitinolytic ability of fungi and their capacity to destroy nematode eggs[7,27]. This may be explained in part by evidence that chitin or a very similar material is a component of the middle layer of the tylenchoid egg shell[3]. Also, there is evidence that chitin may be present in the gelatinous matrix of egg masses of *Meloidogyne* spp.[29]. Several fungal species isolated from soils treated with chitin are able to decompose the polymer and are known colonizers of eggs of *Meloidogyne* spp., or of *Heterodera glycines*[7,27].

Our present knowledge about mode of action of chitinous amendments against nematodes suggests new approaches in the use of organic amendments for the management of phytonematode populations. Chitin studies have suggested the possibility of selecting materials for soil amendment that will stimulate development of a specialized microflora antagonistic to nematodes. A possible approach may be to partially 'match' the chemical composition of an amendment with that of some nematode component exposed to soil, *e.g.,* cuticle, egg shell, gelatinous matrix, *etc.* Addition of these specific compounds to soil would be expected to stimulate development of microbial species capable of degrading similar compounds present in the nematode. Use of these specialized types of organic amendments (*e.g.* chitin) offers the advantage of determining the microbial species that will be present in the amended soil

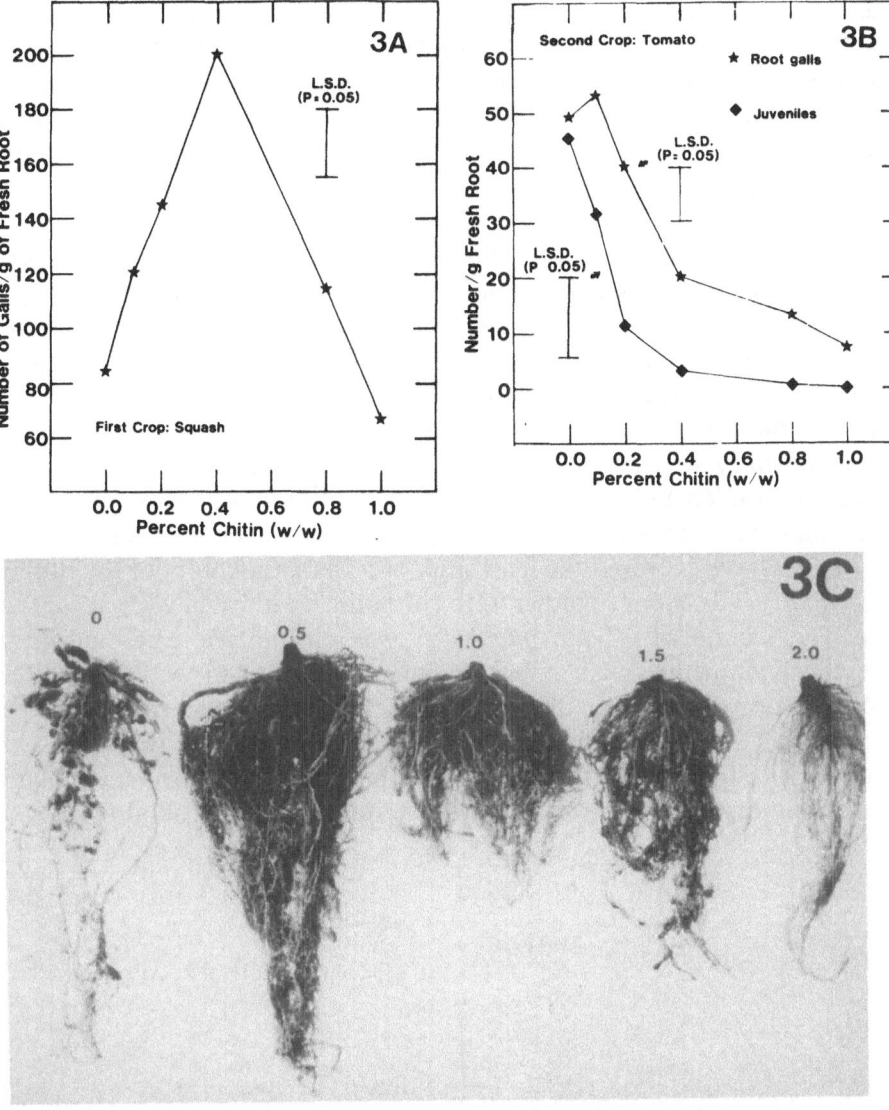

Fig. 3. Effect of chitin amendments on numbers of root galls caused by *Meloidogyne arenaria* on (**A**) squash (first crop) and on (**B**) tomatoes (2nd crop) in soil infested with the nematode; (**C**) effect of increasing dosages (% w/w) of chitin on tomato root systems.

for some time after application of the amendments. Effects of chitin amendments on nematodes may last for several months. Sufficient time must be allowed after addition of the amendment for populations of the specialized organisms to develop to levels adequate for effective nematode control. In our experience, it is not unusual to obtain little or no nematode control in crops soon after chitin amendments are applied to soil, whereas good control often occurs in the second crop following harvest of the first. This phenomenon was well illustrated in a recent study in our laboratory on the long-term effects of chitin amendments on the root-knot nematode *Meloidogyne arenaria*. The incidence of galls caused by the nematode in the roots of squash (*Cucurbita pepo*), which was planted a short time after amending the soil with chitin, was not affected by the amendments or was actually increased (Fig. 3A). However, when tomato (*Lycopersicon esculentum*) was planted after the squash the amendments were very effective in reducing numbers of root galls and curtailing production of juvenile nematodes (Fig. 3B, 3C) in the roots. Numbers of chitinolytic microorganisms, especially actinomycetes and bacteria (Fig. 4), were higher in chitin-amended soil than in unamended soils. A number of the fungal species recovered from the chitin-treated soils were known colonizers of nematode eggs.

A corollary to the use of amendments of specific composition for selection of microbial antagonists is the possibility of modifying organic amendments to make them more specific and effective against nematodes. We have explored the feasibility of adding inoculum of the fungus *Paecilomyces lilacinus* together with chitin to soil infested with *M. arenaria*. The isolate of *P. lilacinus* used was chitinolytic and had been previously shown to destroy eggs of the nematode in *in vitro* tests. The study demonstrated that the fungus inoculum could increase the effectiveness of the chitin amendment against the nematode (Fig. 5). This approach to increasing the selectivity or effectiveness of organic amendments can also be followed using other individual microbial species or

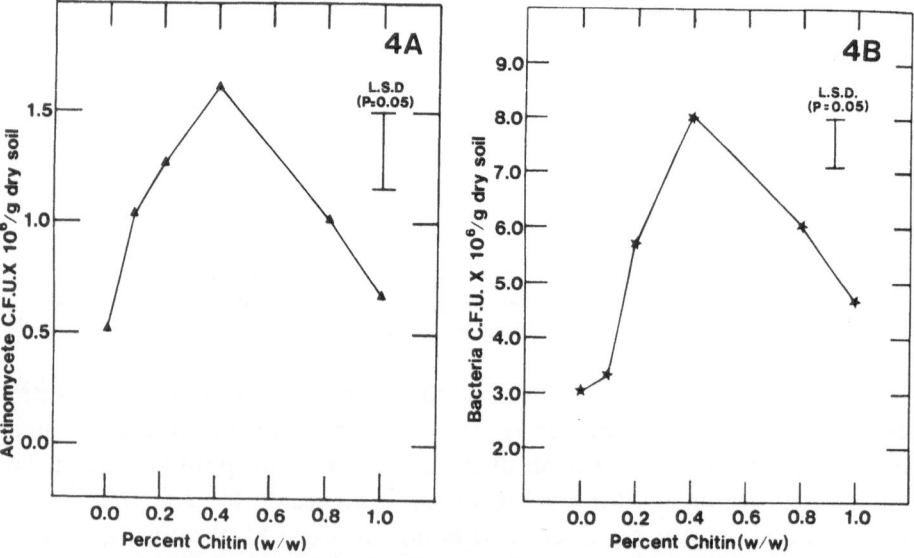

Fig. 4. Numbers of chitinolytic colony forming units (C.F.U.) of actinomycetes (**A**) and bacterial (**B**) in soil infested with *M. arenaria* and treated with several levels of chitin.

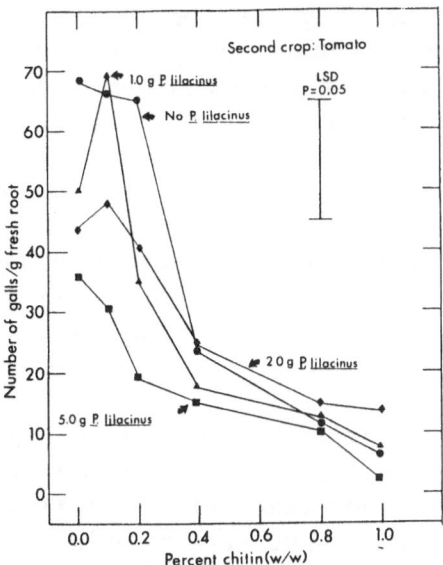

Fig. 5. Effect of chitin amendments and additions of inoculum of the fungus *Paecilomyces lilacinus* on numbers of root galls caused by *M. arenaria* in tomatoes. (Fungus inoculum consisted of oat grains colonized by *P. lilacinus* and was added to soil at rates of 0, 1, 2, or 5 g/kg soil.).

mixtures of species antagonistic to nematodes. Chitin or other appropriate material serves as the substrate or food base for selective development of the biocontrol agent in soil. The approach should permit the design of amendments to control specific target pathogens in soil.

Organic materials may also be applied by mixing them with compounds that will increase the selectivity of the materials for specific components of the soil microflora. This concept has received little attention in studies on the management of phytonematodes with organic amendments. Culbreath *et al.*,[4] demonstrated that addition of ligno-hemicellulosic materials to soil amended with chitin can increase the effectiveness of chitin against nematodes and avoid some of the deleterious effects of chitin when applied at high levels (1.0% (w/w)). It is possible that the reported benefit of the ligno-hemicellulosic supplement may have resulted from increased selectivity of the treatments for nematode antagonists in soil.

Application of organic amendments and the use of commercial nematicides for managing phytonematode populations differ in several respects. The amounts of organic matter added in the case of the amendments is in the hundreds of Kg/ha, whereas nematicides are used at very low levels. Effective organic amendments increase the plant nutrient status of the soil, serving as manures. The nutrient content of the amendments and the large quantities of these materials added to soil result in increased biological activity. Such activity can lead to significant changes in microbial composition. The changes may be deleterious or beneficial to crop plants. For example, considerable phytotoxicity is associated with the addition of low C:N ratio organic matter (oil cake, animal excrement) to soil. This adverse effect may come from accumulation of salts or ionic species (*i.e.* nitrites, nitrates, ammonia) or from development of phytopathogenic microorganisms. These problems in most cases can be avoided by modification of the amendments (*e.g.* widening of C:N ratio) or by the addition of specific organisms antagonistic to phytopathogens that may colonize and develop on the amendment. To effect any of these modifications it is essential to have accurate knowledge of the microbial processes involved in the decomposition of organic matter in soil.

Often, organic amendments are effective against nematodes in some fields but not in others. Physical and chemical soil properties, and influence of amendments on the soil microflora cannot be expected to be the same for all locations. The pH and the buffer capacity of soils can have a direct influence on the effectiveness of chitinous amendments against nematodes. In our experience, the acidic soils typical of southeastern USA (pH 5.5–6.5) require larger amounts of chitin to obtain satisfactory nematode control than do the neutral-alkaline soils of Israel[30].

Unlike commercial nematicides, effects of organic amendments on the soil microflora can last for several months and may affect more than one crop. This phenomenon, if properly understood and exploited, can be of advantage in crop production. The long-term effects of organic amendments against nematodes are in part attributable to the establishment of high antagonistic microorganism populations. It should be possible to maintain a selected microflora through periodic additions of appropriate amendments, thus suppressing nematode populations for sustained periods.

Besides the nutrient value and nematicidal activity, organic amendments may provide other benefits. In our studies with chitin we have shown that soil treated with the amendments became suppressive to *Sclerotium rolfsii*. Germination of sclerotia of this phytopathogen buried in chitin-treated soil was reduced significantly and in direct proportion to the amount of the polymer added to soil (Fig. 6). The ungerminated sclerotia were found colonized by chitinolytic bacterial species. Apparently, the microflora selected by the chitin amendments was effective not only against root-knot nematode but also against a fungus that contains chitin in its mycelium. Therefore, it seems possible, as was suggested by Mitchell and Alexander in 1961[15], that selection of a chitinolytic microflora may result in suppression of organisms that contain chitin in any exposed structure. Equally, one may logically

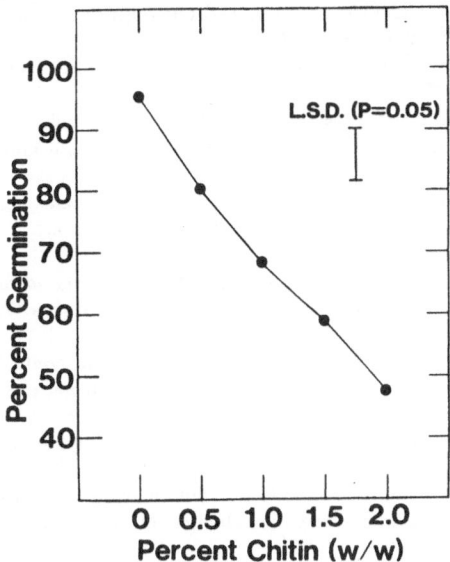

Fig. 6. Relation between the rates of chitin added to soil and germination of sclerotia of *Sclerotium rolfsii* after burial in amended soil for 10 days.

expect that organisms, such as some zoosporic fungi, that do not contain chitin in their mycelia will not be affected.

Present information available on the use of organic amendments for control of phytonematodes in soil indicates that this is a promising area for development of new nematode management practices. There are available throughout the world numerous types of waste materials that could be considered for use as soil amendments. It is now clear that a great deal of knowledge is lacking on the precise mode of action of organic amendments on nematodes and other plant pathogens in soil. Each type of amendment and each environment in which the amendments are used represents a new situation that must be defined in terms of microbial species and biochemical processes involved in decomposition of the amendments, non-target effects of the amendments, and economics. Organic wastes, in many cases, pose problems for finding disposal methods consistent with sound ecological practices. It should be possible to develop technologies for utilization of these waste materials in the manufacture of products that can be used for control of phytonematodes in soil.

References

1   Alam M M, Ahmad M and Khan A H 1980 Effect of organic amendments on the growth and chemical composition of tomato, eggplant and chili and their susceptibility to attack by *Meloidogyne incognita*. Plant and Soil 57, 231–236.

2   Badra T, Saleh M A and Oteifa B A 1979 Nematicidal activity and composition of some organic fertilizers and amendments. Revue Nematol. 2, 29–36.

3   Bird A F and McClure M A 1976 The tylenchoid (Nematoda) egg shell: structure, composition and permeability. Parasitology 72, 19–28.

4   Culbreath A K, Rodriguez-Kabana R and Morgan-Jones G 1985 The use of hemicellulosic waste matter for reduction of the phytotoxic effects of chitin and control of root-knot nematodes. Nematropica 15, 49–75.

5   Godoy G, Rodriguez-Kabana R and Morgan-Jones G 1983 Fungal parasites of *Meloidogyne arenaria* in an Alabama soil. A mycological survey and greenhouse studies. Nematropica 13, 201–213.

6   Godoy G, Rodriguez-Kabana R and Morgan-Jones G 1982 Parasitism of eggs of *Heterodera glycines* and *Meloidogyne arenaria* by fungi isolated from cysts of *H. glycines*. Nematropica 12, 111–119.

7   Godoy G, Rodriguez-Kabana R and Morgan-Jones G 1983 Chitin amendments for control of *Meloidogyne arenaria* in infested soil. II. Effects on microbial population. Nematropica 13, 63–74.

8   Hollis J P and Rodriguez-Kabana R 1966 Rapid kill of nematodes in flooded soil. Phytopathology 56, 1015–1019.

9   Huebner R A, Rodriguez-Kabana R and Patterson R M 1983 Hemicellulosic waste and urea for control of plant parasitic nematodes: effect on soil enzyme activities. Nematropica 13, 37–54.

10   Linford M B, Yap F and Oliveira J M 1938 Reduction of soil populations of root-knot nematode during decomposition of organic matter. Soil Science 45, 127–141.

11   Mankau R and Das S 1969 The influence of chitin amendments on *Meloidogyne incognita*. J. Nematology 1, 15–16.

12   Mian I H and Rodriguez-Kabana R 1982 Soil amendments with oil cakes and chicken litter for control of *Meloidogyne arenaria*. Nematropica 12, 205–220.

13   Mian I H and Rodriguez-Kabana R 1982 Survey of the nematicidal properties of some organic materials available in Alabama as amendments to soil for control of *Meloidogyne arenaria*. Nematropica 12, 235–246.

14   Mian I H, Godoy G, Shelby R, Rodriguez-Kabana R and Morgan-Jones G 1982 Chitin amendments for control of *Meloidogyne arenaria* in infested soil. Nematropica 12, 71–84.

15   Mitchell R and Alexander M 1961 Chitin and the biological control of Fusarium diseases. Plant Disease Reptr 45, 487–490.

16   Morgan-Jones G, Ownley-Gintis B and Rodriguez-Kabana R 1981 Fungal colonization of *Heterodera glycines* cysts in Arkansas, Florida, Mississippi, and Missouri soils. Nematropica 11, 155–164.

17   Morgan-Jones G and Rodriguez-Kabana R 1981 Fungi associated with cysts of *Heterodera glycines* in an Alabama soil. Nematropica 11, 69–77.

18   Muller R and Gooch 1982 Organic amendments in nematode control. An examination of the literature. Nematropica 12, 319–326.

19   Muzzarelli R A 1977 Chitin. Pergamon Press, New York, N.Y. 309 pp.

20   Ownley-Gintis B, Morgan-Jones G and Rodriguez-Kabana R 1983 Fungi associated with several developmental stages of *Heterodera glycines* from an Alabama soybean field soil. Nematropica 13, 181–200.

21   Ownley-Gintis B, Morgan-Jones G and Rodriguez-Kabana R 1982 Mycoflora of young cysts of *Heterodera glycines* in North Carolina soils. Nematropica 12, 295–303.

22   Rodriguez-Kabana R 1986 Organic and inorganic nitrogen amendments to soil as nematode suppressants. J. Nematol. 18, 129–135.

23   Rodriguez-Kabana R and Hollis J P 1965 Biological control of nematodes in rice fields: role of hydrogen sulfide. Science 148, 524–526.

24   Rodriguez-Kabana R, Godoy G, Morgan-Jones G and Shelby R A 1983 The determination of soil chitinase activity. Conditions for assay and ecological studies. Plant and Soil 75, 95–106.

25   Rodriguez-Kabana R and King P S 1980 Use of mixtures of urea and blackstrap molasses for control of root-knot nematodes in soil. Nematropica 10, 38–44.

26   Rodriguez-Kabana R, King P S and Pope M H 1981 Combinations of anhydrous ammonia and ethylene dibromide for conttrol of nematodes parasitic of soybeans. Nematropica 11, 27–41.
27   Rodriguez-Kabana R, Morgan-Jones G and Ownley-Gintis B 1984 Effects of chitin amendments to soil on *Heterodera glycines*, microbial populations, and colonization of cysts by fungi. Nematropica 14, 10–25.
28   Singh R S and Sitaramaiah K 1973 Control of plant parasitic nematodes with organic amendments of soil. Final technical report. Effect of organic amendments, green manures, and inorganic fertilizers on root-knot of vegetable crops. Res. Bull. Expt. Stn. and Coll. Agric. G. B. Plant Univ. Agric. Tech., Pantanagar. 6, 289 pp.
29   Spiegel Y and Cohn E 1985 Chitin is present in gelatinous matrix of Meloidogyne. Revue de Nematologie 8, 184–185.
30   Spiegel Y, Cohn E and Chet I 1986 Use of chitin for controlling plant parasitic nematodes. I. Direct effects on nematode reproduction and plant performance. Plant and Soil 95, 87–95.

Plant and Soil 100, 249–283 (1987).
Ms. 100-21
© 1987 *Martinus Nijhoff Publishers, Dordrecht.*

Practical aspects of mycorrhizal technology in some tropical crops and pastures

R.H. HOWELER*, E. SIEVERDING and S. SAIF
CIAT, Apartado Aéreo 6713, Cali, Colombia

Key words Acid soils Agronomic practices Beans Cassava Coffee Environmental effects Inoculation VA mycorrhiza VAM species effectivity Tea Tropical pastures Yield

Summary Greenhouse and field experiments were conducted on the effect of VA mycorrhiza (VAM) on the growth of cassava, various tropical grass and legume species, as well as beans, coffee and tea. A large number of VAM fungal species were evaluated for effectivity in increasing cassava growth and P uptake in acid low-P soils. The effectivity of VAM species and isolates was highly variable and dependent on soil pH and fertilizer applications, as well as on soil temperature and humidity. Two species, *Glomus manihotis* and *Entrophospora colombiana* were found to be most effective for a range of crops and pastures, at low pH and at a wide range of N, P, and K levels.

At very low P levels nearly all crops and pasture species were highly mycorrhizal dependent, but at higher soil P levels cassava and several pasture legumes were more dependent than grass species.

Mycorrhizal inoculation significantly increased cassava and bean yields in those soils with low or ineffective indigenous mycorrhizal populations. In these soils cassava root yields increased on the average 20–25% by VAM inoculation, both at the experiment station and in farmers' fields. VAM inoculation of various pasture legumes and grasses, in combination with rock phosphate applications, increased their early growth and establishment. Agronomic practices such as fertilization, crop rotations, intercropping and pesticide applications were found to affect both the total VAM population as well as its species composition.

While there is no doubt about the importance of VA mycorrhiza in enhancing P uptake and growth of many tropical crops and pastures grown on low-P soils, much more research is required to elucidate the complicated soil–plant–VAM interactions and to increase yields through improved mycorrhizal efficiency.

Introduction

Vesicular-arbuscular mycorrhizal (VAM) fungi, when associated with higher plants, play an important role in their P nutrition. In this symbiotic association the fungus utilizes carbohydrates produced by the plants, while the plant benefits by the increased uptake of P and some other nutrients through the external hyphae extending from the root surface into the soil[21]. The beneficial effect of mycorrhiza is of special importance for those plants having a coarse and poorly branched root system, since the external hyphae can extend as much as 8 cm away from the roots[26], absorbing nutrients from a much larger soil volume than the absorption zone surrounding a non-mycorrhizal root. This is of particular importance for the absorption of nutrients of low mobility in soil solution such as P, Zn and Cu.

---

* Present address: CIAT Regional Office, Department of Agriculture, Bangkhen, Bangkok 10900, Thailand.

VAM fungi are present in nearly all natural soils and these fungi infect the great majority of plants, including the major food crops. Many tropical crops and pastures are grown on soils that are very deficient in P, and particularly in those soils an efficient mycorrhizal association is of great importance to increase P uptake and crop yields. Since 1980, the Centro Internacional de Agricultura Tropical (CIAT), in collaboration with the German financed Mycorrhiza Project at CIAT, has investigated the efficiency of naturally occurring (or native) VAM species as well as the effect of inoculation with highly efficient isolates of VAM fungi on the yield of various crops and pastures. The objective of this paper is to review the results obtained and discuss the potentials as well as the limitations of mycorrhizal technology in several tropical crops and pastures.

Mycorrhizal dependence of different crops and pastures

Except for a very few plant species, including the Cruciferae, Cyperaceae and Chenopodiaceae[21], most plants form associations with VA mycorrhiza. However, large differences exist between plant species in their dependence on VAM fungi for P uptake and growth. Plant species considered most mycorrhiza dependent are those having thick, fleshy roots with few roots hairs, such as cassava (*Manihot esculenta*, Crantz), citrus (*Citrus* spp), sweetgum (*Liquidambar styraciflua*), grape (*Vitus vignifera*) and many legumes[21]. Yost and Fox[38] reported a greater dependence on VAM fungi for P uptake by cassava and *Stylosanthes* sp. compared with cowpea (*Vigna unguiculata*), onion (*Allium cepa*), soybeans (*Soya max*) and *Leucaena* sp. Bowen[2] reported that all economically important tropical crops are mycorrhizal with the exception of flooded rice (*Oriza sativa*). Upland rice, however, is mycorrhizal and a 40% yield increase was obtained by inoculation of this crop in sterilized Nigerian soils[29].

The mycorrhizal dependence of the crop varies with the P content of the soil as well as the species of VAM fungi used to inoculate. Table 1 shows the dry matter (DM) production of various crops and forage species in a pasteurized Quilichao soil (see Table 2) to which three levels of P had been applied. The crops were either not inoculated or inoculated with fine cassava roots infected with native VAM fungi from Quilichao, containing mainly *Glomus manihotis* and *Entrophospora colombiana*. All crops responded to inoculation except upland rice. If we calculate the relative mycorrhizal dependence of each plant species as the DM yield obtained with inoculation minus the DM yield without inoculation, divided by DM yield with inoculation[24], it is clear that at low levels of applied P, cassava, Stylosanthes guianensis, and *Andropogon gayanus*

Table 1. Effect of inoculation and P application on dry weight (g pot^{-1}) of aerial part of several plant species grown in sterilized soil from Quilichao. Last column shows their relative mycorrhizal dependence calculated from the average yield at three P levels*

Plant species	Non-inoculated			Inoculated			Mycorrhizal dependency**
	P_0	P_{100}	P_{500}	P_0	P_{100}	P_{500}	
Cassava (*Manihot esculenta*)	0.34	0.72	0.54	4.33	14.21	16.36	95
Beans (*Phaseolus vulgaris*)	1.11	3.44	8.29	3.08	18.79	25.01	72
Cowpea (*Vigna unguiculata*)	0.96	0.64	13.65	2.60	20.68	36.32	74
Stylosanthes guianensis	0.08	0.08	2.74	1.25	9.33	12.20	87
Andropogon gayanus	0.15	0.39	34.24	1.26	16.67	32.18	30
Maize (*Zea mays*)	1.19	8.74	59.35	4.84	34.75	53.67	26
Rice (*Oryza sativa*)	3.79	26.63	30.60	3.83	22.36	31.23	−6

* P_0, P_{100} and P_{500} indicate levels of 0, 100 and 500 kg P ha^{-1} applied as TSP.
** Mycorrhizal dependency = dry weight of mycorrhizal plants minus dry weight of non-mycorrhizal plants divided by dry weight of mycorrhizal plants multiplied by 100.

were the most mycorrhiza dependent species. At high levels of applied P cassava was still extremely dependent, the legumes were intermediately dependent, and the grasses were not dependent, mainly because of their finer and more highly branched root system, which makes them more efficient in P uptake; however, this generalization does not always apply.

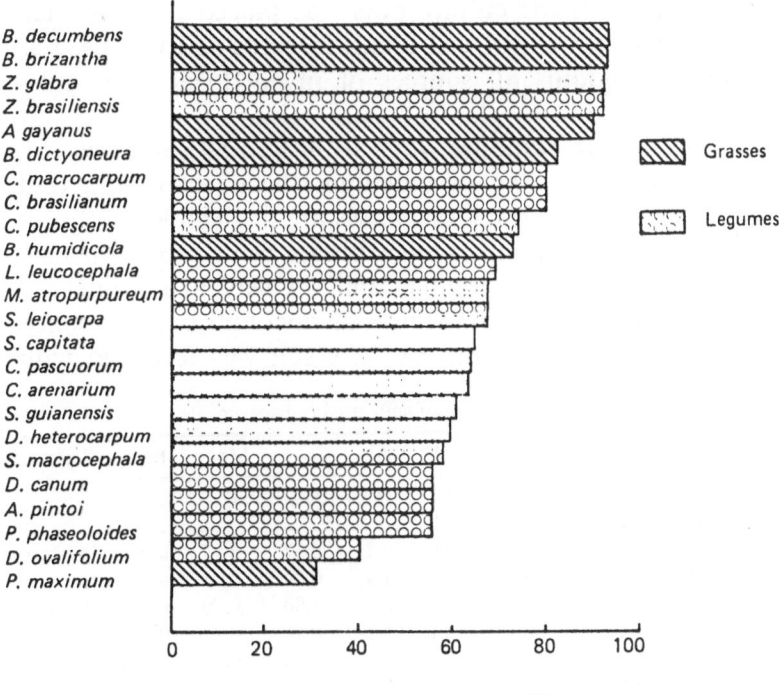

Fig. 1. Mycorrhizal dependency (MD) of twenty four tropical forage species. MD = Dry weight of mycorrhizal plants minus dry weight of non-mycorrhizal plants divided by dry weight of mycorrhizal plants multiplied by 100.

When the mycorrhizal dependence of six tropical grasses and 18 forage legumes were determined in a sterilized Oxisol from Carimagua ($3.5\,\mu g\,P\,g^{-1}$ soil) to which $20\,kg\,P\,ha^{-1}$ were applied as rock phosphate, five of the six grasses were more mycorrhiza dependent than the majority of the legumes (Fig. 1). Thus under conditions of very low levels of available P, as is often the case in tropical soils, the grasses may be as mycorrhiza dependent as the legumes, or more woody species such as cassava.

Collection and evaluation of VAM fungi

In order to identify VAM fungal species or isolates with high efficiency for increasing P uptake and growth, soil samples were collected in various important edapho-climatic zones of Colombia and their VAM spores were isolated. At present this mycorrhizal collection contains 1200 isolates, which are maintained as pure cultures on plants of *Pueraria phaseoloides* (kudzu) in the greenhouse or stored as soil inoculum in the cold room.

To evaluate the isolates for their efficiency to enhance plant growth, 25 g of soil inoculum from the culture pots was placed under rooted cuttings of cassava, which was the test plant. Infected soil was used as inoculum because it contains all sources of mycorrhizal propagules resulting in a rapid infection of plants. In preliminary trials no differences were found in growth response when from 5 to 50 g infected soil was applied; thus, by using 25 g of inoculum the effect of different inoculum concentrations on growth responses are neglegible.

In the first step of evaluation, cassava plants were grown in pasteurized soil from Quilichao. This soil has similar chemical and physical characteristics (see Table 2) as those where cassava and pasture plants are commonly grown in South America. The soil was fertilized with

Table 2. Chemical characteristics and classification of soils used in experiments

Soil	O.M. (%)	P* ($\mu g\,g^{-1}$)	pH	Al	Ca	Mg	K	Classification
				(meq $100\,g^{-1}$ soil)				
Carimagua-Reserva, Meta	3.2	1.4	4.0	3.1	0.37	0.17	0.08	Haplustox
Carimagua-Alegria, Meta	2.4	0.9	4.6	1.4	0.09	0.05	0.04	Haplustox
CIAT-Quilichao, Cauca	7.1	1.8	4.3	2.8	1.80	0.70	0.18	Paleudult
Mondomito, Cauca	4.8	1.6	4.1	5.1	0.45	0.12	0.11	Dystropept
Tres Quebradas, Cauca	8.0	0.9	4.7	2.3	1.26	0.43	0.13	Dystropept
Telecom, Cauca	5.6	0.4	4.2	8.3	0.60	0.12	0.12	Dystropept
San Emigdio, CVC-Valle	8.2	2.1	4.8	3.3	0.69	0.26	0.10	Dystropept
CIAT-Palmira, Valle	4.0	40.0	7.0	–	20.0	10.0	0.50	Calciustoll

* P extracted with Bray II: $0.03\,N$ NH$_4$ F + $0.1\,N$ HCl.

$50 \, \text{kg N ha}^{-1}$ as NH_4NO_3, $25 \, \text{kg P ha}^{-1}$ as $NaH_2PO_4 \cdot H_2O$, $50 \, \text{kg K ha}^{-1}$ as KCl and $5 \, \text{kg Zn ha}^{-1}$ as $ZnCl_2$; no lime was added.

Efficiency of species and isolates of VAM fungi was found to be highly variable, as shown in Table 3. To compare isolates in different trials,

Table 3. Evaluation of some mycorrhizal isolates' effectiviness for cassava in pasteurized Quilichao soil, in the greenhouse

Isolate No.	Species	Dry matter production	P uptake	Root length	Root infection
C-1-1	G. manihotis	xxx[a]	xxx[a]	xx[a]	xxx[a]
C-2-3	A. longula	x	x	x	x
C-2-4	G. occultum	xx	xx	xxx	x
C-3-2	G. occultum	xxx	xxx	xxx	x
C-3-4	A. myriocarpa	xxx	xxx	xxx	xx
C-3-5	E. colombiana	xxx	xxx	–	xx
C-3-10	G. manihotis	xxx	xxx	xx	xxx
C-3-11	E. colombiana	x	x	xx	xxx
C-3-12	A. morrowae	0	0	0	x
C-3-15	A. laevis	x	xx	0	xxx
C-3-17	A. morrowae	x	0	x	x
C-9	Gi. heterogama	0	0	0	0
C-10	E. colombiana	xxx	xxx	–	x
C-12-1	A. longula	xxx	xxx	xxx	x
C-12-2	G. occultum	xxx	xxx	xx	x
C-13-3/4	Glomus sp.	xxx	xxx	x	xxx
C-13-5	E. colombiana	xx	xxx	x	xxx
C-13-6	A. longula	xxx	xxx	xxx	xxx
C-13-8	G. occultum	xx	xx	xxx	xx
C-15-2	A. mellea	xx	xx	xxx	xx
C-17-3	A. myriocarpa	xxx	xxx	x	xxx
C-20-2	G. manihotis	xxx	xxx	xx	xxx
C-78	G. manihotis	xxx	xxx	xx	xxx
C-82	G. manihotis	xxx	xxx	xx	xx
C-85	A. appendicula	xxx	xxx	xxx	xx
C-86-2	A. laevis	xxx	xxx	xx	xxx
C-88-1	A. myriocarpa	x	xx	0	x
C-89-1	Gi. pellucida	x	xx	x	x
C-89-2	A. laevis	0	0	xxx	xxx
C-89-3	A. morrowae	0	0	0	xx
C-90-1	A. appendicula	0	xx	x	x
C-90-2	E. colombiana	xx	xx	xxx	x
C-92-1	G. occultum	x	xx	0	0
C-92-2	G. occultum	xxx	xx	xxx	x
C-93-1	A. mellea	0	0	0	x
C-93-2	A. longula	x	x	0	xxx
C-105	Glomus sp.	xxx	xx	xx	xx

[a] 0 = not effective, plant parameters not different from non-inoculated plants.

x = low effectiveness, plant parameters significantly different from non-inoculated plants, but significantly smaller than trail average.

xx = medium effectiveness, plant parameters not significantly different from trial average.

xxx = high effectiveness, plant parameters significantly better than trial average.

– = lost data.

Table 4. Effect of some mycorrhizal isolates on cassava growth in soil with high mycorrhizal competition (unsterilized soil from Quilichao) in a greenhouse trial

Isolate No.	Species	Shoot dry weight (g plant^{-1})	P content (mg P g^{-1} dry weight)	Root length (m plant^{-1})	Root length infected (%)
Native		5.66^{d}*	1.65^{c}	19.4^{a}	41.0^{bc}
C-1-1	Glomus manihotis	14.05^{a}	1.70^{bc}	11.4^{a}	62.3^{a}
C-3-5	Entrophospora colombiana	8.82^{bcd}	1.78^{bc}	14.4^{a}	35.2^{bc}
C-6-2	Acaulospora longula	7.55^{bcd}	1.63^{c}	17.3^{a}	33.9^{bc}
C-7	A. myriocarpa	11.49^{ab}	1.85^{bc}	15.6^{a}	25.0^{c}
C-10	E. colombiana	9.95^{bc}	1.90^{bc}	13.6^{a}	42.3^{bc}
C-12-1	A. longula	8.19^{bcd}	2.45^{a}	13.3^{a}	48.5^{ab}
C-12-2	G. occultum	6.57^{cd}	2.00^{b}	10.6^{a}	37.9^{bc}
Mean		9.04	1.87	14.45	40.76

* Figures followed by the same letter(s) were not statistically different at the 0.05 level of the Duncan's test.

efficacy was expressed relative to non-inoculated plants, and in relation to the trial average. At present about 150 isolates have been evaluated. It was found that plant growth was strongly related to P uptake, but there was no general relationship between infection ratings and effectiveness for P uptake. However, most isolates effective for P uptake were also highly infective. Using P uptake as an indicator, about 40% of the isolates were not effective or have low effectiveness with cassava[7]. These isolates were not included in the next evaluation under unsterilized soil conditions, where the fungal isolates have to compete with other soil microorganisms including indigenous VAM fungi.

In the second step of evaluation, virgin soil from Quilichao was used. This soil has a large population of very effective VAM fungi[18]. Instead of soluble P, Huila rock phosphate (HRP) was applied at the rate of 50 kg P ha^{-1}. Table 4 shows that in one trial only three out of seven isolates were able to increase significantly shoot production of cassava. Of all the tested isolates about 30% were found to be effective under these conditions and should be further screened in the field.

Similar trials were also conducted with several pasture species, using either pasteurized or natural virgin soil from Carimagua in the Llanos Orientales of Colombia, to which 30 kg P ha^{-1} was applied as HRP. Table 5 shows that of the 13 VAM isolates tested, E. colombiana (C-10 or C-11-2) was most effective for two of the four legumes and for three of the four grasses tested. G. manihotis, Acaulospora longula and A. scorbiculata were most efficient on Desmodium ovalifolium, Zornia glabra and Brachiaria humidicula, respectively.

Table 5. Percent increase in shoot dry weight of forage legumes and grasses caused by inoculation with best-yielding mycorrhizal fungi. NM = not inoculated; IMF = indigenous mycorrhizal fungi (Source: CIAT[10])

Plant species	Most effective mycorrhizal fungi	CIAT Isolate No.	Percent dry weight increase over	
			NM	IMF
Legumes				
Centrosema macrocarpum	E. colombiana	C-10	242	173
Stylosanthes capitata	E. colombiana	C-11-2	1005	238
Desmodium ovalifolium	G. manihotis	C-20-2	893	221
Zornia glabra	A. longula	C-12-1	1087	240
Grasses				
Andropogon gayanus	E. colombiana	C-11-2	4950	300
Brachiaria decumbens	E. colombiana	C-11-2	1281	212
B. brizantha	E. colombiana	C-11-2	571	203
B. humidicola	A. scorbiculata	C-76-1	499	178

Other greenhouse trials with beans (*Phaseolus vulgaris*) showed greatest effectivity of *G. manihotis* (C-1-1), and *E. colombiana* (C-10 and C-3-5)[13]. Thus, it was concluded that those isolates that are most efficient for cassava are also quite efficient for other crops grown under similar edapho-climatic conditions.

Responses to mycorrhiza in greenhouse trials

To determine the specific effect of mycorrhiza on plant growth and nutrient uptake it is often necessary to eliminate the native VAM fungi by soil sterilization, and reinoculate the plant with a specific VAM isolate. At CIAT, greenhouse trials are generally conducted using virgin soil from CIAT-Quilichao or Carimagua. The chemical characteristics and classification of these and other soils are listed in Table 2. When necessary the soil was steam pasteurized for two 4 hour periods. This was found adequate to eliminate completely the native VAM fungi population[8]. Soils were generally fertilized with P as TSP or HRP, N as urea, K as KCl, Zn as $ZnSO_4 \cdot 7H_2O$ and Ca and Mg as dolomitic lime. Cassava trials were conducted with rooted tip cuttings, while other species were tested using sexual seed. Plants were inoculated by placement of a soil-root inoculum or root inoculum, usually produced with kudzu as host, under the plant, seedling or seed at time of planting.

Since the efficiency of isolates of VAM fungi often depends on the chemical and physical environment, some factors that might influence this efficiency were studied in the greenhouse.

A. *Effect of P levels and sources*

1. *Cassava.* Cassava is highly mycotrophic and its response to P application depends entirely on the mycorrhizal association. Howeler *et al.*[17] reported that non-mycorrhizal cassava had an external P requirement of 190 ppm (Bray II) compared with 15 ppm for mycorrhizal plants. In that pot trial, application of about 1600 kg P ha^{-1} were required for non-mycorrhizal plants to give the same growth as that of mycorrhizal plants without P.

Many researchers[1,12,37] have reported that the efficiency of VAM fungi decreases as the level of plant available P of the soil increases. However, at very low levels of soil P the VAM association cannot be effective either because VAM fungi are absorbing P from the same labile pool as the plant root[21]. Moreover, VAM species vary in their relative effectiveness at different P concentrations in the soil solution. Figure 2 shows that when isolates of 5 VAM fungi were evaluated in pasteurized Quilichao soil, to which 5 levels of P were applied, none of the isolates was very effective without P application; only *G. manihotis* (C-1-1 from Quilichao and C-20-2 from Carimagua) was highly effective at 25 kg P ha^{-1}, *E. colombiana* (C-10) was effective mainly at 50 to 200 kg P ha^{-1}, while

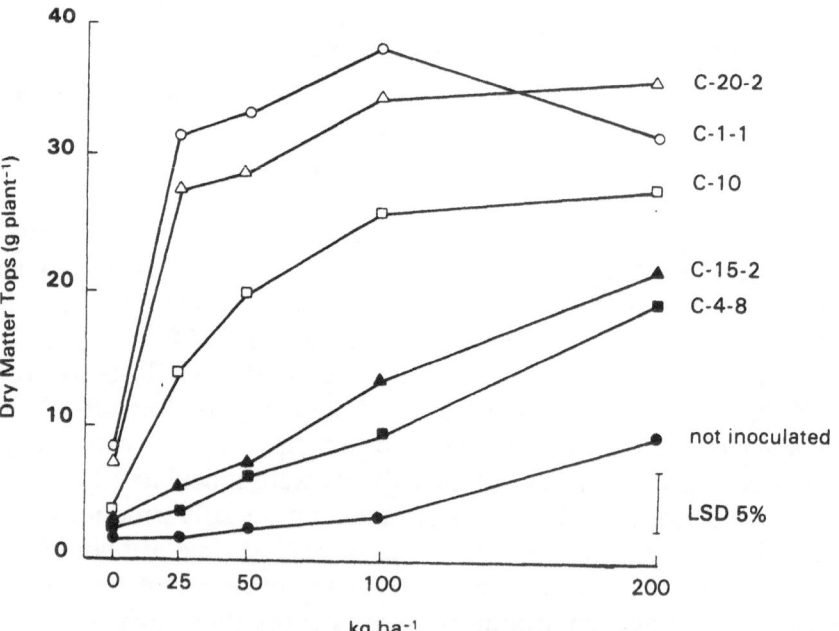

Fig. 2. Dry matter response of cassava, M Ven 77, to application of various levels of P and inoculation with five mycorrhizal strains in sterilized Quilichao soil.

A. mellea (C-15-2) and *Acaulospora* sp. (C-4-8) were only intermediately effective at high levels of $200 \, kg \, P \, ha^{-1}$. On the average, root infection increased slightly from 53% without P to 66% with $200 \, kg \, P \, ha^{-1}$, while spore numbers increased from 39 to 93 spores g^{-1} dry soil with the highest P rate. However, in case of *G. manihotis* (both C-1-1 and C-20-2) spore numbers declined above $50 \, kg \, P \, ha^{-1}$, while total P uptake increased up to $200 \, kg \, P/ha$ (data not shown).

2. *Pastures.* The effect of P, applied either as HRP (9.6% P) or basic slag (7.4% P), on the growth of *Pueraria phaseoloides, Centrosema macrocarpum* and *Brachiaria decumbens*, was investigated using unsterilized Carimagua soil in the greenhouse[27]. Plants were either not-inoculated or inoculated with a mixture of *G. manihotis* (C-1-1), *E. colombiana* (C-10) and *A. longula* (C-12-1). Figure 3 indicates that both the legumes and the grass responded markedly up to the highest level of P application, that there were no significant differences between the two P sources, and that VAM inoculation increased DM production significantly in all species and at each P level, although the response was greater in the legumes than in the grass, and relatively greater at lower levels of P application. VAM inoculation also increased significantly total P absorption in the two legumes, but this effect was not significant in *B. decumbens*, except at the highest P application rate. Thus, it may be surmised that in the extremely P deficient soils of the Llanos Orientales of Colombia pasture establishment can be aided by a combination of relatively insoluble (and cheap) P sources and VAM inoculation with highly effective fungi.

B. *Effect of soil pH*
Soil acidity and related problems of Al toxicity and low levels of Ca and Mg can affect the efficiency of VAM species, either directly, or indirectly through the deficient growth of plants. Tropical Oxisols and Ultisols are generally characterized by very low pH (4.0–5.5) and only those VAM fungi that tolerate this low pH and the corresponding high levels of Al can be expected to be effective in stimulating plant growth and P uptake.

To study the effect of pH on the efficiency of four VAM species, two soils, from Carimagua and Quilichao, were incubated with 1 or $2 \, t \, ha^{-1}$ of elemental S to decrease pH, and with 1 and $5 \, t \, ha^{-1}$ of calcitic lime to increase pH. With these soil amendments the pH ranged from 3.7 to 6.4 in the Carimagua soil and from 4.0 to 5.4 in the Quilichao soil. Rooted

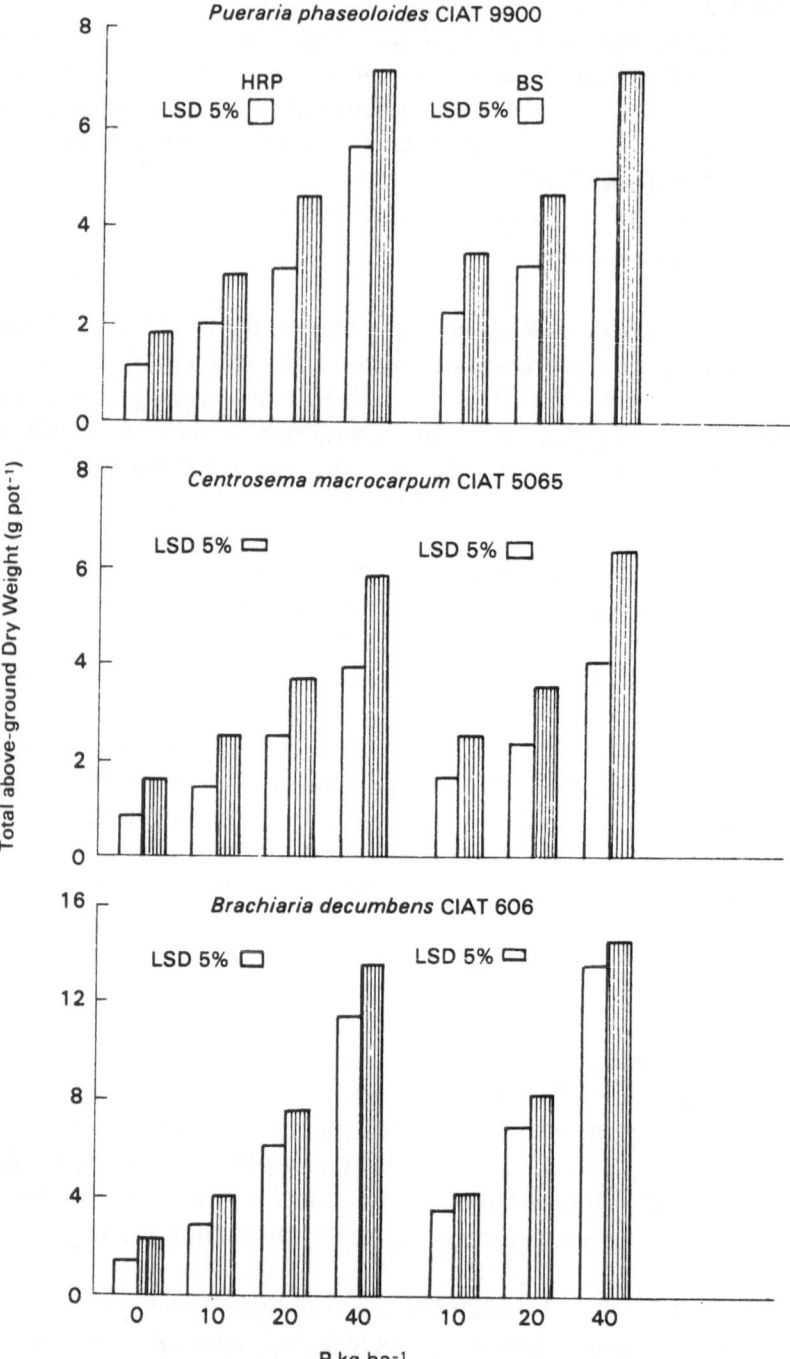

Fig. 3. Total dry weight of Pueraria, Centrosema and Brachiaria grown in unsterilized Oxisol in pots. HRP = rock phosphate Huila: BS = basic slag; □ = non-inoculated; ■ = inoculated with mycorrhiza (mixture of C-1-1, C-10 and C-12-1).

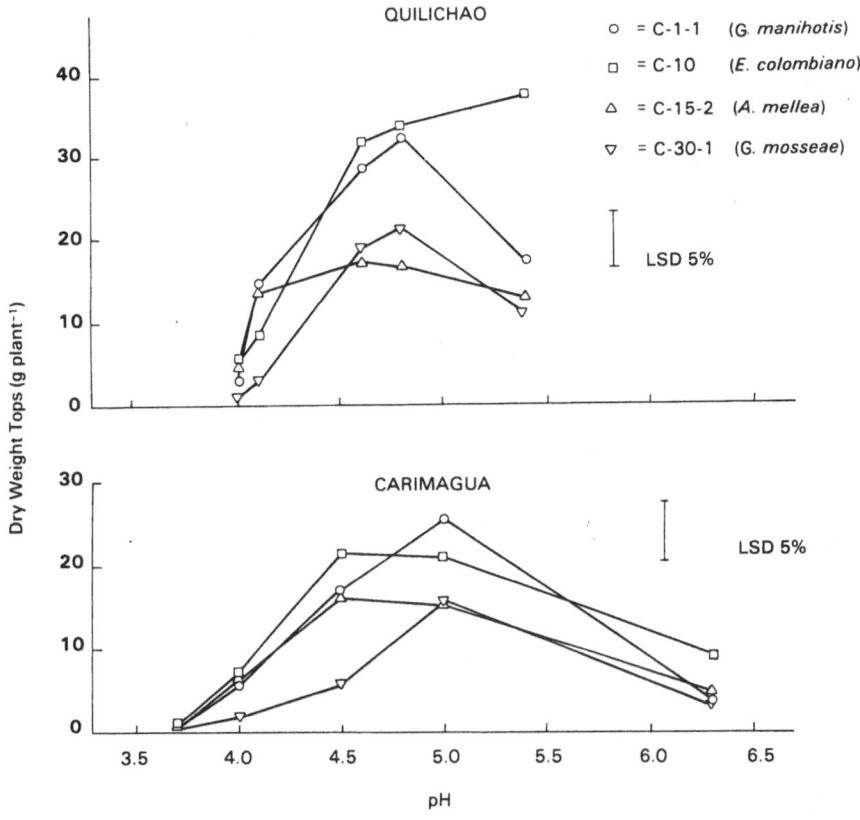

Fig. 4. Effect of soil pH on dry matter production of tops of two month old cassava, M Ven 77, inoculated with four VAM species and grown in sterilized soil from Quilichao and Carimagua.

tip cuttings of cassava were inoculated with four species of VAM fungi and planted in the amended and sterilized soils. Figure 4 shows the dry matter production of tops after two months of growth in both soils in relation to soil pH. Dry matter production was extremely low when S had been added to decrease pH, probably due to Al toxicity (6–8 meq $100\,g^{-1}$), and also when high levels of lime were added, due to induced Zn deficiency. However there were large differences among species of VAM fungi: *G. manihotis* (C-1-1), *E. colombiana* (C-10) and *A. mellea* (C-15-2) were quite effective even at a pH of 4.1, while *G. mosseae* (C-30-1) was only effective above pH 4.5. While all four species produced maximum DM production at pH between 4.5 and 5.0, *G. manihotis* and *E. colombiana* were overall most effective and best adapted to low soil pH. *A. mellea* was effective at low pH but relatively ineffective at higher pH, while *G. mosseae* was quite ineffective at low pH and only intermediately effective at higher pH. Figure 5 shows that for three of the four VAM fungi root infection was well related with soil pH in both soils,

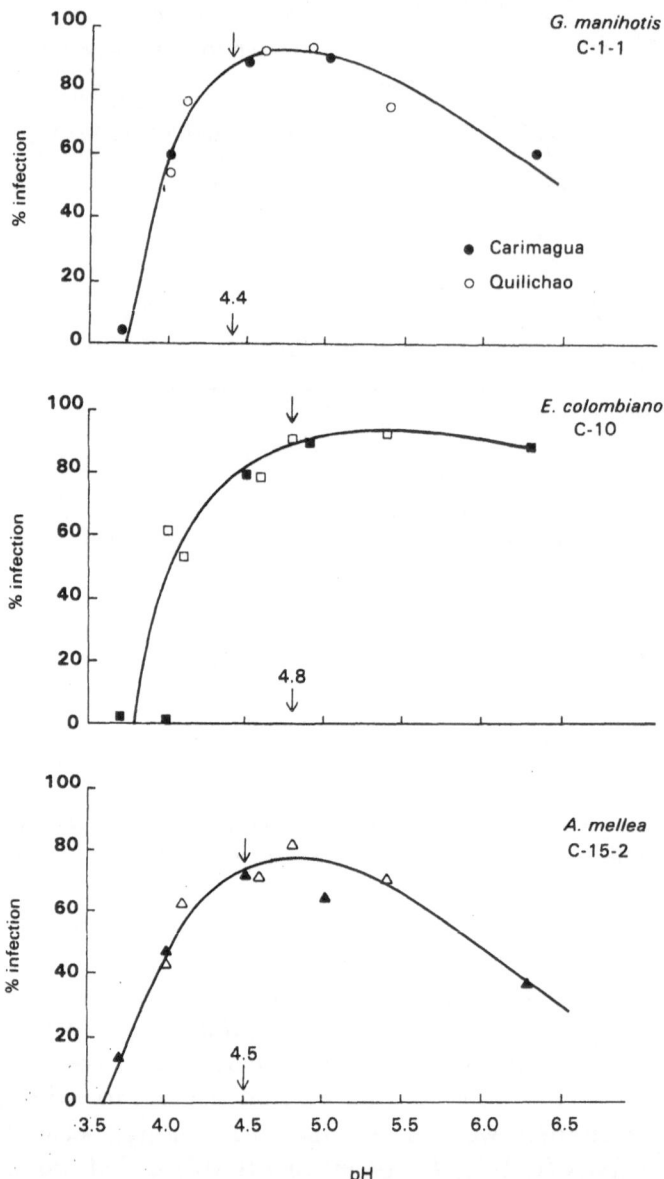

Fig. 5. Relation between percent VAM root infection and soil pH determined for three mycorrhizal fungi in sterilized soils from Carimagua and Quilichao. Arrows indicate the critical pH, corresponding to 95% of maximum infection.

increasing from nearly zero at pH below 4 to a maximum between pH 4.5 and 5.0. A critical pH, corresponding to 95% of maximum infection, was determined to be 4.4 for C-1-1, 4.5 for C-15-2, and 4.8 for C-10. Infection ratings around 80% were obtained between pH 4.5 and 5.5 which can explain the high effectiveness of these species in acid soils.

C. *Effect of N and K*

Several reseachers[3,11,14] have shown that mycorrhizal infection and effectiveness not only decrease with high levels of soil P, but may also decrease at high levels of inorganic N. Since cassava production on highly infertile tropical soils requires high levels of fertilization, the effect of N and K levels was studied by applying different levels of NH_4NO_3 and KCl, respectively, in solution to a virgin Quilichao soil. Rooted cassava tip cuttings were inoculated at time of planting with three isolates of VAM fungi in a pasteurized soil; their effectiveness was compared with those of native strains in a non-sterilized soil. All treatments received 100 kg P ha^{-1} as ground TSP. Figure 6 shows that in the absence of K there was a marked DM response to high levels of applied N, except in those plants inoculated with C-20-2. A similar response was observed when K was applied at the rate of 200 kg K ha^{-1}. Again C-20-2 showed a negative response to the highest level of N, while the native VAM was also negatively affected. *G. manihotis* (C-1-1) and *E. colombiana* (C-10) were not affected by even the highest levels of N and K in the soil solution. *G. manihotis* (C-20-2), collected from the very infertile soils in Carimagua, was the most effective species at low and medium levels of N and K application, but was most seriously affected by high levels of these elements. Growth of non-mycorrhizal plants was extremely poor and was not improved by N or K fertilization, even in the presence of 100 kg P ha^{-1}.

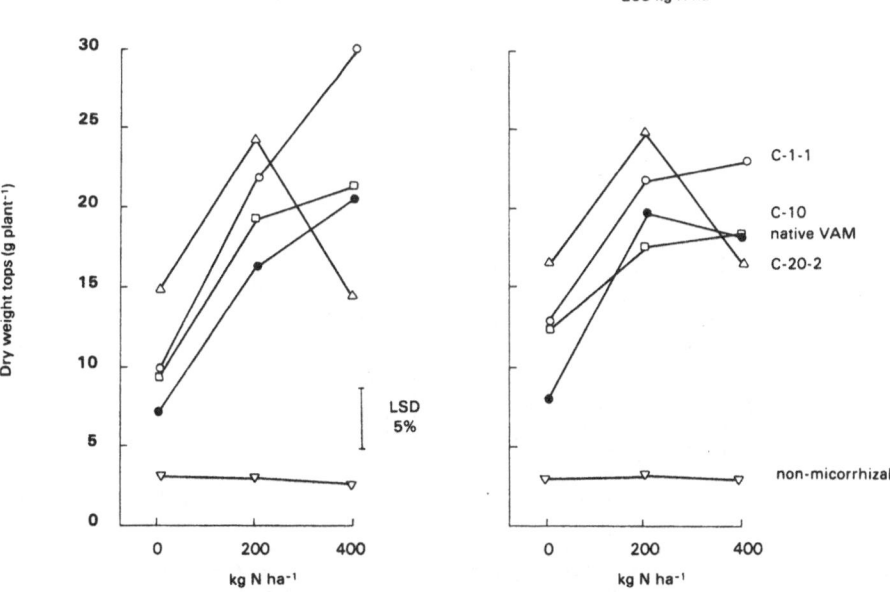

Fig. 6. Effect of different levels of N and K application and inoculation with several VAM fungal isolated on the dry weight of 2½ months old cassava plants grown in Quilichao soil in the greenhouse.

In two tropical pastures, *Centrosema macrocarpum* and *B. decumbens*, there were no significant effects of N application, up to 100 kg N ha^{-1}, on the response to VAM inoculation in sterilized Carimagua soil. However, there was a strong P level × mycorrhiza interaction, with the greatest growth and P uptake response at 20 kg P ha^{-1} combined with 30 kg N ha^{-1} for *Centrosema* and 20–40 kg P with 100 kg N ha^{-1} for *Brachiaria*[10].

D. *Effect of soil temperature and humidity*

Soil temperature and humidity not only affect nutrient uptake and plant growth but may also have a differential effect on the effectiveness of species of VAM fungi. Sieverding[30] determined the adaptation of several species to either high (30°C) or low (20°C) soil temperature, by growing cassava in sterilized Quilichao soil, maintained at these temperatures in a waterbath. Figure 7 shows that compared with the uninoculated check four isolates were very effective at 30°C, but only two were effective at 20°C. Even within the same species of *E. colombiana* only the isolate C-3-5, collected at high altitude in Popayan (1800 masl) was effective at low soil temperature, while the isolate C-10, collected in Carimagua at 175 masl, was not adapted to low soil temperature. On the

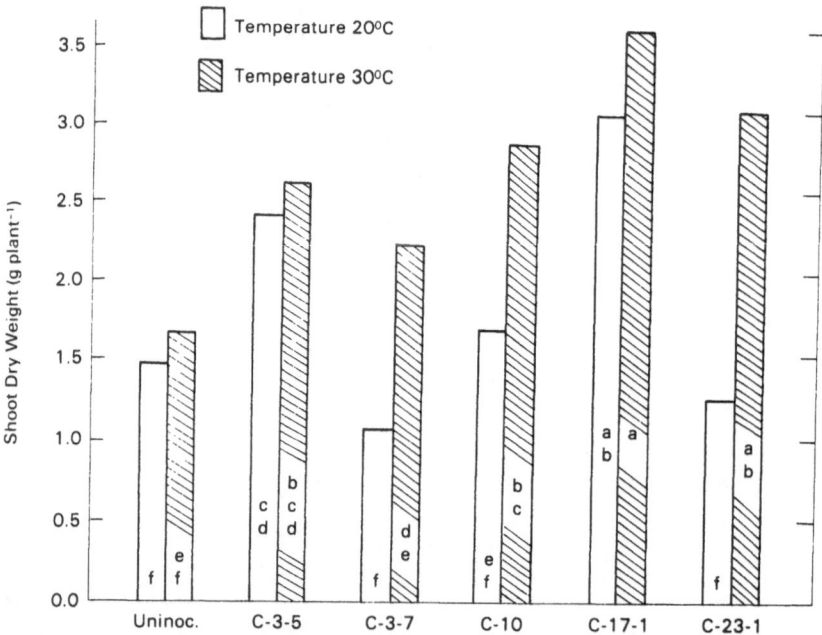

Fig. 7. Effect of different soil temperatures on growth of cassava M Col 113 without mycorrhiza or inoculated with different mycorrhizal isolates (C-3-5, *E. colombiana* from Popayán; C-3-7, *Gi. pellucida* from Popayán; C-10, *E. colombiana* from Carimagua; C-17-1, *G. manihotis* from the North Coast of Colombia; C-23-1, *A. longula* from Antioquia).

Table 6. Effect of soil water conditions on the effectivity of mycorrhizal fungi with the cassava variety CM 342-170 in pasteurized Quilichao soil

Isolate number	Mycorhizal fungi	Total DM (g plant^{-1})	Spore number 100 g^{-1} soil
Wet conditions			
	Not inoculated	8.9^h	0^e
C-1-1	G. manihotis	84.3^a	2714^{cd}
C-18-1	G. fasciculatum	12.6^{gh}	150^d
C-19-3	G. occultum	43.2^c	790^d
C-37-8	Gi. heterogama	14.1^{gh}	12^d
C-6-2	A. longula	29.8^{de}	21208^b
C-7	A. myriocarpa	35.1^{cd}	3900^{cd}
C-10	E. colombiana	71.4^b	31265^a
Dry conditions			
	Not inoculated	8.8^h	0^e
C-1-1	G. manihotis	21.6^{efg}	350^d
C-18-1	G. fasciculatum	12.9^{gh}	83^d
C-19-3	G. occultum	19.2^{fgh}	325^d
C-37-8	Gi. heterogama	9.7^h	48^d
C-6-2	A. longula	9.5^h	292^d
C-7	A. myriocarpa	14.9^{fgh}	94^d
C-10	E. colombiana	25.0^{def}	5600^c

other hand, G. manihotis, isolate C-17-1, collected in the hot climate of the Colombian north coast, was the most effective at both high and low soil temperatures. Thus, when VAM fungi are to be introduced into areas with cooler climates at higher altitudes, their adaptation to low temperature should be evaluated.

To study the effect of waterstress on VAM effectiveness seven fungal isolates were evaluated in sterilized Quilichao soil that was either maintained at maximum water holding capacity ("wet" treatment) or left to dry until first signs of wilting occurred before being brought back to maximum water holding capacity ("dry" treatment). One 5 cm cassava stake was planted and inoculated with a pure culture of each VAM isolate. Table 6 shows that under "wet" conditions G. manihotis (C-1-1) and E. colombiana (C-10) were the most effective species. Under "dry" conditions DM production decreased markedly, but the above two species remained the most effective. On the other hand, A. longula (C-6-2), which was intermediately effective under "wet" conditions, was shown to be quite ineffective under conditions of water stress. Under the "dry" conditions there was a marked reduction in spore production of all fungi.

E. *Effect of native VAM populations*

To what extent the introduction of highly efficient VAM isolates will improve P uptake and plant growth in a particular crop and soil depends on the mycorrhizal dependence of the crop and the quantity and quality of the native mycorrhizal population[18]. In the same soil the native VAM fungal population can vary considerably due to cropping patterns, agronomic practices and climatic conditions. Also, the population of the virgin soil may depend to a large extent on soil chemical characteristics, especially the organic matter content[7]. Mycorrhizal inoculation of a crop will only increase yields if the native population is low or ineffective and the introduced isolate is highly effective, well-adapted to the particular soil and climatic conditions and applied in large enough quantitites to compete with the indigenous VAM population.

Figure 8 shows the effect of inoculation on DM production of cassava in six unsterilized soils (see Table 2) with different populations of native VAM fungi, as well as the plant growth of non-mycorrhizal plants. Non mycorrhizal plants grew very poorly in all soils, but those in the CIAT-Palmira soil with a high level of $40 \mu g P g^{-1}$ soil, grew best compared with those plants growing in P deficient $(2 \mu g P g^{-1})$ soils. Still, even in the Palmira soil, cassava plants infected with native VAM fungi produced over three times the DM as non-mycorrhizal plants; however, inocula-

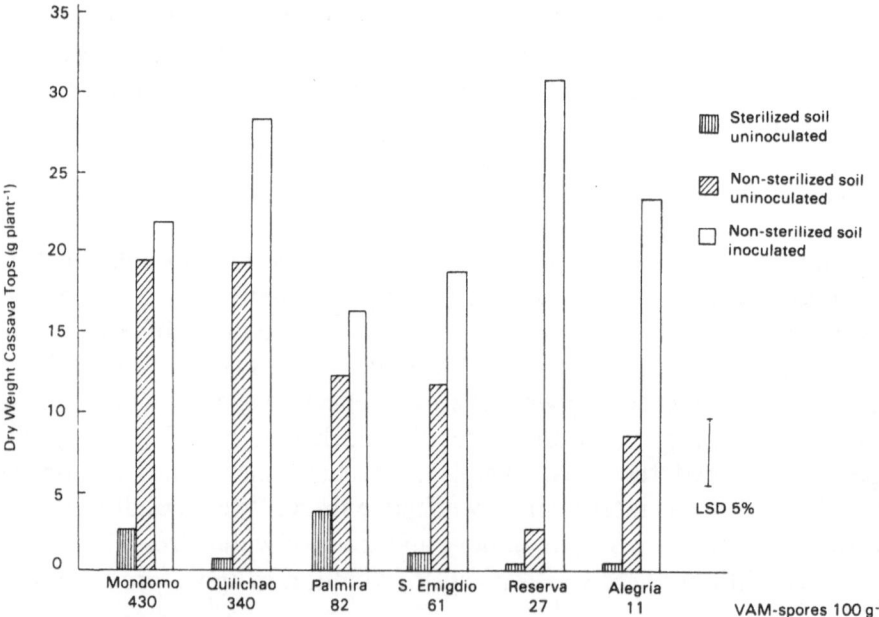

Fig. 8. Effect of soil sterilization and VAM inoculation (average of plants inoculated with C-1-1 and C-10) on the top dry weight of cassava grown in six soils with different indigenous VAM populations (indicated by number of VAM spores in non-sterilized soil).

tion did not further increase yields significantly, as this soil had an intermediate number of VAM spores. In the Monodomo soil, with the highest population of native VAM fungi, inoculation had no significant effect while in Quilichao soil, with also high spore numbers the effect of inoculation was intermediate. In the two Carimagua soils (Reserva and Alegria), with very low spore numbers, inoculation had the greatest effect, more than doubling the DM production of tops. These results indicate that the growth responses to inoculation depend on the P status of the soil as well as the quantity and quality of indigenous VAM fungi. Spore numbers give an indication but can not always be used to determine accurately plant growth respones. Low spore numbers, however, indicate field sites where a response to inoculation may be expected.

Mycorrhizal responses under field conditions

To be of any practical significance mycorrhizal technology has to be tested in the field. Compared with pot-bound plants, those growing in the field have a more disperse root system. Also climatic conditions and agronomic practices greatly affect the number of VAM spores, the fungal species composition and their distribution in the soil profile. It was shown[8] that during a long dry season the VAM population decreased and that this population was concentrated at deeper layers. Also, the growth cycle of a crop in the field until final harvest is generally much longer than in the greenhouse, where plants are generally harvested prematurely. Thus, in the field, even a low population of native VAM fungi may have enough time to develop a highly effective mycorrhizal association, and therefore responses to inoculation tend to be much smaller under natural field conditions than in the greenhouse.

1. *Cassava*

Yost and Fox[38] reported that elimination of the native VAM population in the field by methyl bromide sterilization greatly reduced the DM production of cassava, soybean, cowpea, *Stylosanthes guianensis* and *Leucaena leucocephala*. They noted that plants at borders of fumigated plots recovered from P deficiency once their roots reached the non-fumigated soil. A similar effect was reported by Howeler et al.[17]; when cassava plants were grown on previously fumigated miniplots they recovered completely at 6 months from an initial very severe P deficiency and at harvest, at 12 months, soil sterilization had no significant effect on root yields. However, in a similar experiment using larger plots (Figure 9), plants recovered more slowly from an initial P deficiency and final root yields in sterilized plots were only about 50 and 60% of those

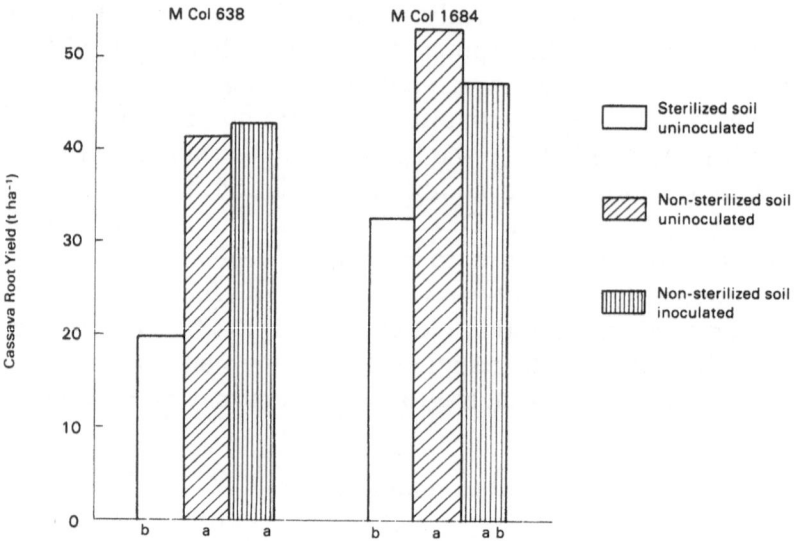

Fig. 9. Effect of soil sterilization and mycorrhizal inoculation on the fresh root yield of two cassava cultivars grown in Quilichao, Colombia. Treatments with same letter under bars are not significantly different (5%) within each cultivar.

in non-sterilized plots for the two cultivars tested. In this case inoculation with VAM fungi in the non-sterilized plots had no significant effect on yield, due to the highly effective population of native VAM fungi in Quilichao, dominated by *E. colombiana* and *G. manihotis*. Other inoculation trials of cassava at this site have also resulted in nonsignificant yield increases[17,18].

In the Monodomo (Cauca) region of Colombia cassava is grown on steep and often eroded slopes. In this area native VAM populations are extremely variable and dominated by less effective species like *G. fasciculatum*, *A. morrowae* and *A. longula*. Sieverding[31] showed that inoculation of cassava with three highly effective species of VAM fungi in three farmer's fields in the area (Table 7) generally increased cassava root

Table 7. Effect of field inoculation and P sources (50 kg P ha^{-1} as triple superphosphate, TSP, or Huila rockphosphate, HRP) on fresh root yields (t ha^{-1}) of cassava cv. CMC 92, at three sites in Cauca Department of Colombia

Inoculation*	Pescador		Mondomito		Agua Blanca		Average	
	TSP	HRP	TSP	HRP	TSP	HRP	TSP	HRP
Not inoculated	18.5	I1.3	7.0	6.2	13.1	12.7	12.9	10.1
C-1-1	22.9	17.3	7.2	8.0	18.1	15.5	16.1	13.6
C-10	19.7	18.7	7.8	9.5	16.0	11.8	14.5	13.3
C-33-1	21.1	20.4	8.2	9.2	14.0	14.6	14.4	14.7

* Inoculated with 500 g soil substrate infected with *Glomus manihotis* (C-1-1), *Entrophospora colombiana* (C-10) or *Glomus occultum* (C-33-1).

yields, especially when rock phosphate ($50\,kg\,P\,ha^{-1}$) was used as the P source. On the average, the isolate C-1-1 was most effective with TSP, while C-33-1 (*G. occultum*) was most effective with rock phosphates.

In Carimagua, in the Colombian Llanos, populations of native VAM fungi are low and consist mainly of *A. longula, E. colombiana, Gigaspora* spp. *G. fasciculatum* and *G. manihotis.* It has been reported[18] that cassava responded significantly to inoculation when $100\,kg\,P\,ha^{-1}$ were applied, but the response was not significant at low levels of 0 and $50\,kg\,P\,ha^{-1}$ or at high levels of $200\,kg\,P\,ha^{-1}$. There were no significant differences among five P sources.

Table 8 shows a summary of responses of cassava to VAM inoculation in several field experiments in Carimagua and the Mondomo-Agua Blanca-Pescador region. In nearly all trials there was a significant response to inoculation with at least some mycorrhizal isolates, especially to C-1-1 and C-10, which are highly effective isolates for cassava over a wide range of environmental and acid soil conditions. Best responses were generally obtained with the application of $50\,kg\,P\,ha^{-1}$, either as banded TSP or incorporated rock phosphate.

Table 8. Effect of field inoculation with selected mycorrhizal strains on cassava fresh root yields after one year of growth at different soil sites with the application of different sources and levels of P fertilizer (Means of four replications at each site comparing not inoculated control with treatment inoculated with most effective strain). (Source: CIAT[7])

Soil sites*	P** source	P level (kg/ha)	Root yields (t ha^{-1})		Most effective mycorrhizal strain No.
			Not inocul.	Inocul.	
Mondomito I		0	26.1	27.3	C-4-2
Carimagua-Yopare		0	9.8	9.3	C-1-1
Mondomito I	TSP	50	29.9	36.7	C-19-1
Mondomito II	TSP	50	7.0	8.2	C-33-1
Agua Blanca I	TSP	50	13.1	18.1	C-1-1
Pescador	TSP	50	18.5	22.9	C-1-1
Carimagua-Alegria	TSP	50	15.9	18.3	C-19-1
Carimagua-Alegria	TSP	100	16.4	19.9	C-4-2
Carimagua-Yopare	TSP	100	11.6	17.6	C-1-1
Carimagua-Alegria	BS	50	18.0	18.6	C-10
Mondomito II	HRP	50	6.2	9.5	C-10
Agua Blanca I	HRP	50	12.9	16.1	C-1-1
Agua Blanca II	HRP	50	21.2	27.1	C-1-1/C-10
Agua Blanca III	HRP	50	15.6	18.3	C-1-1/C-10
Agua Blanca IV	HRP	50	24.7	31.1	C-3-5
Tres Quebradas	HRP	50	17.7	19.1	C-1-1/C-10
Pescador	HRP	50 ·	11.3	20.4	C-33-1
Carimagua-Alegria	HRP	50	15.9	19.8	C-10
Carimagua-Yopare	HRP	100	11.7	19.2	C-1-1

* Carimagua sites are Oxisols; all others are Inceptisols.
** TSP: Triple superphosphate; BS: Basic slag; HRP: Huila rock phosphate.

Table 9. Effect of field inoculation with *G. manihotis* (C-1-1) on dry grain yields of beans in a cassava/beans intercropping trial in Mondomito

Fertilization*	Grain yield beans (kg ha^{-1})	
	Beans not inoculated	Beans inoculated
0	378	524
+P	620	539
+Lime	608	801**
+P + Lime	908	1112**

* 0 = Basic fertilization: 100 kg N ha^{-1}, 100 kg K ha^{-1}, 10 kg Mg ha^{-1}, 5 kg Zn ha^{-1}, 1 kg B ha^{-1}, 1 t Huila rock phosphate ha^{-1}.
+P = Basic fertilization plus 100 kg P ha^{-1} as TSP, broadcast and incorporated.
+Lime = Basic fertilization plus 3 t ha^{-1} lime. All fertilizers broadcast and incorporated.
** Response to inoculation was significant at $P = 0.05$.

2. Beans

Common beans have a high P requirement[5] and have been shown in greenhouse trials to be intermediately dependent on an effective VAM association. A field trial in Mondomo showed (Table 9) that uninoculated beans responded markedly to applications of lime and P; only when lime or lime and P were applied did the beans respond significantly to inoculation with isolate C-1-1. In this case the inoculum was applied under the seed at a rate of 200 g per linear meter (4 t ha^{-1}).

Another trial with a bean-cassava intercrop showed a positive response of beans to inoculation only when both crops were inoculated. Cassava responded to inoculation when either one or both crops were inoculated[9].

3. Coffee (Caffea arabica) and tea (Camellia sinensis)

Being transplanted crops with a coarse and woody root system, coffee and tea could potentially benefit from a mycorrhizal association, especially in acid low P Andepts where these crops are often grown. Moreover, seedlings are often produced in sterilized soil in nurseries to control nematodes and pathogenic fungi. Thus, reinoculation with VAM isolates are expected to greatly increase seedling establishment and early vigor.

Lopes et al.[19] reported a highly significant shoot dry weight increase of coffee seedlings when inoculated with four VAM species; *Gi. margarita* and *G. mosseae* were the most effective species. Sieverding and Toro[36] showed that both plant height and transplant survival-rate increased markedly when coffee seedlings were raised in Quilichao soil infected with several VAM isolates (Table 10). Best results were obtained with *G. occultum, G. manihotis, E. colombiana* and *A. myriocarpa*, both

Table 10. Effect of planting coffee in soil containing different species of VAM fungi on plant height at 7 months in the nursery and on plant survival at 3 and 7 months after transplanting to the field

Treatment	Plant height (cm)	Plant survival after transplanting (%)	
		3 months	7 months
Non-mycorrhizal	9.27[c]	60	47
Acaulospora longula	9.95[c]	71	61
Acaulospora myriocarpa	14.63[a]	95	93
Entrophospora colombiana	13.42[b]	92	85
Gigaspora heterogama	9.87[c]	78	68
Glomus fasciculatum	9.47[c]	72	63
Glomus manihotis	13.45[b]	97	88
Glomus occultum	14.70[a]	84	81

Data followed by the same letter are not statistically ($P = 0.001$) different.

in terms of plant growth in the nursery and survival in the field after transplanting.

The results of two experiments conducted by the same authors[36] with tea grown in sterilized Quilichao soil and inoculated with ten isolates of VAM fungi are shown in Table 11. Ten to eleven months after inoculation, plant height, DM of tops, and root length markedly increased by inoculation with some species, with *G. manihotis, Glomus* sp. (C-22-5), *G. occultum, A. scrobiculata, A. spinosa* and *Acaulospora* sp. (C-179-3) being the most effective. Thus, in both coffee and tea, when seedlings are raised

Table 11. Effect of the inoculation of tea (*Camellia sinensis*) with different VAM isolates on plant growth at 11 (first trial) and 10 months (second trial) after inoculation. Plants were grown in sterilized Quilichao soil

Inoculation treatment	CIAT isolate No.	Plant height (cm)	Dry weight tops (g plant^{-1})	Total root length (m plant^{-1})
First trial				
Not inoculated	–	23.1	3.72	30.0
E. colombiana	C-3-5	50.5	14.28	94.7
C. manihotis	C-1-1	51.0	31.82	151.9
Glomus sp.	C-22-5	68.3	27.86	150.5
LSD 5%		22.0	10.28	101.1
Second trial				
Not inoculated	–	21.8	3.78	30.1
A. appendicula	C-179-2A	34.0	6.92	32.4
A. scrobiculata	C-179-6	38.8	9.68	43.4
A. spinosa	C-179-5	41.5	11.36	40.6
Acaulospora sp.	C-179-3	40.5	10.36	25.0
Gi. pellucida	C-179-7	18.6	1.88	8.5
G. occultum	C-43-4B	38.8	9.81	22.6
G. versiformis	C-141	28.3	8.57	46.6
LSD 5%		10.4	3.86	22.2

in sterilized soil, inoculation with highly effective VAM isolates is recommended to improve seedling establishment and early vigor upon transplanting.

4. *Pastures*

Pasture establishment in acid low P Oxisols and Ultisols is often slow due to inadequate P nutrition, especially in case of legumes, which require relatively high levels of P for N fixation. To study the effect of P fertilization and inoculation with VAM fungi on pasture establishment in an Oxisol in Carimagua, *Andropogon gayanus* and *Pueraria phaseoloides* were seeded with four treatments: 1) control without P or inoculation, 2) VAM inoculation only, 3) $20 \, kg \, P \, ha^{-1}$ as rock phosphate, and 4) rock phosphate + VAM inoculation.

Figure 10 shows that both the grass and legume responded markedly to rock phosphate application; they also responded to VAM inoculation, but only in the presence of the rock phosphate. Maximum responses were observed with the second cut at 6 months after sowing, while in the third cut of the grass at 9 months, and the fourth cut of kudzu at 15 months, the mycorrhizal effect had disappeared, possibly due to other growth limiting factors. The early response to VAM inoculation combined with low levels of rock phosphate, can be beneficial by shortening the time in which recently established pastures are available for grazing.

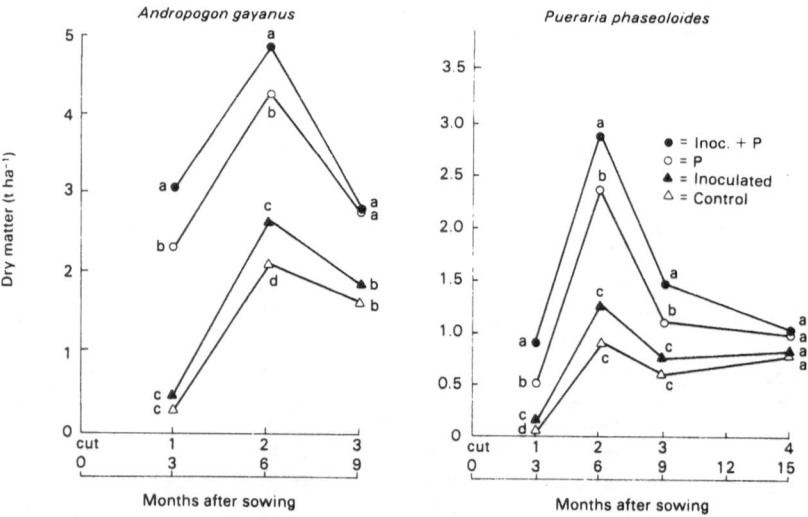

Fig. 10. Effect of mycorrhizal inoculation and rock phosphate application on the dry matter production of *A. gayanus* CIAT 621 and *P. phaseoloides* CIAT 9900 grown under natural field conditions in Carimagua. (Different letters indicate statistical significance at each cut).

Inoculum production and conservation

Unlike *Rhizobium* which can be produced relatively easily on artificial media, VAM fungi have not been successfully reproduced in axenic culture. Thus VAM fungi reproduce only in the presence of living roots and inoculum production requires a host plant, and a growth medium, usually soil, with optimum conditions for growth and reproduction of the fungus.

1. *Host plant*

Being rather promiscuous, VAM fungi can be reproduced on a wide range of host plants[6]. Grass species like maize and *Andropogon* produce more roots, but legumes tend to produce more spores in the same volume of soil (Figure 11). For choosing the right host plant one has to consider the edaphic and climatic adaptation of the species as well as the fact that some host plants may harbor pathogenic organisms detrimental to the crop to be inoculated. Also, some host plants may be effective in the reproduction of only certain VAM fungi, as has been shown in the case of *Canavalia ensiformis* having preference for *Gi. heterogama*[8].

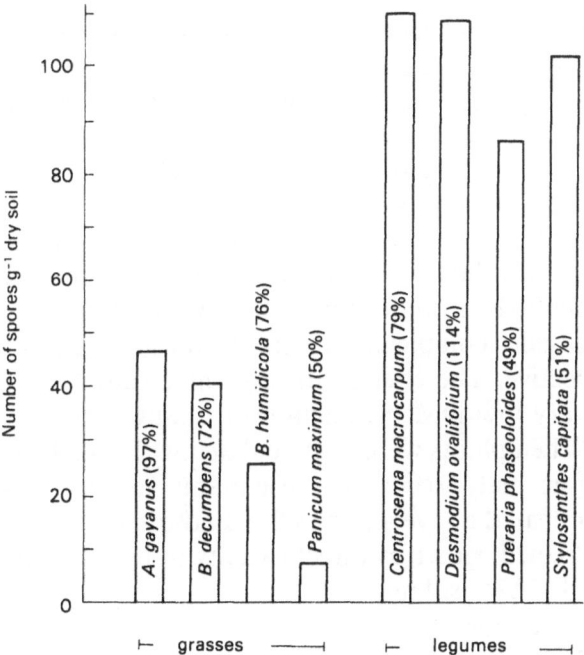

Fig. 11. Effect of host plants on mycorrhizal spore production; in parenthesis the coefficient of variability due to six different levels of P applications. (Sieverding, 1985b).

Table 12. Effect of different soil sterilization methods on the percent infection of kudzu roots at 1 and 2 months after planting, and on the top growth of cassava grown for 3 months in the treated Quilichao soils

Treatment	Kudzu—root infection (%)		Cassava top growth (g plant^{-1})
	1 MAP	2 MAP	
Untreated	49.1a	55.0a	28.4a
Sun dried	43.0a	44.1a	19.3bc
Solar heated under black PVC plastic	43.7a	40.1ab	31.8a
Pasteurized	0.0b	0.0c	2.1f
Sterilized with methyl bromide	0.0b	0.0c	1.6f
Autoclaved	0.0b	0.0c	1.2f
Heated in can above open fire	0.0b	0.0c	2.5df
Vapam ($150 \, cm^3 \, m^{-2}$)	0.0b	43.2ab	6.7d
Basamide ($50 \, g \, m^{-2}$)	0.0b*	3.1c	2.5df
Di-Trapex ($60 \, cm^3 \, m^{-2}$)	6.6b	33.8ab	14.7c
Formol ($150 \, cm^3 \, m^{-2}$), uncovered soil	0.0b	30.7b	22.7b
Formol ($150 \, cm^3 \, m^{-2}$), covered	0.4b	0.7c	19.2bc

MAP: Months after planting.
* Plants were affected by toxicity of the product.
Note: Vapam = Metam-sodium = sodium-N-methyldithiocarbamate. Basamid = Dazomet = tetrahydro-3, 5-dimethyl-2H-1,3,5-thiadiazine-3-thione. Di-Trapex = Vorlex = mixture of methyl isothiocyanate and chlorinated C3-hydrocarbons. Formol = formaldehyde.

2. Growth media

While nutrient solution or nutrient-sand culture media have been proposed[22,23] for inoculum production, the most common medium is soil, previously sterilized and then inoculated with a pure stock culture or spores of a specific isolate. For production of pure cultures complete elimination of all native VAM propagules is obviously required. Table 12 shows that several methods of sterilization were effective in eliminiating native VAM propagules, as indicated by a bio-assay using kudzu as the test plant; however, only autoclaving, pasteurization for 8 hours, or methyl bromide application were effective in eliminating VAM propagules completely. Of the chemical fumigants, only Basamid* was effective. The effectiveness of these fumigants is also indicated by the very low DM production of cassava grown in these fumigated soils. When plants were re-inoculated with C-1-1, the growth of cassava was normal, except for those plants grown in soil treated with Vapam, or heated above an open fire; in those treatments growth was seriously affected, either by direct toxicity or by a residual effect on the introduced VAM isolate.

For practical field application, it may be advantageous to produce inoculum directly on the farm, in order to eliminate high transport

* Note: Commerical names in this article are used as examples and do not imply an endorsement by CIAT of any product or company. Chemical composition of products are indicated in Table 12.

Table 13. Spore production of mycorrhizal fungi in inoculum of *B decumbens* soil-root mixture collected four months after application of various soil fumigants and planting of *B. decumbens* in a farmer's field in Mondomito, Cauca (Source: CIAT)

Soil treatment	Inoc. treatment*	No. of spores 100 g⁻¹ dry soil		Percent of *G. manihotis* in total VAM population
		G. manihotis	Others**	
Untreated	Not inoc.	86	996	8
	Inoc.	506	252	67
Formol***	Not inoc.	15	313	5
	Inoc.	4338	132	97
Basamid	Not inoc.	38	83	31
	Inoc.	7930	70	99
Di-Trapex	Not inoc.	16	549	3
	Inoc.	3335	181	95
Methyl-bromide	Not inoc.	8	102	7
	Inoc.	12803	83	99

* Inoculated with *G. manihotis* (C-1-1); 10 g of infected soil was applied to each of 25 planting holes per square meter.

** Other species of mycorrhizal fungie were: *G. fasciculatum, A. myriocarpa, A. mellea, E. colombiana, A. foveata*, two not identified *Acaulospora* spp., *G. occultum* and two *Gigaspora* spp.

*** For chemical composition of products see Table 12.

costs. In that case, the soil in a small area should be fumigated to eliminate the native VAM fungi, before inoculation with a small amount of a pure culture and planting of a host species. Table 13 shows that when soil in a farmer's field in Mondomo was fumigated with methyl bromide or Basamid, planted with *Brachiaria decumbens*, and inoculated with *G. manihotis*, C-1-1, at four months after planting about 99% of spores encountered in the soil were of the introduced species, compared with 67% in case of the unfumigated soil. While methyl bromide is highly toxic and should not be recommended for use by small farmers, the granular Basamid can be handled rather easily and safely under these conditions. Research of the effectiveness of this on-farm produced inoculum is still in progress, but it is believed that this methodology can be used to produce inoculum cheaply even in far removed places.

Field inoculation can be done with either infected roots or soil containing pieces of roots, spores and hyphae. It has been shown[4] that root inoculum containing the VAM fungus *G. manihotis* stored for 30 days at room temperature was as effective as fresh inoculum in increasing growth of cassava plantlets grown in sterilized Quilichao soil. Soil inoculum maintains its effectiveness for even longer periods, and when stored in a slightly moist state in a cold room it was found to be effective even after 3 years of storage.

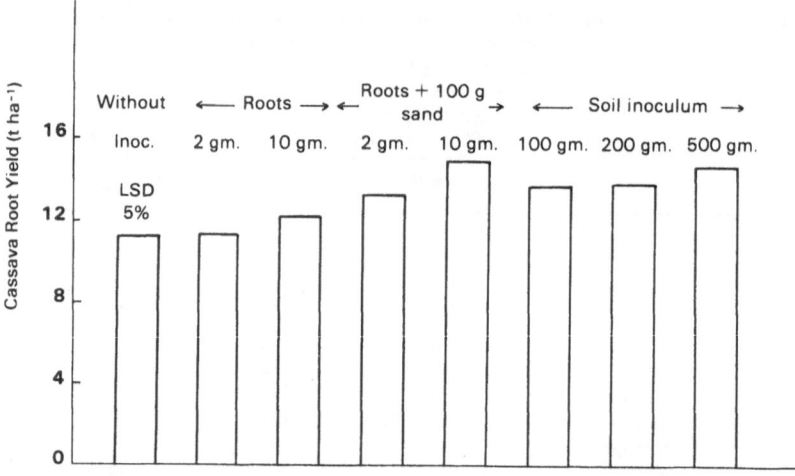

Fig. 12. Effect of different levels and methods of inoculation with mycorrhizal isolate C-1-1 on the yield of cassava, M Ven 77 grown with $100\,kg\,P\,ha^{-1}$ as band applied triple superphosphate or incorporated Huila rock phosphate in Carimagua. Data are the average of the two P-sources.

Methods of inoculation

The best method of inoculation depends on the crop, its planting system and growth cycle. In short season, row planted crops like beans good results have been obtained by applying soil inoculum in a continuous band under the seed at time of planting, *i.e.* placing the seed on top of the inoculum in a small furrow. Fertilizers can be side banded at time of planting or after germination.

Small seeded pastures are inoculated in a similar way or seeds are pelleted with inoculum[25]. For cassava or tree crops the inoculum is best placed under the stake or seedling at time of planting and reinoculated in side bands after a prolonged dry season[33]. It appears that once the introduced strain is well established there is no need to reinoculate the second year[8].

Figure 12, shows that for cassava, 2 or 10 g root inoculum per stake was less effective than 100–500 g soil inoculum. However, when this small amount of root inoculum was mixed with 100 g of sand to increase its spacial distribution below the plant, the root inoculum was as effective as soil inoculum in increasing cassava yields in the field. Using this method, only small amounts of root inoculum have to be transported, while sand (or sterilized soil) can be available at the site of inoculation.

Effect of agronomic practices on VAM population

The beneficial effect of mycorrhizal associations can be increased, either by inoculation with highly effective VAM fungi, or by the use of those cultural practices that stimulate the population of indigenous VAM fungi, especially the more efficient components of that population. This population can be altered by the selection of crop species, by crop rotations, by fertilization, the use of pesticides or by crop residue management.

A. *Plant species and cropping systems*

The introduction of fast growing crops or pasture species that are strongly mycorrhizal can greatly enhance the population of native VAM fungi. Figure 13 shows that the mycorrhizal inoculum potential of an Oxisol in Carimagua, as measured in a bioassay by the percent root infection of tropical kudzu, was markedly increased by the introduction of improved grasses and pasture legumes, compared with the native savanna. Most effective were the grass-legume associations of *A. gayanus*

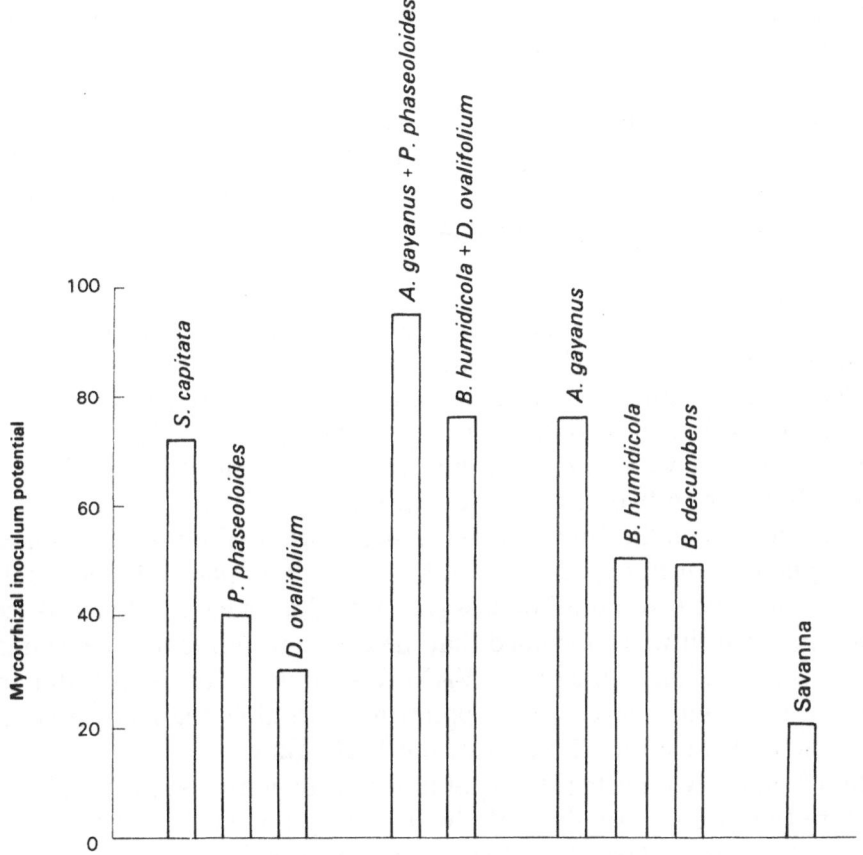

Fig. 13. Effect of vegetation cover on VA mycorrhiza inoculum potential of an oxisol (0–10 cm depths) as measured by per cent root infection of *P. phaseoloides* in a bio assay at 4 weeks after planting.

with *P. phaseoloides* or *B. humidicola* with *D. ovalifolium*; this resulted in a higher inoculum potential than either of the grasses or legumes alone. Thus, the introduction of more productive plants species, with their recommended cultural practices (including fertilization), markedly increased the quantity of the population of indigenous VAM fungi.

Similar results were obtained with the planting of cassava in an Inceptisol in Mondomo. While the soil under natural pasture contained only 103 propagules of native VAM fungi $100 g^{-1}$, this increased to 213 under bush fallow, and to 823 and 1717 with one or three continuous years of cassava cropping, respectively. Cassava grown on these plots with different cropping histories responded positively to inoculation in all soils, except those with the 3 year cassava history, probably because of the very high population of native VAM fungi in this soil[7]. Similarly, spore counts in the sandy soils of Media Luna, on the north coast of Colombia, indicated the highest population of native VAM fungi in plots with 2 years continuous cassava (with fertilization), compared with cassava intercropped with legumes or cassava grown in rotation with grain legumes[7].

Thus, it appears that strongly mycorrhiza-dependent plants like cassava or certain forage plants markedly increase the population of native VAM fungi. The effect of this on other crops grown in association or in rotation should be further investigated.

B. *Fertilization*

The application of organic or chemical fertilizers not only affects the total population of VAM fungi but may also stimulate some species while reducing others. Sieverding and Howeler[34] reported that the application of P from 0 to $200 kg P ha^{-1}$ to cassava grown in 3 locations in Colombia decreased the total mycorrhizal root infection at all locations. However, in two of the 3 sites, application of P actually increased root infection with the fungal species *G. manihotis*. Only in these two sites the population of this species was enhanced by P application and resulted in a positive P response of cassava up to $200 kg P ha^{-1}$. From these and other trials it was concluded that the crop's response to P depended on the composition of the native VAM fungi population. This is due to the fact that higher levels of P application can alter the relative composition, and thus, the efficiency of the VAM fungi.

Saif[28] similarly reported that root infection by native VAM fungi in 3 pasture legumes was significantly increased by application of K, up to $40 kg K ha^{-1}$, in an Oxisol at Carimagua; however, this effect was much less in grasses, and only in *Brachiaria humidicola* was root infection increased by application of $20 kg K ha^{-1}$. At the same location, lime

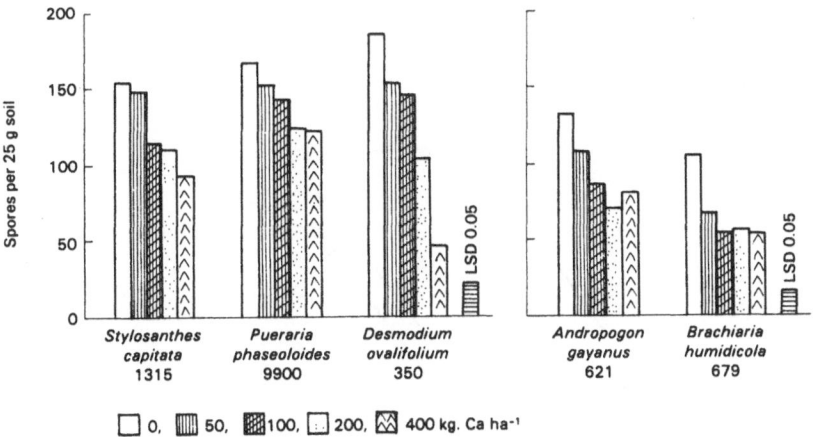

Fig. 14. Effect of calcium ($CaCO_3$) on the native mycorrhizal spore population in five tropical pasture plants grown in an Oxisol at Carimagua, two years after planting.

application had no consistent effect on root infection of *A. gayanus, B. humidicola* and *P. phaseoloides*, but significantly increased that of *S. capitata* and *D. ovalifolium* up to the level of 200 kg Ca ha^{-1} (0.5t $CaCO_3$ ha^{-1}). However, in all pasture species tested, lime application significantly decreased the concentration of native VAM spores in the soil, two years after planting and lime application (Figure 14). Greenhouse trials with cassava (Figure 5) also showed that lime application to two very acid soils increased per cent root infection up to a pH of 4.5–5.0, but higher application rates decreased root infection as well as spore concentrations in Carimagua soil.

C. Pesticides

Menge[20] reported that many chemicals used to control insects, diseases and weeds in crops also affect the VAM, either positively or negatively. Determinations of root infection or spore numbers as influenced by pesticides are often inconsistent or contradictory, as many pesticides may temporarily lower the population of mycorrhizal fungi as well as that of other competing microorganisms until, with time, a new balance in the population is reached[35].

Figure 15 shows that when the application of three herbicides (or mixtures) were compared with hand weeding in cassava, the herbicides had no significant effect on the VAM spore population during the first three months. However, in the herbicide treated plots the build up in spore population was delayed about 3–4 months compared with the handweeded plots, but eventually reached the same level at 50 weeks after planting. Thus, the herbicides had only a temporary effect on the population of VAM fungi, which did not affect final root yield.

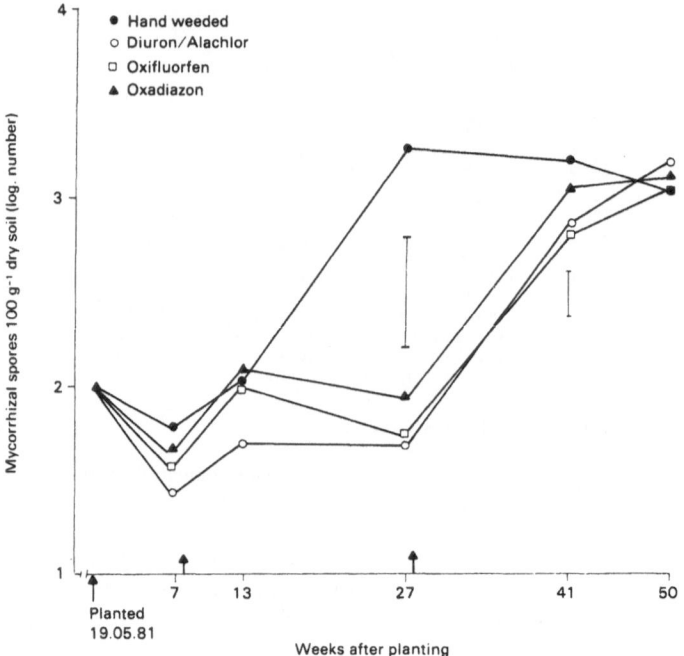

Fig. 15. Effect of hand weeding and herbicides on native mycorrhizal spore production in a cassava field with cv. MCol 638 at CIAT Quilichao. (Bars indicate LSD 5%. Arrows indicate dates of hand weeding for checks without herbicide application).

Similarly, it was found[6] that the fungicide–insecticide–zinc sulphate mixture, that is sometimes used to treat cassava stakes before planting, initially affected the build up of spores of *G. manihotis*, but actually enhanced that population at 78 days after planting. Surely, different pesticides have different effects on different VAM fungi. These complicated interactions will have to be studied in greater detail before any recommendations can be made. Pesticides may not only depress the total population of VAM fungi, but their continued use may cause a change in the composition of the population. If the most efficient VAM species are depressed this may result in reduced yields[35].

Potentials and limitations of mycorrhizal technology

Recent research on the mycorrhizal dependence of crops have clearly indicated that some important tropical crops and pastures are highly mycotrophic and will not grow or produce well in low P soils without an effective mycorrhizal association. Crops like cassava may even be mycorrhizal in some very high P soils. These highly mycorrhiza dependent crops have a very high external P requirement when grown in the absence of mycorrhiza[16,17], but have a low P requirement when associated with highly effective VAM fungi. Thus, cassava and some tropical pastures have a much lower critical level for available soil P than less mycorrhizal dependent crops like maize, sorghum, beans and soybean[15]. When the highly efficient species of the native VAM fungi in Quilichao was enhanced by continuous cassava production with adequate N and K fertilization, cassava yields were extremely high even without P application in a soil with only $5 \mu g\,g^{-1}$ Bray II-P[9]. The same crop required high P applications in a soil with a low or inefficient population of VAM fungi in Carimagua. Thus, the critical soil P level and the response to P application varies greatly according to the efficiency of the native VAM fungi. Since nearly all natural soils contain VAM fungi, these highly mycotrophic crops become infected sooner or later and grow and produce in very low P soils, where other crops may not grow at all without heavy P applications. These crops are therefore grown on very poor soils under low input conditions, as is often the case in tropical agriculture or pasture production systems.

Since mycotrophic crops depend on a VAM association when grown under low external P conditions, their yield can be increased by increasing the efficiency of the VAM fungi, either by inoculation, *i.e.* the introduction of more efficient fungal species or isolates to the rhizosphere, or by manipulation of the population of native VAM fungi through the use of agronomic practices that enhance the most efficient species in that population. Recent field trials with cassava, pastures and some transplanted crops like tea and coffee have shown that even under very rustic small farmer's conditions, yields can be significantly increased by inoculation with highly effective VAM isolates. In the Mondomo–Pescador region of Colombia cassava yields on the average increased 25% by introduction of more efficient VAM isolates, through inoculation. Similarly in the very acid and infertile Oxisols of Carimagua, cassava yields increased 23%, while the rate of establishment of most tropical grasses and legumes was greatly increased by mycorrhizal inoculation in combination with low applications of rock phosphates.

Two species of VAM fungi, *G. manihotis* and *E. colombiana*, were consistently among the most efficient isolates tested, but the most efficient species under any particular condition varied with different crops, different fertilizers regimes, and different edapho-climatic conditions.

Further research is still required to study these complicated interactions, in order to recommend the most efficient isolate for a particular situation. In the mean time, the two species mentioned above, can be used as they were found to be highly effective on a large range of crops and pastures grown under highly variable conditions in acid, infertile tropical soils.

The beneficial effect that can be expected from inoculation depends largely on the efficiency of the native VAM fungal population. Soils that have low concentrations of total VAM propagules are likely to give a crop response to mycorrhizal inoculation; however, soils with high total propagule numbers are not necessarily non-responsive to VAM inoculation, because the indigenous population of VAM fungi may contain many ineffective fungal species. In that case the composition of the population of native VAM fungi has to be determined as well as the efficiency of each species identified; alternatively, the efficiency of the population is determined by a bio-assay using a highly mycotrophic test plant like kudzu or cassava. Although total spore numbers were found to correlate positively with high organic matter content of the soil, the efficiency of the VAM fungi is not necessarily related to it, or to any other easily determined chemical or physical soil characteristic. Thus, preliminary greenhouse testing and bio-assays may be necessary to identify those soils where positive responses to inoculation can be expected.

Mycorrhizal inoculum will have to be produced on living roots. This is rather laborious and carries the danger of introducing phytosanitary problems with the inoculum. Production of high quality soil inoculum was not found to be particularly expensive, but the cost of transport of large amounts of soil inoculum may be prohibitively expensive, depending on the distance and road conditions. In that case small amounts of root inoculum can be transported and mixed with locally available sand or sterile soil; or alternatively, inoculum can be produced on farm using safe and effective soil fumigants to partially sterilize the soil before inoculating a host plant with a small amount of base inoculum of a highly effective isolate. This locally produced inoculum was found to contain as much as 99% of the introduced *G. manihotis* species. The efficacy of this method for inoculum production of other VAM species still needs to be tested. It may be concluded that VAM inoculation has great potential for increasing yields of highly mycorrhiza dependent crops, grown on low P soils, but the methodology has its limitations when the soil has a highly effective or competitive native population, when the introduced VAM fungi is not well-adapted to the local edaphoclimatic conditions, or when the area to be inoculated is far away from the site of inoculum production. The production and distribution of high quality inoculum, free of pathogenic organisms, will also require considerable institutional organization.

The enhancement of the efficiency of the native VAM population through agronomic practices may be a practical solution as long as this native population contains at least some highly efficient VAM fungal species. However, this alternative requires a great deal more research to determine the specific effects of many agronomic practices on the different VAM species. In this respect, the use of chemical or organic fertilizers, green manures, crop associations and rotations, residue management and soil conservation practices will have to be studied under different edapho-climatic conditions, to determine their effect on the most efficient components of the population of native VAM fungi. It was found that for a highly mycotrophic crop like cassava, continuous production in the same soil may be more productive than crop rotations, because cassava stimulates and maintains a high population of VAM fungi in the soil, which greatly enhances the crop's ability to absorb P. Similarly, while highly soluble P fertilizers give the greatest responses in crops like maize, beans and soybeans, the use of less soluble basic slag or rock phosphates may be more effective in very acid soils to stimulate the population of native VAM fungi. Thus, many traditionally, recommended practices should be reevaluated in the light of their effect on specific components of the population of native VAM fungi.

While much has been learned about the importance of mycorrhiza for increasing the productivity of many tropical crops and pastures, the practical utilization of the present technology will still require more research and more public expenditure before maximum benefits can be obtained from this naturally occurring symbiosis between plants and mycorrhizal fungi.

References

1   Abbott L K and Robson A D 1977 Growth stimulation of subterranean clover with vesicular-arbuscular mycorrhizas. Aust. J. Agric. Res. 28, 639–649.
2   Bowen G D 1980 Mycorrhizal roles in tropical plants and ecosystems. *In* Tropical Mycorrhiza Research. Ed. Mikela. Clarendon Press, Oxford. pp 165–189.
3   Bowen G D and Smith S E 1981 The effect of mycorrhizas on nitrogen uptake by plants. Ecol. Bull. 33, 237–247.
4   Burckhardt E A and Howeler R H 1984 Efecto de micorrizas en el crecimiento de la yuca, estudiado en ensayos de invernadero. Suelos Ecuatoriales 14, 158–165.
5   Centro Internacional de Agricultura Tropical (CIAT), 1978 Annual Report 1977.
6   Centro Internacional de Agricultura Tropical (CIAT) 1982 Annual Report 1981. Cassava Program.
7   Centro Internacional de Agricultura Tropical (CIAT) 1985 Annual Reports for 1982 and 1983. Cassava Program. 519 p.
8   Centro Internacional Agricultura Tropical (CIAT) 1985 Annual Report 1984, Cassava Program. 270 p.
9   Centro Internacional de Agricultura Tropical (CIAT) 1986 Annual Report 1985. Cassava Program (*in press*).
10 Centro Internacional de Agricultura Tropical (CIAT) 1986 Annual Report 1985. Tropical Pastures Program (*in press*).
11 Chambers C A, Smith S E and Smith F A 1980 Effects of ammonium and nitrate ions on mycorrhizal infection, nodulation and growth of *Trifolium subterraneum*. New Phytol. 85 47–62.
12 Daft M J and Nicolson T H 1969 Effect of endogeny mycorrhiza on plant growth. II. Influence of soluble phosphate on endophyte and host in maize. New Phytol. 68, 945–952.

13   Haanschoten J 1984 Response of beans (*Phaseolus vulgaris* L.) growing in phosphorus deficient Colombian soils to introduction of vesicular-arbuscular mycorrhizae. MSc. Thesis. Univ. of Florida, Gainsville, Fl. 88 p.

14   Hays R, Ried C P P, St. John T V and Coleman D C 1982 Effects of nitrogen and phosphorus in blue grama (*Boutelona gracilis*) growth and mycorrhizal infection. Oceologia (Berl.), 54, 260–265.

15   Howeler R H 1985 Mineral nutrition and fertilization of cassava., *In* Cassava: Research, Production and Utilization. UNDP-CIAT Cassava Program. Cali, Colombia, 745 p.

16   Howeler R H, Asher C J and Edwards D G 1982 Establishment of an effective mycorrhizal association on cassava in flowing solution culture and its effect on phosphorus nutrition. New Phytol. 90, 229–238.

17   Howeler R H, Cadavid L F and Burckhardt E A 1982 Response of cassava to VA mycorrhizal inoculation and phosphorus application in greenhouse and field experiments. Plant and Soil 69, 327–339.

18   Howeler R H and Sieverding E 1983 Potentials and limitations of mycorrhizal inoculation illustrated by experiments with field grown cassava. Plant and Soil 75: 245–261.

19   Lopes F S, Oliveira F, Neptune A M L and Morais F R P 1983 Efeito da inoculacao do cafeeto com diferentes especies de fongos micorrizicos vesicular-arbusculares. Revista Brasileira Ciencia Solo 7, 137–141.

20   Menge J A 1982 Effect of soil fumigants and fungicides on vesicular-arbuscular fungi. Phytopathology 72, 1125–1131.

21   Mosse B 1981 Vesicular-arbuscular mycorrhiza research for tropical agriculture. Research Bul. 194. Hawaii Institute of Trop. Agric. and Human Resources. Univ. of Hawaii. 82 p.

22   Mosse B and Thompson J P 1981 Nutrient film culture of vesicular-arbuscular mycorrhiza for mass production of inoculum: Exploratory experiments with beans (*Phaseolus vulgaris*). J. Exp. Bol. volume?

23   Ojala J C and Jarrell W M 1980 Hydroponic sand culture systems for mycorrhizal research. Plant and Soil 57, 297–303.

24   Plenchette C, Fortin J A and Furlan V 1983 Growth response of several plant species to mycorrhizae in a soil of moderate P fertility. I mycorrhizal dependency under field conditions. Plant and Soil 70, 199–209.

25   Powell C L 1984 Field inoculation with VA mycorrhizal fungi. *In* VA mycorrhizae. Eds. C L Powell and D G Bagyaraj. CRC Press, Boca Raton, Florida. pp 205–222.

26   Rhodes L A and Gerdemann J W 1975 Phosphate uptake zones of mycorrhizal and non-mycorrhizal onions. New Phytol. 75. 555–561.

27   Saif S R 1984 Respuesta de plantas forrajeras tropicales a las aplicaciones de roca fosfórica y micorriza en un Oxisol no esterilizado. Proc. 1st. Latinamerican Conference on Rock Phosphate. Cochabamba, Boliva. Oct. 10–15, 1983.

28   Saif S R 1986 Vesicular-arbuscular mycorrhizae in tropical forage species as influenced by season, soil texture, fertilizers, host species and ecotypes. Angew. Botanik 60, 125–139.

29   Sanni S O 1976 Vesicular-arbuscular mycorrhiza in some Nigerian soils. The effect of *Gigaspora gigantea* on the growth of rice. New Phytol. 77; 673–674.

30   Sieverding E 1984 Posibilidades de aumentar la producción de yuca en suelos acidos de regiones montañosas con el uso de hongos micorrizicos. Suelos Ecuatoriales 15, 190–198.

31   Sieverding E 1985 Yield response of cassava to field inoculation with VA mycorrhiza in acidic soils. *In* Proceedings. 6th North American Conf. Mycorrhizae. Ed. R Molina June 25–29, 1984. Bend, Oregon. 241 p.

32   Sieverding E 1985 Selección de hospederos, substrato de suelo, nivel y fuente de P para la producción de esporas formadores de micorriza vesiculo-arbusculares. *In* Investigaciones sobre micorrizas en Colombia. Memorias Primer Curso Nacional sobre Micorrizas. Eds. E. Sieverding *et al.* Feb. 7–10, 1984. Palmira, Colombia. pp 237–250.

33   Sieverding E 1985 Influence of method of VA mycorrhizal inoculum placement on the spread of root infection in field grown cassava. J. Agron. Crop Sci. 154, 161–170.

34   Sieverding E and Howeler R H 1985 Influence of species of VA mycorrhizal fungi on cassava yield response to phosphorus fertilization. Plant and Soil 88, 213–222.

35 Sieverding E and Leihner D E 1984 Effect of herbicides on population dynamics of VA-mycorrhiza with cassava. Angew. Botanik 58, 283–294.

36 Sieverding E and Toro S 1986 Efecto de la inoculación de hongos micorrizicos VA en plantulas de cafe (*Caffea arabica* L.) y de te (*Camellia sinensis* (L) O. Kuntze). Memorias Seminario sobre Micorrizas. March 19–21, 1986. Medellin, Colombia (in press).

37 Tinker P B 1980 Role of rhizosphere microorganisms in phosphorus uptake by plants. *In* The Role of Phosphorus in Agriculture. Eds. F E Khasawuch, et al. Am. Soc. Agronomy, Tennessee Valley Authority. pp 617–654.

38 Yost R S and Fox R L 1979 Contribution of mycorrhizae to the P nutrition of crops growing on an Oxisol. Agron. J. 71, 903–908.

Plant and Soil 100, 285–295 (1987).   Ms. 100-31

Effect of spore concentration on germination and autotropism in *Trichoderma hamatum*

LEA NOL and YIGAL HENIS

The Hebrew University of Jerusalem, Department of Plant Pathology and Microbiology, Faculty of Agriculture, P.O. Box 12, Rehovot 76 100, Israel

Key words   Competition   Germtube emergence   Inhibitory compounds   Nutrients

Summary   The effect of spore concentration on spore germination and germtube growth of *Trichoderma hamatum* on water agar and on potato dextrose agar (PDA) was studied. Increasing inoculum size up to 10^9 spores/plate on PDA and up to 10^7 spores/plate on water agar shortened the incubation period required for germtubes emergence and increased germination rate. However, on water agar germination was inhibited at 10^8 and was completely arrested at 10^9 spores/plate. Inhibition in germination of 10^7 spores/plate was observed on water agar when the plates were preincubated with 10^9 spores/plate for 5 h or more. Addition of glucose and ammonium nitrate to the water agar medium allowed only 25% of the spores to germinate at 10^9 as compared to 78% at 10^7 spores/plate after 8 h of incubation. Addition of polysaccharides to the C + N supplemented medium, significantly increased germination up to 84% as compared to 100% on PDA, after 8 h of incubation. Germlings of *Trichoderma hamatum* phialospores exhibited positive autotropism and anastamosis on both media. The phenomenon was positively related to inoculum size, being most pronounced at 10^7 spores/plate.

Introduction

Spores of many fungi germinate poorly or not at all when in dense suspensions or crowded upon a surface. Inhibition in these fungi is directly proportional to the size of spore population[3]. Macroconidia of *Fusarium solani* require exogenous carbon and nitrogen for germination when incubated at high densities, while they are fully independent of exogenous nutrients at low densities. These findings suggest that conidia of *F. solani* elaborate on endogenous reserves self-inhibitors of germination, which at high conidial densities must be counteracted by exogenous carbon and nitrogen, to allow germination to occur[1]. Following germination, germtubes of some fungi are either attracted to, or repelled from each other, exhibiting positive or negative autotropism, respectively[4,5]. Robinson *et al.*[4] reported on the absence of positive autotropism between germinating spore pairs of *Trichoderma viride* on agar surface.

In a study of the aggregation of germinating conidia of *T. hamatum* in shake, it was found that the polysaccharides, amylopectin and polytran N (produced by *Sclerotium rolfsii* Sacc) prevented aggregation and increased germination rate considerably (Carlos Spitz and Yigal Henis, Unpublished data, 1985).

The present investigation was undertaken to determine the effect of spore concentration, external nutrients, amylopectin and polytran N on germination and germtube growth in *T. hamatum*.

Materials and methods

The isolate of *Trichoderma hamatum* (Tri-4) used in this study was obtained from G.C. Papavizas, Soilborne Disease Laboratory, Beltsville MD. USA[2]. The fungus was grown on a synthetic agar medium of the following composition (g/L): glucose, 15; NH_4NO_3, 1.0; KH_2PO_4, 0.9; $MgSO_4 \cdot 7H_2O$, 0.2; microelements, 1 mL of a stock solution containing (mg/mL): $FeSO_4 \cdot 7H_2O$, 2; $MnSO_4 \cdot 7H_2O$, 2; $ZnSO_4 \cdot 7H_2O$, 2, and Bacto agar (Difco), 20; in distilled water. Inoculated agar plates were incubated at 30°C. Phialospores were harvested from 21-days-old cultures with sterile distilled water and washed twice with sterile distilled water by centrifugation in sterile capped centrifuge tubes for 10 min at 10,000 g.

Germination on agar media

Plates containing 10 ml agar were inoculated by spreading spores in a volume of 0.1 ml per plate on their surface, by means of a sterile glass rod. The plates were incubated at 30°C, and germination and growth were stopped by adding a few drops of cotton blue-lactophenol solution.

Phialospores of *Trichoderma hamatum* were incubated on PDA, and on water agar at concentrations of 10^5–10^7 spores/plate for 10 h and at concentrations of 10^7–10^9 spores/plate for 9 h.

Spores of *T. hamatum* (10^9/plate) were incubated for 0–24 h on cellophane sheets placed on water agar. The cellophane sheets with the spores were removed from the water agar plates at intervals, and the plates were re-inoculated with 10^7 spores/plate. Germination was examined after 8 h of incubation at 30°C.

Quantitative estimation of germination rate was done by direct microscopic observation of stained inoculated agar discs with cover glasses. Germination was defined by development of a germtube 0.125 μm length or longer.

In some experiments 27 × 76 mm glass slides were placed in empty sterile petri plates, covered with 0.5 ml PDA, and inoculated with 0.05 ml spore suspension per slide, using the edge of another glass slide.

Spore counts

Spores were counted with haemacytometer. Densities of 10^8–10^5 spores/ml were obtained from suspensions of 10^9 spores/ml by dilution in sterile distilled water.

Preparation of cellophane sheets

Cellophane round sheets of 9.0 cm in diameter were boiled in distilled water which was changed 3 times until turning clear. Then the cellophane sheets were placed in a glass petri dish and sterilized in the autoclave. Sterile cellophane discs were placed on the surface of water agar plates, one sheet per plate.

Media

The following media were used to study spore germination and germtube growth:

1. Potato Dextrose Agar (PDA, Difco).

2. Water agar (WA): Bacto agar (Difco) 20 g in (1 L) distilled water.

3. Washed-water agar (WWA). Bacto agar (Difco) powder (20 g) was suspended in distilled water (1 L) and incubated overnight at 4°C. Water was decanted and replaced with fresh distilled water 15–20 times for 2 days until agar and water turned clear. Finally distilled water was added to 1 L volume and the medium was sterilized in the autoclave.

4. Mineral agar (MA). This medium consisted of (g/L) K_2HPO_4, 0.9; $MgSO_4.7H_2O$, 0.2; 1 mL of a stock microelements solution, same as for synthetic agar, and washed agar, 20. Final pH-6.0.

5. Synthetic medium (SM). This medium consisted of (MA) supplemented with (g/L) NH_4NO_3,

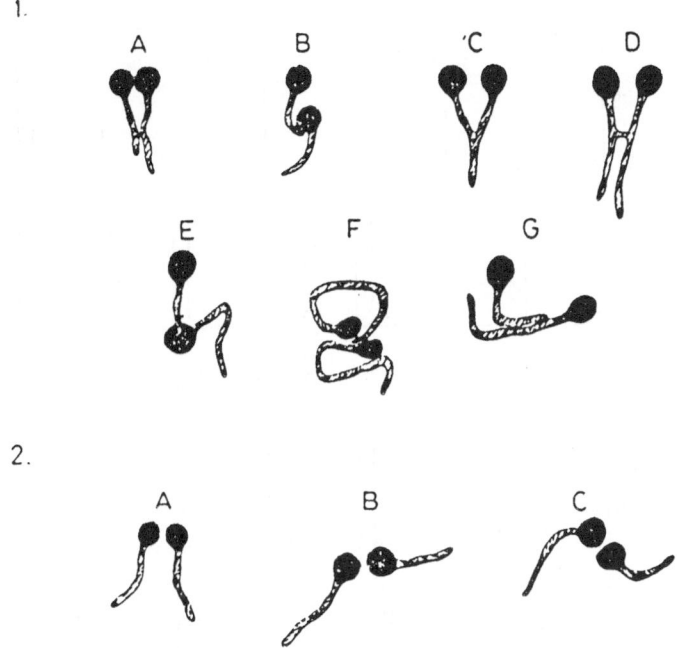

Fig. 1. Autotropism of *Trichoderma hamatum* germlings on potato dextrose agar (PDA) after 10 h at 30°C. 1. Positive autotropism. 2. Negative autotropism.

1; and glucose, 10. In some experiments, this medium was supplemented with (g/L) dialysed amylopectin, 10 or with polytran N, 10.

Polysaccharides

Amylopectin (from corn, Sigma Chemicals Co. U.S.A., Cat. No. A-7780) and polytran N (an exopolysaccharide from *Sclerotium rolfsii*, Jetco Chemicals Co. U.S.A. Cat. No. L-6-2) were suspended in water, placed in a dialysis tube, dialyzed against streaming distilled water for 24 h and added to the agar medium before autoclaving.

Autotropism

Autotropism was estimated according to Robinson et al.[5] by recording under the microscope the mutual growth response of the emerging germtubes; however instead of the four forms of autotropism used by these authors, we have adopted two types: positive and negative (Fig. 1).

Statistical analysis

Data expressed in percentage unites were transformed to the arcsine before analysis. All experiments repeated and analyzed follow standard procedures for analysis of variance; differences between means were evaluated for significance ($P = 0.05$) with Duncan's multiple range test[6].

Results

Germination

In PDA and WA germtubes of *Trichoderma hamatum* emerged after 6 to 8 h of incubation at concentrations of 4×10^6 and 4×10^7 spores/

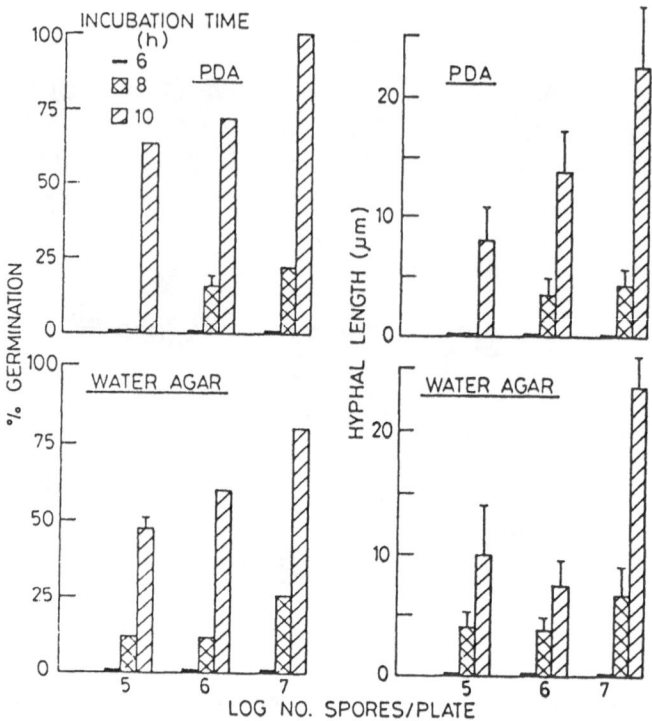

Fig. 2. Spore germination and hyphal length as related to spore concentration and growth medium. Values of 10^5, 10^6, and 10^7 spores/plate are equivalent to 6.9×10^2, 6.9×10^3 and 6.9×10^4 spores/cm², respectively. Vertical bars represent standard error of the mean.

plate. At 4×10^5 spores/plate germination commenced earlier on WA than on PDA, whereas at 4×10^7 spores/plate all spores germinated on PDA and 80% germinated on WA after 10 h (Fig. 2).

Germination rate and hyphal length on PDA and on water agar increased with inoculum size up to 10^7 spores/plate. To find out whether increased hyphal length was due to increased growth rate of the germ-tubes or to a shorter incubation time necessary for germination, the behaviour of individual spores on glass slides covered with PDA was followed. Germtubes growth rate increase and their emergence occurred earlier, as spore concentration was increased up to 3.5×10^7 spores/slide (Table 1). However, the difference between the average growth rate at 3.5×10^7 and 3.5×10^6 spore/slide was not significant.

Increasing spore concentration above 10^7 up to 10^9/plate stimulated germination of *Trichoderma hamatum* spores on PDA but inhibited it on water agar. Germination was completely arrested on WA at 10^9 spores/plate. However, up to a concentration of 10^7 (10^5 to 10^7) spores/plate germtubes emergence on WA was faster than on PDA (Fig. 3).

Table 1. Growth rate of *Trichoderma hamatum* individual germtubes on potato dextrose agar as affected by spore concentration

Incubation time (h)	3.5 × 10⁵ᵃ				3.5 × 10⁶					3.5 × 10⁷				
	I	II	III	IV	I	II	III	IV	V	I	II	III	IV	V
Hyphal length (μm) of individual germtubesᵇ														
6	0	0	0	0	0	0	0	0	0	0	0	0	0	0
7	0	0	0	0	0	0	0	0	0	6.1	6.1	3.0	3.0	0
8	0	0	0	0	6.1	3.0	3.0	3.0	0	12.2	12.2	9.1	9.1	6.1
9	6.1	6.1	0	0	18.3	18.3	12.2	12.2	9.1	24.4	24.4	30.5	24.4	12.2
10	9.1	9.1	6.1	0	24.4	24.4	24.4	30.4	18.3	48.7	42.6	48.7	42.6	21.3
11	12.2	18.3	9.1	0	36.5	46.7	42.6	54.8	30.4	66.9	66.9	97.4	85.3	48.7
Germtubes growth rate (μm/h)														
Individual	3.0	6.1	3.0	—	10.2	14.2	13.26	21.3	10.6	15.2	15.2	23.6	20.5	14.2
Average	4.1 ± 1.4				13.9 ± 4.0					17.8 ± 3.7				
During first hour	3.0				10.9					6.1				
During first 2 hours	4.5				10.9					11.3				
During first 3 hours					14.7					15.1				
During first 4 hours										18.6				

ᵃ 2.8 × 10² spores/cm²
ᵇ Roman numbers for individual spores
± Standard deviation.

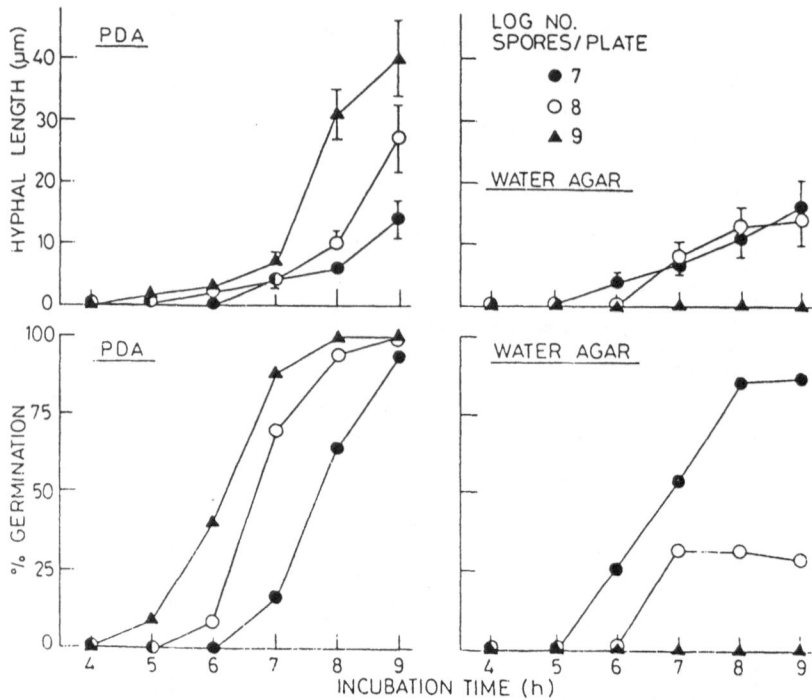

Fig. 3. Effect of spore concentration and growth medium on germination and germtube elongation. Values of 10^7, 10^8 and 10^9 spores/plate are equivalent to 1.72×10^4, 1.72×10^5 and 1.72×10^6 spores/cm^2, respectively. Vertical bars represent standard error of the mean. Standard errors for means of germination percent are negligible.

Possible role of nutrients

There was a decline in germination per cent in plates which had been preincubated for 5 h or more with 10^9 spores/plate. No germination could be observed in plates which had been preincubated with 10^9 spores/plate for 24 h (Fig. 4). To find out whether this inhibition was due to competition for nutrients, WWA plates were supplemented with either glucose, ammonium nitrate or both and inoculated with spores at concentrations of 10^7 and 10^9 spores/plate. Whereas at 10^7 spores/plate addition of glucose or ammonium nitrate either did not affect (63–70%) or slightly affected (+15%) germination, both carbon and nitrogen sources were required to allow for some germination (25%) at 10^9 spores/plate. No germination was observed in the absence of these nutrients.

Addition of 1% (final conc.) of either amylopectin or polytran N to synthetic agar significantly increased germination (Table 2)., Germination at 10^9 spores/plate on SM supplemented with amylopectin almost equalled that observed on PDA (Fig. 5).

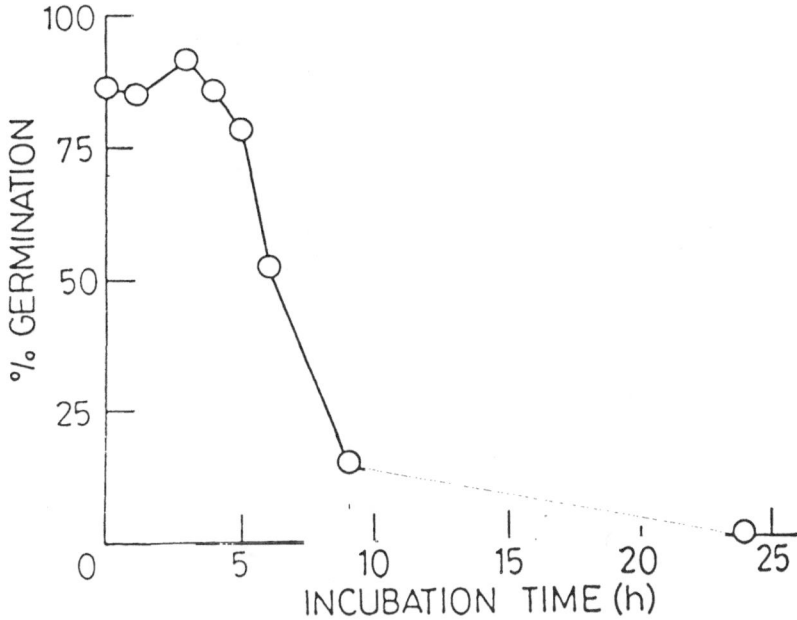

Fig. 4. Germination of 10^7 spores/plate incubated for 8 h on water agar which had been preincubated with 10^9 spores/plate, spread on a cellophane sheet. Values 10^7 and 10^9 spores/plate are equivalent to 1.55×10^4 and 1.55×10^6 spores/cm^2, respectively. Standard errors for means of germination percent are negligible.

Autotropism

As shown in Fig. 1, germtubes of germinating *Trichoderma hamatum* spores grew towards adjacent spores or towards other germtubes in several patterns. The frequency of the observed positive autotropism varied with medium composition and spore concentration (Table 3). Positive autotropism was most prominent at spore concentration of 10^7/plate on PDA, WA and WA supplemented with 1% glucose. No

Table 2. The effect of amylopectin and polytran on germination of *Trichoderma hamatum*[1]

Supplements (1%) to Synthetic Medium[2]	Spore on Cellophane	Germination (%)
None	±	10.9d
Amylopectin	−	43.2bc
Amylopectin	+	38.9c
Polytran N	−	49.4b
Polytran N	+	59.5a

[1] 10^9 Spores/plate equivalent to 1.44×10^6 spores/cm^2 incubated for 9 h at 30°C.
[2] Basic medium containing minerals, 0.1% NH_4NO_3 and 1% glucose.
Values followed by the same letter do not differ significantly ($P = 0.05$).
[+,−] Addition or absence of cellophane sheets, respectively.

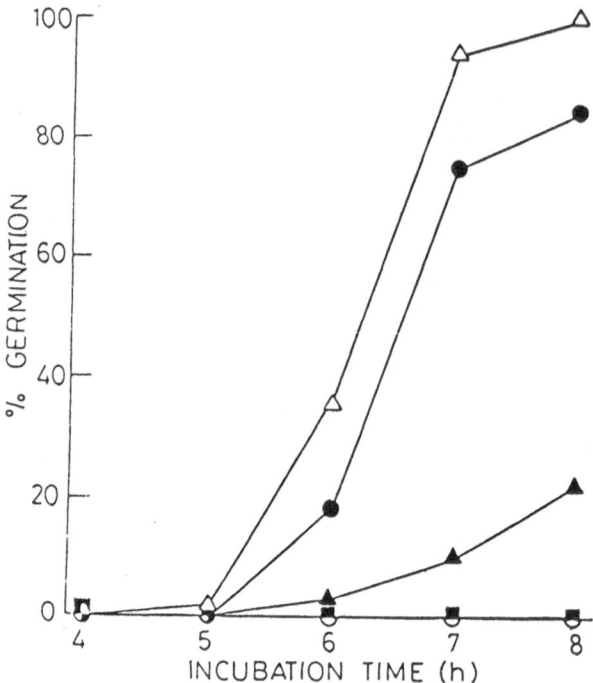

Fig. 5. Effect of amylopectin on germination of 10^9 spores/plate as related to incubation time. A concentration of 10^9 spores/plate is equivalent to 1.6×10^6 spores/cm². (O) Mineral agar (MA) (■) MA supplemented with amylopectin 1%. (▲) Synthetic medium (SM) (●) SM supplemented with amylopectin 1%. (△) PDA. Standard errors for means of germination percent are negligible.

positive autotropism was revealed at 10^5 spores/plate. As the inoculum level was increased, positive autotropism occurred between pairs of spores situated at greater distances.

Discussion

The results presented in this work demonstrate that germinating phialospores of *Trichoderma hamatum* behaved differently on PDA and on WA according to spore concentration. Three variables have been examined: (i) Length of the incubation time preceding the onset of germination, (ii) Germtubes elongation rate and (iii) positive autotropism.

Whereas on PDA increasing spore concentration at all levels resulted in an improvement in these growth parameters, on WA only inoculum levels up to 10^7 spores/plate shortened incubation time and brought germination to reach its highest level. At 10^9 spores/plate germination was completely blocked but inhibition was partially reversed when glucose and ammonium nitrate were added. These results differ from the

Table 3. Positive autotropism in *Trichoderma hamatum* as affected by incubation medium and spore concentration[1]

	Incubation medium								
	Potato Dextrose Agar			Water Agar			Water Agar + 1% Glucose		
	spores per plate[2]								
	10^5	10^6	10^7	10^5	10^6	10^7	10^5	10^6	10^7
Spores per[3] field	3.7 ± 1.5	14.0 ± 4.5	56.3 ± 12.4	1.8 ± 0.7	8.2 ± 2.0	52.7 ± 11.3	4.2 ± 1.7	23.4 ± 4.9	66.0 ± 4.5
% spores in positive autotropism	0 e	8.82 d	56.9 a	0 e	34.2 b	45.9 ab	0 e	18.5 c	29.7 bc
Distance between spores in positive autotropism (μm)	–	7.7 3.5	12.9 ± 4.3	–	14.5 ± 4.4	20.8 ± 7.9	–	8.2 ± 2.8	12.2 ± 4.5
Hyphal length (μm)	14.9 ± 4.1	20.5 ± 12.0	35.3 ± 7.5	31.4 ± 12.0	37.6 ± 8.4	41.3 ± 7.2	16.9 ± 5.5	18.7 ± 4.1	17.2 ± 2.7

[1] As recorded after 10 h of incubation at 30°C.
[2] Equivalent to spore concentration of 4.1×10^2, 4.1×10^3 and $4.1 \times 10^4/cm^2$ for 10^5, 10^6 and 10^7 spores/plate, respectively.
[3] Each microscopic field under × 400 magnification. Each value represents a mean of 10 fields.
± Standard error (values at 5% level).
Means followed by the same letter do not differ significantly ($P = 0.05$).
Positive autotropism follows Fig. 1.

findings of Griffin[1] regarding *Fusarium solani*, who found that inhibition at high spore densities was completely reversed by a combination of carbon and nitrogen sources. Inhibition in germination of *T. hamatum* spores was almost completely reversed on synthetic agar supplemented with amylopectin or polytran N. The effect of the dialyzed polysaccharides on germination was exerted through a cellophane membrane. Contribution of nutrients must therefore be excluded. The presence of germination stimulators and self-inhibitors in fungal spores and their chemical structure have been evaluted in many fungi[3,8,9]. However, the mechanism(s) of inhibition are not clear yet and they seem to be different in different fungi[3]. Possibly, the polysaccharides tested may act as binding agents of inhibitory factors excreted by the spore. Indeed Spiegel *et al.*[7] have demonstrated a leakage of ^{14}C-labelled compounds during germination of *T. viride*. Some of thse compounds could be involved in germination initiation and/or inhibition.

Another possible mechanism which may affect spore germination at high densities is competition for nutrients. In the case of *T. hamatum*, however, glucose and ammonia probably play a partial role only, and other nutrients should be supplemented to support a high rate of germination. Indeed phialospores of *T. viride* have been shown to adsorb amino acids during germination[7]. These observations support both possibilities, *i.e.* that spores of *T. hamatum* excrete both stimulatory and self-inhibitory substances during the early stages of germination. The levels of these substances in the agar increases as a function of inoculum size and time. It appears that one or several components in the PDA medium neutralize(s) the self-inhibitory compounds. Therefore, on this medium there is only a stimulatory effect on both germination and germtube elongation. As the self-inhibitory substance(s) accumulate(s) in the WA medium, they reach a critical concentration which inhibits germination, although there are still stimulatory substances present (as deduced from spore behaviour on PDA). Our results imply that on WA the concentration of the hypothetical inhibitory substances is a critical factor for germination, even in the presence of the spore-stimulatory substances.

Direct evidence for the presence of stimulating and inhibiting factors active on germination and autotropism in *T. hamatum* spores is still missing. Evaluation of the chemical nature and mode of action of these factors awaits further studies.

Germtubes of *T. hamatum* show positive autotropism: they grow towards adjacent spores or adjacent germtubes, in various patterns. Robinson *et al.*[5] reported on the absence of autotropism in *T. viride* germinating on agar surface. The extent of positive autotropism in *T. hamatum* varies with medium composition and inoculum size, being most prominent at concentration of 10^7 spores/plate on both PDA and water agar. At high spore densities positive autotropism occurs between pairs of spores situated at greater distances, probably reflecting the faster growth rate of the germtubes at higher spores concentration.

References

1 Griffin G J 1970 Carbon and nitrogen requirements for macroconidial germination of *Fusarium solani*: dependence on conidial density. Can. J. Microbiol. 16, 733–740.

2 Henis Y, Lewis J A and Papavizas G C 1984 Interactions between *Sclerotium rolfsii* and *Trichoderma* spp: Relationship between antagonism and disease control, Soil Biol. Biochem. 16, 391–395.

3 Macko V 1981 Inhibitors and stimulants of spore germination and infection structure formation in fungi. *In* The Fungal Spore: Morphogenetic Controls. Eds. G Turian and H R Hohl. Academic Press. pp 565–584.

4 Robinson P M 1978 Spore germination. *In* Practical Fungal Physiology. W F Wiley and Sons. N.Y. pp 1–19.

5 Robinson P M, Park D and Graham T A 1968 Autotropism in fungal spores. J. Exp. Bot. 19, 125–134.

6 Snedecor G W and Cochran W G 1967 Statistical Methods. Iowa State University Press, Ames. Iowa. 593 pp.

7 Spiegel S, Henis Y, Chet I and Messer G 1975 The effect of heat shock on differentiation of germinating conidia of *Trichoderma viride*. Can. J. Bot. 53, 2274–2281.

8 Sussman A S 1876 Activators of fungal spore germination. *In* The Fungus Spore, Form and Function. Eds. D J Weber and W M Hess. J Wiley and Sons Publishers, N.Y. 895 p.

9 Weber D K and Hess W M 1975 Diverse spores of fungi. *In* Spores *VI*. Ed. P Gerhardt. R N Costilow and H L Sadoff Publishers, ASM. pp 97–111.

Plant and Soil 100, 297–322 (1987).
© 1987 Martinus Nijhoff Publishers, Dordrecht.

Ms. 100-22

The effect of fertilization on litter decomposition in clearfelled spruce stands

B.D. TITUS* and D.C. MALCOLM
Department of Forestry and Natural Resources, University of Edinburgh, Mayfield Road, Edinburgh EH9 3JU, UK.

Key words Clearfelling Leachate Litterbags Litter decomposition Logging residue NPK-fertilizer Nutrient dynamics Peaty gley Sitka spruce

Summary   The influence of NPK-fertilizer on decomposition of litter layers and deposited logging residues (brash) on a clearfelled Sitka spruce stand was followed during two years by zero-tension lysimetry and litterbags. Root development of second rotation planted trees on this peaty gley soil are restricted to the litter layers (LFH) and without fertilizer are dependent on its decomposition for release of nutrients. A comparison of fertilized and control plots showed few site differences and similar hydrological properties.

Fertilizer addition (urea-N, rock-P and KCl) immediately raised leachate concentrations of NH_4-N, K and PO_4-P, the last remaining high over the period. NO_3-N increased dramatically in the second year leachates from the fertilized area. Within two years 45, 60 and 75% of added NPK respectively were recovered in leachates.

Two-year-old litter in litterbags lost weight significantly more slowly after fertilization. The calculated $\bar{k}$ values were 0.28 (control plot) and 0.15 (fertilized plot). Mean $\bar{k}$ values were derived from individual regressions and allowed microsite variation to be assessed. The difference in $\bar{k}$ is attributed to available C limitation although N concentration of needles increased on both plots, P concentration increased slowly on the control plot while K and Mg decreased on both. Ca concentrations doubled with fertilization and remained constant over two years.

The rapid loss of fertilizer and slight response from planted trees indicate delayed application of fertilizers may be desirable.

Introduction

Almost forty percent of recent afforestation in the British uplands has been on peaty gley soils[33] which developed post-glacially by the accumulation of organic matter over glacial tills of low permeability. The organic horizons (5–45 cm) have low hydraulic conductivity[7] which, combined with the inadequate pore-size distribution of the mineral horizons, prevents lowering of the water table by artificial drainage. Initial afforestation was achieved by planting on raised turves cut or ploughed from the organic horizon but this method is not practicable for establishing second rotation stands. Thus on clearfelled sites the water table rises to within 15 cm of the surface and remains at this depth almost all year[25]. To facilitate extraction of timber a 'bench' felling system is used so that all brash (logging residue) is banded into 8 m wide swathes

* Present address: Newfoundland Forestry Centre P.O. Box 6028 St. John's Newfoundland Canada A1C 5X8

which act as a roadbed for extraction machinery. These swathes are separated by 4 m wide clear strips with are brash-free. After several years delay, to allow the brash swathes to partially break down, planting takes place, the trees being located close to an old stump with their nursery roots mainly in the original organic horizon. Anaerobic conditions kill many of these roots and survival is dependent on the formation of adventitious roots which can only ramify through the litter layers (LFH) derived from the previous stand. Until the planted trees have extensive root systems there is no 'sink' to trap nutrients released from the LFH horizons as these sites are very slow to revegetate. Leachates from the LFH cannot readily infiltrate the original peaty material[7] so tend to flow over it to the nearest drainage channel and may be lost from the site. Release of nutrient to leachate from around the planting position is thus importance to the early growth of planted seedlings.

The aggregation of brash into swathes affects the distribution of nutrient return to the soil after clearfelling and thus alters the nutrient dynamics of the litter layer. It also affected the microclimate of newly planted seedlings and intercepts water flow into the LFH[40]. As the needle fraction of the brash has the highest initial nutrient and lowest lignin content it will decompose before the woody fraction and release available nutrients for newly planted trees.

The potential for nutrient loss by the delay in replanting and the dependence of the transplants on the uppermost horizons for nutrition suggest a possible role for fertilizer applications to ensure adequate early growth rates[23]. Applied fertilizer could be expected to suffer leaching losses and to alter the decomposition rate of litter through its influence on microbial populations.

To investigate these questions the dynamics of nutrient elements in the LFH horizons of a series of comparable sites (0, 2 and 5 years after felling) were followed over two years, during which both leachate release and litter decomposition rates were determined[40]. Fertilizer additions were made to the two-year-old site and it is the results of this part of the study that are reported here.

Materials and methods

Site description
    The study site was located in compartment 720 of Kielder Forest, Northumberland (National Grid Reference NY 657928). The soil was a uniform peaty gley[33] which had developed post-glacially on Scremerston Coal Group Sandstones overlying Carboniferous Limestone. The gently-sloping site (6–7°) was at an elevation of 300–310 m above sea level and had a southerly aspect. The climate is cool and moist, with mean monthly temperatures ranging between 0°C and 15°C, and 1300 mm annual rainfall. The previous crop of pure Sitka spruce *Picae sitchensis* (Bong. Carr.) was planted in 1939 at a density of about 4,115 stems ha^{-1}. In 1975 it had attained a top height of 15 m and was

assessed as Yield Class 10–12 (maximum mean annual increment, $m^3ha^{-1}a^{-1}$). The stand was harvested in November 1979 using a bench felling method that leaves all residue (brash) in distinct swathes 8 m wide, separated by clear strips devoid of brash 4 m wide. For this study these were considered separate treatments[40]. Two adjacent 0.1 ha rectangular plots which were part of a Forestry Commission experiment (Falstone 7/81) were selected. Plot 2 received no fertiliser and pl ot 2F received the equivalent of $150 kg N ha^{-1}$ as urea, $50 kg P ha^{-1}$ as rock phosphate and $100 kg K ha^{-1}$ as muriate of potash. The fertiliser was hand broadcast on 19 June 1981. These plots (2 and 2F) also represented sites two years after clearfelling in a larger time series experiment designed to examine litter layer nutrient dynamics for the first seven years after clearfelling[40].

Soil sampling

Fifteen randomly located soil cores were taken in late November 1981 from both the brash swathes and clear strips of both plots, using a corer of $25 cm^2$ cross-sectional area that minimizes compaction in organic horizons[11] thus allowing an accurate assessment of bulk density to be made. Cores were air dried and ground after division into LFH (derived from spruce litter) and 0 (organic horizon pre-dating planting) horizons. Subsamples were then oven dried to constant weight at 95°C prior to digestion and chemical analysis.

Lysimeters

Zero-tension lysimeters* were constructed of polystyrene trays of $881 cm^2$ surface area. Excised LFH horizons were placed on a nylon screen, supported by a 1.25 cm high grid of PVC, in the bottom of each tray. This allowed for vertical drainage of leachate from any part of the surface of the enclosed litter layer, and free horizontal drainage beneath the screen to buried outlet pipes connecting the lysimeters to 25 litre collection vessels, which had light excluded to discourage algal growth. The sides of the lysimeters protruded several millimetres above the surrounding litter layer to prevent surface flow, and the lysimeters were adjusted so that the surface of the litter layers within the lysimeters was level with that outside. Lysimeters were installed in the usual planting position beside ten randomly chosen stumps in each treatment of both plots, giving a total of forty lysimeters in all. Leachate collections were made every three weeks, although both freezing conditions and drought occasionally led to longer collection periods. Volumes of leachate were measured in the field, and subsamples were returned to the laboratory for immediate chemical analysis.

Precipitation inputs were sampled with four plastic rain-gauges (15.2 cm diameter plastic funnels in 5 litre collection vessels). One was used to calculate volume inputs, and the other three had fibreglass wool wads inserted to exclude insects and debris and were used for chemical determinations. All precipitation samples were collected at the same time as leachate samples.

Litterbags

Litterbags (10 cm × 10 cm) were constructed of nylon curtain material with a rectangular pore size of 0.3 mm × 1.0 mm, which was small enough to prevent the loss of needles. Two-year-old needles were collected from under the brash swathes on plots 2 and 2F on 14 May 1981, combined, and air dried. 3.1 g of the prepared needles were accurately weighed out and enclosed in each litterbag. Completed litterbags were stored at − 20°C until laid out in the field on 17 June 1981, two days before fertilization.

An estimate of initial oven-dry weight was obtained by oven-drying ten litterbags at 105°C at the outset of the experiment. Weight loss, and hence total nutrient content of litter could then be calculated for each collection date.

Seven litterbags were randomly arranged in a 0.5 × 0.5 m subplot or collection station, in the planting position by ten randomly chosen stumps in the brash swathes of each plot, and buried under a thin cover of litter to approximate a position halfway down the thin layer of brash litter that overlay the original LFH horizons. Any branches that were removed during the laying out of litterbags were returned to their original position.

Collections of litterbags were made approximately every 3.5 months. Bags were randomly chosen from each of the ten collection stations per plot and returned to the laboratory. On attaining

* Detailed design available from the authors

a constant weight at 105°C, litter was removed from the bags, weighed, and then ground and analysed after digestion.

Decomposition rate constants (k) were obtained from the negative slope of a regression of ln (percent weight remaining) over time[30]. This "best fit of a straight line" method allows determination of both $\bar{k}$ and r^2. As litterbags were retrieved from distinct collection stations rather than from individual random locations for every litterbag, the weight remaining in each litterbag was a dependent variable with respect to collection station, rather than an independent variable with respect to site. Thus, regressions were calculated for each of the 10 individual collection stations per plot, rather than for either the entire data set (which assumes independent variables) or the mean weight per plot on each collection date (which does not take into account the variation inherent in each mean). This allows the calculation of a mean decomposition rate constant for a given plot ($\bar{k}$) and mean descriptors of the regression ($\bar{a}$, $\bar{b}$, $\bar{r}^2$) for which standard errors may be calculated. It is thus possible to obtain a measure of variation in 'weight remaining' for each collection station (r^2) as well as variation across a site ($\bar{k} \pm$ SE). Further, differences in k between treatments may then be tested by analysis of variance rather than analysis of covariance.

Chemical analysis

Leachate and precipitation samples were analysed colorimetrically on an autoanalyser for concentrations of soluble NH_4-N[10], NO_3-N[20] and PO_4-P[27]. An atomic absorption spectrophotometer was used to determine concentrations of total K and Na by flame emission and Ca and Mg by atomic absorption after the addition of La.

Soil and litterbag samples were likewise analysed for total N, P, K, Na, Ca and Mg after digestion using a modified micro-Kjeldahl technique[1].

Results

Soil sampling

Comparison of brash swathes and clear strips. While the needle input from brash led to an increase of about 2 cm in LFH depth on both plots, the increase in dry weight was not significant (Table 1). Nutrient concentrations in the LFH horizons on plots were always greater under the brash than in clear strips, although this was not significantly so for N and Mg. On plot 2F, presumably due to the addition of the fertilizer, only Mg was significantly greater under the brash swathe. Brash swathes also showed a higher element capital than the clear strips on plot 2, but again this was not consistent for plot 2F. Apart from Mg the organic horizons (O) beneath brash swathes and clear strips did not differ in concentration or total content (Table 1).

Comparison of treatments on plots 2 and 2F. The two plots showed no differences in the physical characteristics of LFH horizons (Table 2) except for a higher water content (%) in the brash swathes of plot 2F. This was also the case for the O horizon in 2F which had lower bulk density and greater depth as well. Total element levels in the LFH were enhanced by fertilization although not significantly so for N and K when sampled four months after treatment. Comparison of concentrations (Table 2) showed some differences which were not always consistent with

Table 1. Comparison of clear strips (C) and brash swathes (B): (a) Soil charactistics; (b) element concentration (mg g^{-1}); (c) element totals (kg ha^{-1} on equal area basis)

LFH horizons				Organic horizons			
Plot 2		Plot 2F		Plot 2		Plot 2F	
C	B	C	B	C	B	C	B
(a)							
Depth (cm)							
5.7	7.5***	6.3	8.5***	16.3	14.2*	21.7	23.7
Bulk density (g cm^{-3})							
0.117	0.106	0.118	0.101*	0.225	0.241	0.190	0.181
o.d.wt. (t ha^{-1})							
66.8	78.6	73.6	86.7	361.9	342.1	407.6	431.5
Water content %							
451	413	473	466	342	315	406	435
(b)							
Nitrogen							
13.22	13.49	12.61	13.21	16.78	16.25	17.18	18.57
Phosphorus							
0.61	0.72***	0.82	1.03	1.14	1.21	1.10	1.22
Potassium							
0.72	0.90*	1.23	0.95	1.71	1.96	1.25	1.67
Sodium							
0.17	0.63***	0.27	0.22	0.25	0.21	0.22	0.21
Calcium							
1.00	1.75***	2.12	2.79	0.19	0.16	0.26	0.20*
Magnesium							
0.54	0.59	0.52	0.65	0.12	0.03***	0.09	0.02***
(c)							
Nitrogen							
873	1059	935	1157	5979	5458	6969	7910
Phosphorus							
40	57	65	87	407	441	447	525
Potassium							
51	71	96	83	636	708	521	769
Sodium							
11	47***	21	19	90	71*	89	92
Calcium							
68	137***	171	238	69	51	105	83
Magnesium							
36	47	38	58**	45	12***	38	11***

Significant differences between C and B ($P < 0.05*$, $< 0.01**$, $< 0.001***$).

with the expected higher status of the fertilized plot. Except for Ca there was little evidence of downward movement of fertilizer into the O horizons.

Lysimeters

Precipitation input and leachate volumes. The values of precipitation input, expressed as volume expected per lysimeter, are compared with

Table 2. Statistical comparison of plots 2 and 2f for differences in a) soil characteristics, b) element concentration, and c) element totals by treatments. Units as in Table 1

	LFH horizons		Organic horizons	
	Clear strip	Brash swathe	Clear strip	Brash swathe
a) *Soil characteristics*				
Depth	ns	ns	***	***
Bulk density	ns	ns	**	***
o.d.wt.	ns	ns	ns	*
w.c. %	ns	**	**	***
b) *Concentrations*				
Nitrogen	ns	ns	ns	*
Phosphorus	ns	**	ns	ns
Potassium	**	ns	*	ns
Sodium	**	***	ns	ns
Calcium	**	***	***	ns
Magnesium	ns	ns	ns	ns
c) *Totals*				
Nitrogen	ns	ns	*	*
Phosphorus	ns	**	ns	ns
Potassium	*	ns	ns	ns
Sodium	*	***	ns	ns
Calcium	*	***	***	ns
Magnesium	ns	ns	ns	ns

T-test significant differences $P > 0.05$ ns, $P < 0.05*$, $< 0.01**$, $< 0.001***$.

leachate volumes collected in Fig. 1. It is clear that evaporative losses can be considerable particularly where the brash has an interception effect. Differences between plots 2 and 2F were minimal but brash swathes yielded significantly less leachate than clear strips on 61 and 37 percent of collection dates respectively[40]. There appears thus to be no hydrological differences in the LFH layers between plots.

Element concentrations

Leachate concentrations of NO_3-N, NH_4-N, PO_4-P and K were greatly increased by the addition of NPK fertilizer (Fig. 2). Levels of Ca, Na and Mg were initially increased (Fig. 3) but rapidly fell to plot 2 levels apart from Ca, which increased again in the second year after fertilization. The addition of fertilizer decreased the hydrogen ion concentration.

Nitrate concentrations only increased slightly on plot 2F in the first year following fertilization. However, in the early summer one year after fertilization, NO_3-N concentrations dramatically increased in both treatments on both plots. While the concentrations on plot 2 rapidly declined, those on plot 2F remained high. By contrast, NH_4-N concentrations were greatly increased with the addition of NPK, but the high concentrations had dropped substantially by the first winter, and for the brash treatment the plots did not differ by the end of the second winter.

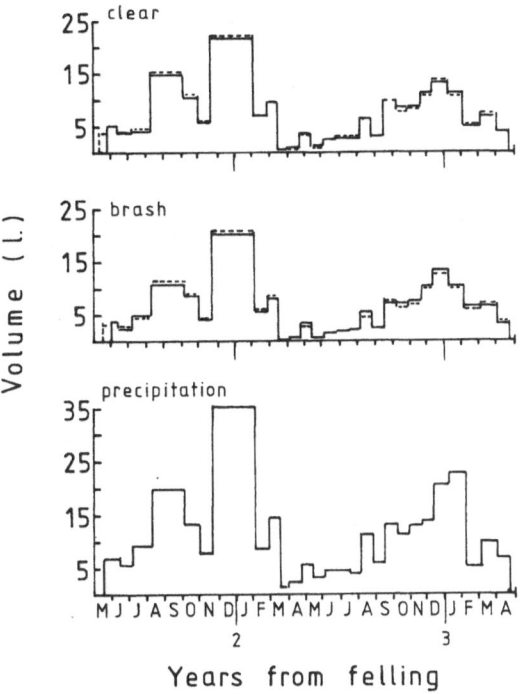

Fig. 1. Leachate volumes. Precipitation expressed as expected volume (l) per lysimeter and differences between plots 2 (*solid line*) and 2F (*dashed line*) for both clear strips and brash swathes.

Orthophosphate concentrations in leachate from plot 2F remained greatly increased throughout the study, although they diminished considerably over the first six months following fertilization. Potassium concentrations decreased dramatically after the initial high pulse. Initial pulses were observed also for Na, Ca and Mg. While Na and Mg concentrations from plot 2F immediately returned to plot 2 values, the differences being insignificant, calcium concentrations rose again one year after fertilization, and remained higher than in plot 2.

Element flux

By taking into account the leachate volumes the measurement concentrations were converted to periodic fluxes thus removing differences between infiltration rates in the brash swathes and clear strips and variation in leachate volume between collections[40]. These showed a similar pattern of events to that found for the concentrations. The cumulative fluxes of elements through the LFH horizon were then combined proportionally for the brash swathes and clear strips (2:1) to give total values ($kg\,ha^{-1}$) for estimating gross leaching losses (Table 3). Net outputs from LFH horizons were derived by subtraction of precipitation input.

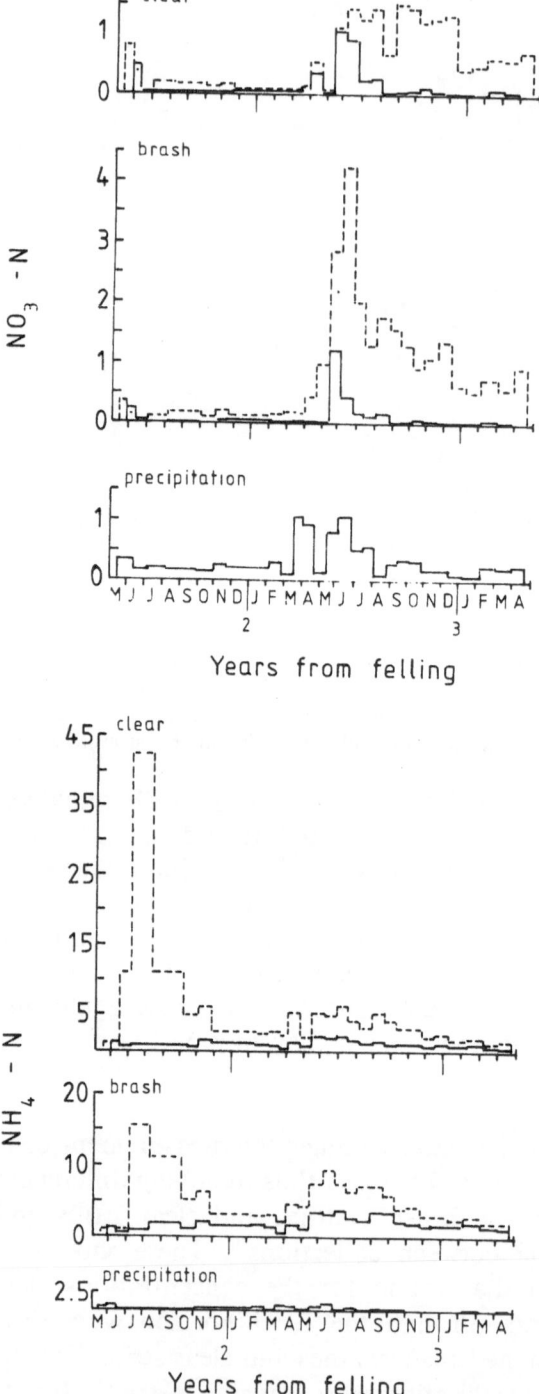

Fig. 2. Concentrations (mgl^{-1}) of (i) NO$_3$-N, (ii) NH$_4$-N, (iii) PO$_4$-P and (iv) K in leachates from plots 2 (*solid line*) and 2F (*dashed line*) after fertilization on 19.6.81.

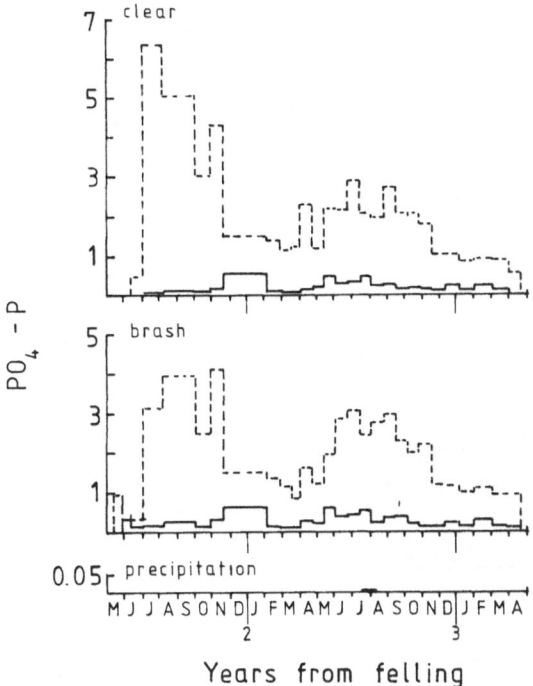

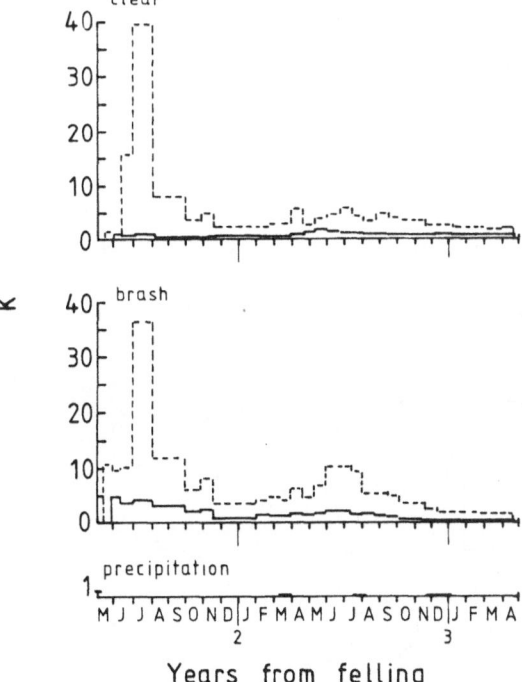

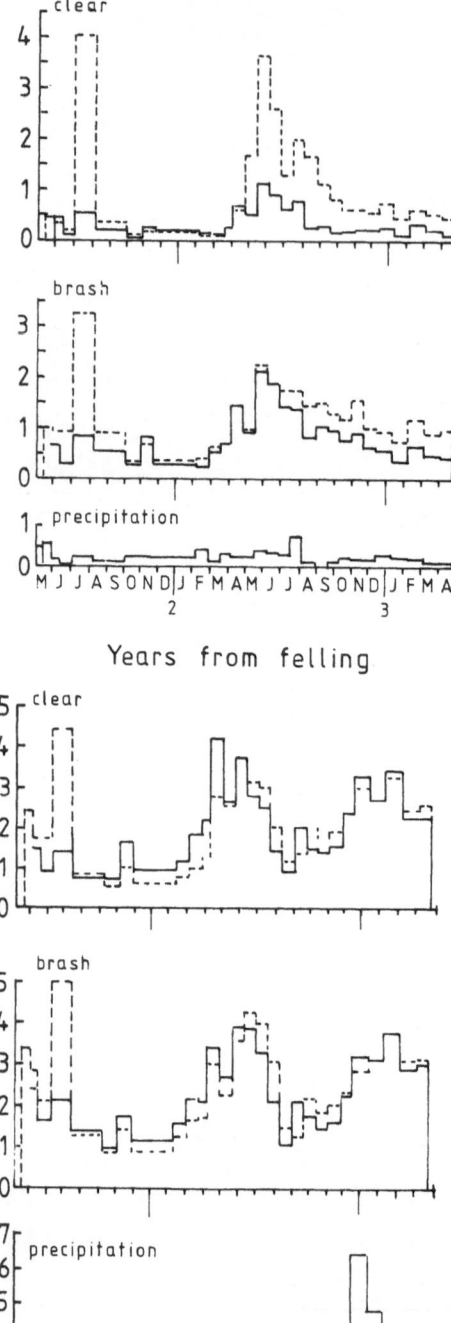

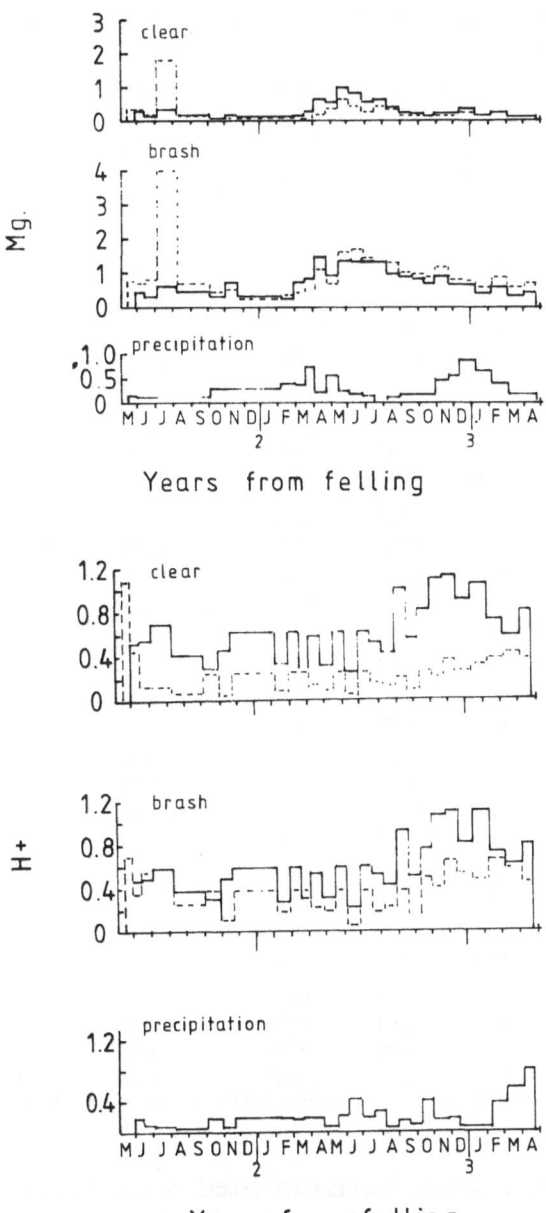

Fig. 3. Concentrations (mg l^{-1}) of (i) Na, (ii) Ca, (iii) Mg and (iv) H in leachates from plot 2 (*solid line*) and 2F (*dashed line*) after fertilization on 19.6.81.

Comparison of gross outputs from 2 and 2F show (Table 3) that fertilization led to enhanced outputs of 9, 58, 30 and 75 kg ha^{-1} of NO_3-N, NH_4-N, PO_4-P and K over two years. The patterns of early or late release of these elements, shown by the concentrations in leachate,

Table 3. Estimated cumulative inputs and outputs (kg ha^{-1}) and differences after NPK fertilization shown by six month periods over two years

Element	Period	Precipitation input	Gross output		Difference	Net output[1]	
			2	2F		2	2F
NO$_3$-N	1	1.06	0.43	0.75	0.32	+0.63	+0.31
	2	2.76	0.64	1.49	0.85	+2.12	+1.27
	3	4.35	1.17	5.75	4.58	+3.18	-1.40
	4	6.79	1.38	10.00	8.62	+5.41	-3.21
NH$_4$-N	1	1.14	5.29	39.41	34.12	-4.15	-38.27
	2	2.86	12.35	54.71	42.36	-9.49	-51.85
	3	4.57	19.41	71.18	51.77	-14.84	-66.61
	4	6.86	28.82	86.47	57.65	-21.96	-79.61
PO$_4$-P	1	0.03	1.07	12.88	11.81	-1.04	-12.85
	2	0.06	2.58	20.40	17.82	-2.52	-20.34
	3	0.08	3.44	28.13	24.69	-3.36	-28.05
	4	0.12	4.08	34.57	30.49	-3.96	-34.45
K	1	1.25	9.70	50.91	41.21	-8.45	-49.66
	2	1.56	13.94	67.88	53.94	-12.38	-66.32
	3	2.50	18.18	83.03	64.85	-15.68	-80.53
	4	3.75	20.61	95.15	74.54	-16.86	-91.40
Na	1	6.67	5.56	7.50	1.94	+1.11	+0.83
	2	23.33	13.06	13.33	0.27	+10.27	+10.00
	3	31.11	20.00	20.83	0.83	+11.11	+10.28
	4	71.11	38.33	38.06	-0.27	+32.78	+33.05
Ca	1	0.84	1.82	4.30	2.48	-0.98	-3.46
	2	2.52	3.64	6.28	2.64	-1.12	-3.76
	3	4.20	6.28	11.24	4.96	-2.08	-7.04
	4	6.72	9.26	16.69	7.43	-2.54	-9.97
Mg	1	0.83	1.21	3.41	2.20	-0.38	-2.58
	2	3.33	2.97	4.62	1.65	+0.36	-1.29
	3	4.33	5.50	7.38	1.88	-1.17	-3.05
	4	10.33	8.04	10.46	2.42	+2.29	-0.13
H	1	1.30	1.87	1.27	-0.60	-0.57	+0.03
	2	2.38	4.33	2.63	-0.70	-1.95	-0.25
	3	3.24	6.53	3.48	-3.05	-3.29	-0.24
	4	4.76	12.81	6.19	-6.62	-8.05	-1.43

[1] + Retention - Loss.

are summarised in Table 4 where annual loss is expressed as a percentage of the two year total. Over 90 percent NO$_3$-N was leached in the second year compared to 74 percent NH$_4$-N in the first year. PO$_4$-P showed only a 20 percent reduction in leachate in year 2 while K and Mg were mainly lost in the first year.

Litterbags

Litter weight loss. Although the litter used in litterbags had been decomposing for two years, the effect of fertilization on weight loss can

Table 4. Element losses per year due to the addition of fertilizer (plot 2F gross totals—plot 2 gross totals), expressed as a percent of the total loss after two years

Time from fertilization (years)	NO$_3$-N	NH$_4$-N	PO$_4$-P	K	Na	Ca	Mg	H
0–1	9.9	73.5	58.5	72.4	100	35.5	68.2	+ 10.6
1–2	90.1	26.5	41.5	27.6	+ 200	64.5	31.8	+ 89.4

+ denotes retention.

still be examined by assuming a starting weight of 100 percent. The distinct difference in $\bar{k}$ shown in Fig. 4 for the two plots (0.15 for 2F; 0.28 for 2) was highly significant ($P < 0.001$) although the $\bar{r}^2$ values (0.91 for 2F; 0.94 for 2) did not differ. The mean litter weights and standard error bars only indicate variation observed on any given collection date and were not used in calculating the regression. Individual regressions for

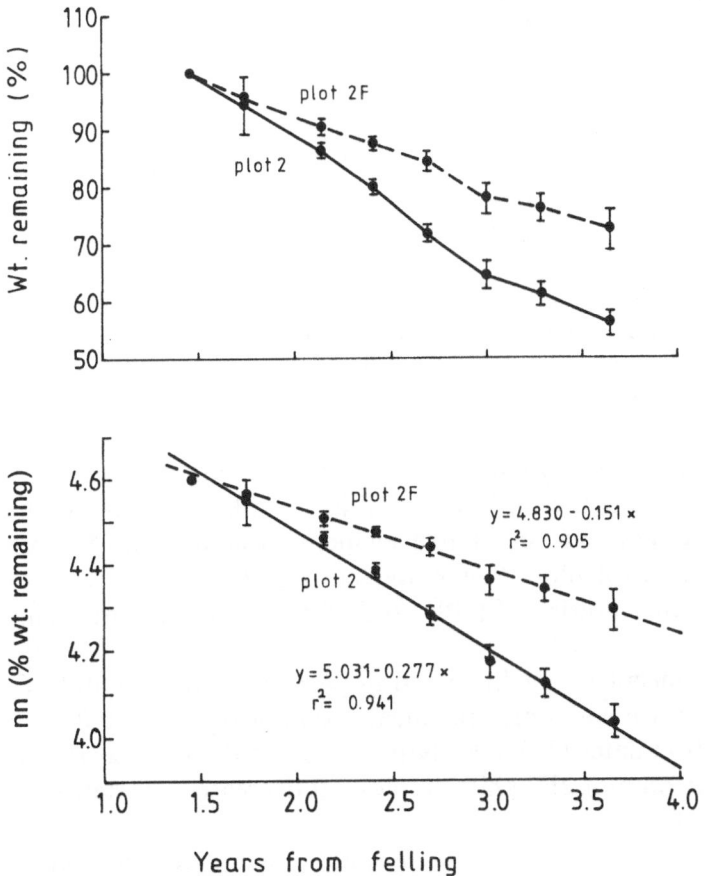

Fig. 4. Percent 'weight remaining' and ln (percent weight remaining) of litterbags against time from plots 2 (*solid line*) and 2F (*dashed line*).

Table 5. Regression equations for each collection station on plots 2 and 2F in the form ln (percent weight remaining) against time ($k = -b$)

	Station	a	b	r^2	n
Plot 2	1	5.045	−0.268	0.966	8
	2	4.899	−0.195	0.966	8
	3	4.968	−0.258	0.968	8
	4	5.183	−0.366	0.923	8
	5	5.104	−0.307	0.978	8
	6	5.063	−0.307	0.986	8
	7	5.055	−0.286	0.877	8
	8	5.066	−0.290	0.956	8
	9	5.029	−0.292	0.846	8
	10	4.901	−0.197	0.948	8
	Mean	5.031	−0.277	0.941	
	SE	0.028	0.016	0.015	
Plot 2F	1	4.663	−0.077	0.791	8
	2	4.798	−0.115	0.882	8
	3	4.684	−0.077	0.893	8
	4	4.850	−0.159	0.867	8
	5	4.833	−0.163	0.945	8
	6	4.757	−0.109	0.956	8
	7	5.028	−0.264	0.878	8
	8	5.066	−0.282	0.966	8
	9	4.812	−0.125	0.879	8
	10	4.810	−0.134	0.988	7
	Mean	4.830	−0.151	0.905	
	SE	0.041	0.022	0.019	

each collection station are given in Table 5 and show a range of k of 0.20–0.37 and 0.08–0.28 for plots 2 and 2F respectively. However r^2 was never less than 0.79 for any individual collection station on either plot and ranged up to 0.99.

Litter element concentrations and contents. From Figures 5 and 6 it can be seen that in general the addition of NPK fertilizer increased the concentration and content of all five elements studied above those observed on the control plot. This could be due to (1) a decrease in weight loss due to the addition of NPK, and (2) the addition of nutrients to the litter.

The nitrogen concentrations increased in parallel on plots 2 and 2F. However, the increase was only statistically significant ($P < 0.05$) on half of the collection dates (Table 6). Nitrogen content barely decreased over the two years on plot 2F, while on plot 2 it decreased constantly to 75 percent of its starting content.

Phosphorus concentrations increased sharply with the addition of fertilizer relative to the control, and then decreased for the next 1.5 years until they were not significantly greater than the control. At that point

(3 years after felling; 1.5 after fertilization) the concentrations on both plots increased slightly. The same pattern was observed for the P content of the litter, with the addition of fertilizer raising the content initially, followed by a fairly rapid decrease over the following two years.

Potassium concentrations were also initially greatly increased on the fertilized plot, and then fell off exponentially over the following year until they were approximately the same as or slightly less than control concentrations. The pattern for potassium content was similar, with the difference due to the addition of fertilizer rapidly declining, so that within a little over a year the content of K in litterbags from both plots was virtually identical.

With calcium, the concentration nearly doubled with the addition of fertilizer, and then remained approximately constant, as did the control concentration (with the exception of one date). When litter weight loss was taken into account, the calcium contents of litterbags from both plots decreased gradually, being less for plot 2F than plot 2. The content of plot 2F was virtually twice that of plot 2.

Both magnesium content and concentration decreased on plots 2 and 2F, and although levels on plot 2F were always higher, this was the only nutrient where fertilization did not cause an immediate increase in either content or concentration.

Discussion

Lysimetry

The use of zero-tension lysimeters placed under the LFH horizons to determine potential nutrient loss from the rooting zone of newly planted seedlings, depends on several assumptions. Firstly, leachate movement down through the LFH horizons is assumed to be largely due to saturated flow, and is best sampled using a "low tension, low resistance" instrument[17]. Secondly, it is assumed that leachate flows laterally on reaching an impervious horizon, so that soluble nutrients do not readily infiltrate the O horizon. The general similarity of element concentrations in the O horizon beneath the brash swathes and clear strips of plots 2 and 2F, as well as the low hydraulic conductivity of the peat[7] supports the latter assumption. Further, the similarities in the physical characteristics of the litter layers of the two plots (bulk density, depth, oven dry weight) indicates that physical leaching processes on plots 2 and 2F were comparable. However, leachate losses from the planting position are not necessarily equatable with total losses from the site. As the seedling roots of replanted trees grow laterally, they increasingly will intercept and take up soluble nutrients in the leachate before it flows into drainage ditches.

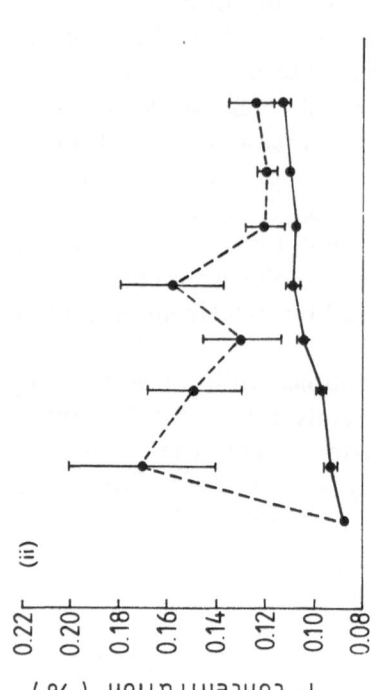

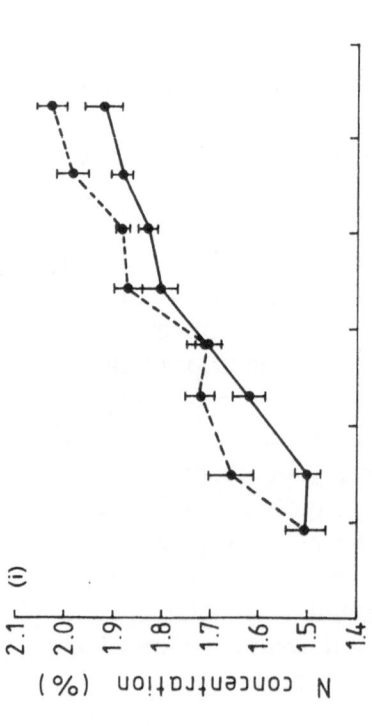

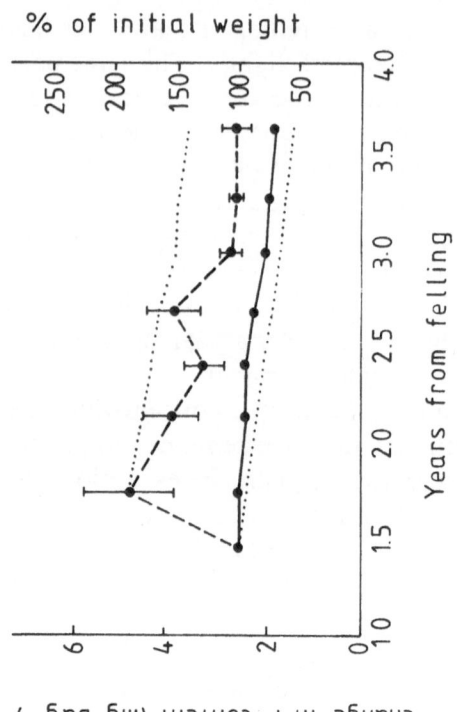

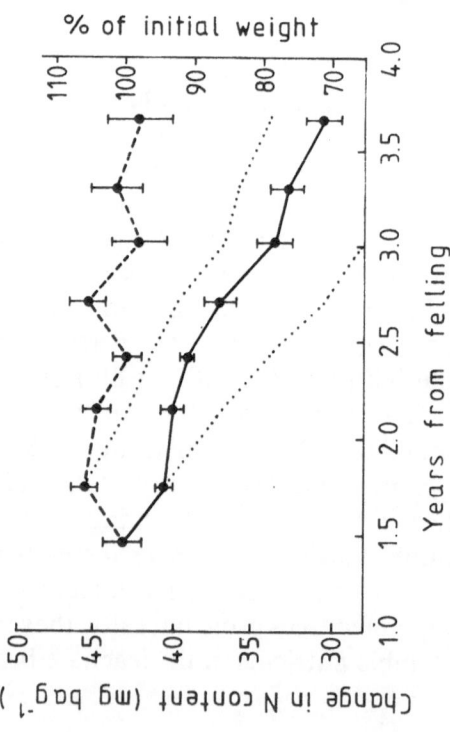

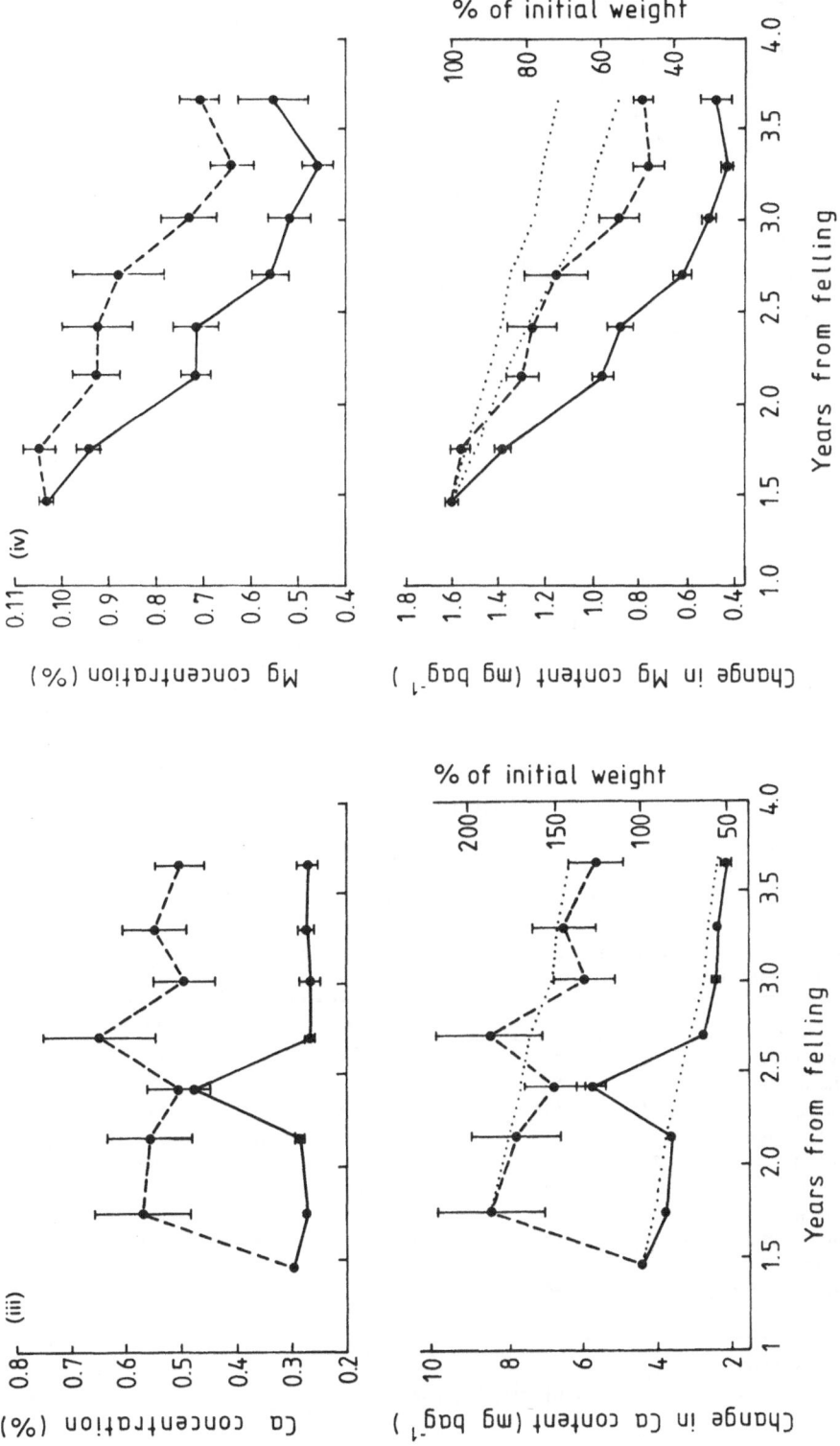

Fig. 5. Changes in concentration (% wt.) and content (mg bag⁻¹) and as percent of initial content of (i) N, (ii) P, (iii) Ca and (iv) Mg for plots 2 (*solid line*) and 2F (*dashed line*). Weight of litter remaining shown as *dotted line*.

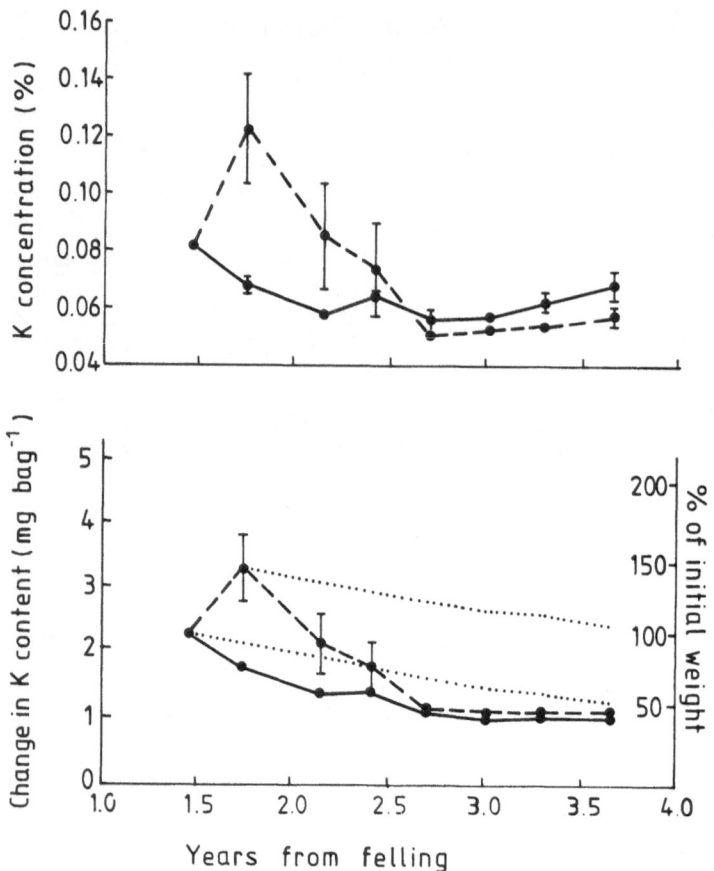

Fig. 6. Changes in concentration (% wt.) and content (mg bag^{-1}) and as percent of initial content of K for plots 2 (*solid line*) and 2F (*dashed line*). Weight of litter remainng shown as *dotted line*.

Table 6. Comparison of element concentration (%), content (mg bag^l) and litter weight remaining (%) between plots 2 and 2F (Student's t-test)

Date[1]	640	787	884	988	1101	1206	1388
Wt. remaining	ns	*	***	***	**	***	***
N concentration	**	*	ns	ns	ns	*	*
Content	***	***	**	***	***	***	***
P concentration	*	*	ns	*	ns	ns	ns
Content	*	*	ns	*	*	***	*
K concentration	*	ns	ns	ns	ns	*	ns
Content	*	ns	ns	ns	ns	ns	ns
Ca concentration	**	**	ns	**	**	***	***
Content	**	**	ns	**	**	***	***
Mg concentration	*	**	*	**	*	**	ns
Content	**	***	**	**	**	***	***

[1] Day from felling
ns $P > 0.05$; * $P < 0.05$; ** $P < 0.01$; *** $P < 0.001$.

The magnitude, timing and duration of increases in nutrient flow in leachate at Kielder were in keeping with results of other fertilizer studies. NH_4-N concentrations at Kielder rose sharply immediately after fertilization, while NO_3-N levels showed little response for the first year. NO_3-N production increased rapidly one year after fertilization, as has been observed in soil incubation[32], lysimeter[31] and watershed studies[39]. However, results on nitrate production following addition of urea are not consistent[32]. At Kielder the increase in nitrate production on plot 2F coincided with a similar increase on plot 2 which was of much shorter duration, and happened at the same time as a further release of NH_4-N. The same NO_3-N and NH_4-N response in the spring of the year was also observed on plots studied 0 and 5 years after felling[40].

The increased levels of NH_4-N and NO_3-N observed on plot 2F could have arisen from a combination of the added urea (via ureolysis and nitrification) and from organic N mineralized by increased decomposer activity in the litter layer. However, the reported effects of urea on micro-organisms and hence decomposition is contradictory. Respiration rates have been observed to both increase[35,34] and decrease[32] with the addition of urea. Similarly, the size of fungal and bacterial populations can increase[36] or decrease[2] after fertilization with this compound. Soil meso-faunal populations will vary mainly through the indirect effects of urea fertilizer on their food webs[21]. The reasons for changes in decomposer population sizes and activities are complex and could include changes in carbon availability, nutrient deficiencies or fertiliser-induced deficiencies[19]. The increase in pH due to ureolysis[13] may also cause changes in soil micro-organism community dynamics[22]. In forests, the overall microbiological shifts can lead to increased rates of litter[16] and forest floor[24] weight loss, as a result of changes to the organic chemistry of LFH horizons[29].

Orthophosphate losses following fertilisation at Kielder were large, and similar losses following fertilisation with rock phosphate have been well documented for forest sites[18] and peat soils[26].

Carbon mineralisation has been observed to decrease and mineral nitrogen accumulation to increase with the addition of phosphate[18] and the rate of forest floor weight loss to decrease[23]. However, combinations of urea and phosphate can increase forest floor weight loss above that of urea alone suggesting that P has an important role to play in the N cycle in some ecosystems[24]. Further, combined applications of P and K to peat can reduce leaching losses of both elements, due to increased vegetation uptake[24]. This mechanism of nutrient retention was not applicable to Kielder, however, due to the pronounced lack of vegetation on site.

Potassium levels in leachates decreased sharply immediately following fertilisation, with most (72%) being lost in the first year. Similar rapid leaching of K has been observed elsewhere[26], due to the high mobility of this ion[6].

Addition of urea has also been observed to increase leaching of potassium[9] and may have contributed to K losses at Kielder.

Calcium losses remained elevated on the fertilized plot, presumably as a result of calcium input to the site in the form of rock phosphate, and fluctuated seasonally with the PO_4-P losses. Calcium losses may have been less than for P because of its greater adsorption by organic matter.

The results for Na and Mg suggest that these elements were largely unaffected by additions of NPK fertilizer, while the pH increased due to the addition of rock phosphate and ureolysis[13].

The reaction of the forest soil system to combined NPK inputs is clearly complex and variable, especially as regards microbial dynamics[34]. However, it is the overall availability of nutrients in the soil solution, from whatever source, that is of importance to young planted seedlings. Results for Kielder indicate that addition of NPK at planting will increase availability of NO_3-N, NH_4-N, PO_4-P, K and Ca. However, the timing and duration of the increased availability will vary. Of the 150, 50 and 100 kg ha^{-1} of added NPK fertilizer, 67, 30 and 75 kg were recovered in lysimeters in the first two years as NH_4-N and NO_3-N, PO_4-P and K, respectively. Losses of organically bound nutrients and volatilization of urea were not included, suggesting that at a maximum only 55, 40 and 25 percent of added nutrients remained on site, in the planting position, after two years. While it may by feasible to fertilize with N and P, K losse were rapid and high, and thus this fertilizer should only be used judisciously.

Seedling growth, however, is dependent on a number of variables as well as nutrient availability. As part of experiment Falstone 7/81, seedlings were planted in clear strips and brash swathes of both plots in May 1981, at the time of lysimeter installation. After three summers' growth seedling heights were measured in the winter of 1983/84 (Forestry Commission, pers. com.), and had mean heights in the brash swathes of 78 cm and 83 cm in plots 2 and 2F respectively while the clear strip mean heights were 66 cm and 69 cm in plots 2 and 2F.

T-tests indicated that for both plots 2 and 2F, seedling heights in brash swathes were significantly greater than heights in the clear strips. However, for both the clear and brash treatments, heights in plot 2F were not significantly greater than those in plot 2, suggesting that for at least the first three years after planting, the retention of brash had a greater influence on seedling height than did the addition of fertilizer. Microclimatic differences may be more important to early seedling growth than greater availability of nutrients.

It may be that delaying fertilization until the microclimatic effects of the brash are less pronounced and seedling root systems are more developed would provide a more economic manipulation of the complex plant/soil interactions to increase forest productivity.

Litter weight loss

Generally, litterbags are either laid out at a series of collection stations and one bag randomly retrieved from each on every sampling occasion[4,12] or all bags are laid out randomly across the site and sampled randomly on each collection date[28]. The use of collection stations implies that litter weight loss is treated as a dependent variable, whereas for randomly laid bags litter weight loss is independent. The distinction is not important if the decomposition rate constant k is derived arithmetically from only initial and final weights:

$$k = - \frac{\ln \left(\frac{x_t}{x_0} \right)}{t} \qquad \text{(after Olson}^{30}\text{)}$$

In both experimental designs initial litter weight is recorded for the same bags at the start and end of the study. Each collection station then effectively consists of only one bag, as regards weight loss. This method of calculation of k allows for a measure of spatial variability in litter weight loss because a standard error of the mean k can be calculated[14]. However, temporal variability is not considered. The negative slope of the best-fitting line through percent ln litter weight remaining over time can also be used to calculate $\bar{k}$[30]. In this case, r^2 values will indicate the magnitude of variability.

When using best-fit regressions to derive $\bar{k}$, however, it is important to distinguish between litter weight loss as a dependent or independent variable. Where collection stations have not been used, and all variables are independent, the r^2 value is a measure of a combination of both spatial and temporal variation in weight loss. If a regression is fitted through only mean weight loss data (either dependent or independent) then any measure of spatial variability is lost. Fitting a regression line for every collection station will give a measure of temporal variation at each station (r^2), and calculating a mean decomposition rate constant ($\bar{k}$) will give a measure of spatial variation across the study site ($\pm$ SE).

While this does not affect the actual value of k[40], the way in which the data is handled will alter both the amount of information that is gained about decomposition rate constants in the field, and the way in which the data can be statistically analysed. The use of collection stations means that it is possible to characterise microsite decomposition rate differences across a site, as well as to describe the mean rate for the whole site.

At Kielder the r^2 values for individual collection stations on both plots were very high, regardless of the value of k (Table 5). This suggests that the temporal variation at each collection station was small, and that the microsite properties of each were constant over time, relative to each other.

The high r^2 values for plot 2F also indicate that the observed slower mean decomposition rate ($\bar{k} = 0.15$) with the addition of NPK fertiliser as compared with plot 2 ($\bar{k} = 0.28$) was a long term effect. Further, analysing the data on an annual basis yielded $\bar{k} = 0.15$ for the first and $\bar{k} = 0.14$ for the second year following fertilisation[40].

It might have been expected that fertiliser amendments would increase decomposition rates, as the initial immobilisation of N and occasionally P and S observed in other coniferous studies[38] suggests that shortages of these nutrients with respect to carbon can limit microbial activity and hance litter weight loss. However, results from fertilisation studies are often contradictory, with additions of N sometimes increasing and sometimes decreasing[37] soil microbial activity, depending on the form of the added nitrogen[2]. In another litterbag study in a western hemlock stand, k increased from 0.50 and 0.60 to 0.69 and 0.76 with the addition of urea-N[16]. In experiments where urea-N has increased microbial activity or decomposition rates, combinations of N with P and/or K increased rates of activity or respiration even further[24,34], as excessive addition of only one nutrient may induce shortages of others. The addition of P and K with N at Kielder should have ensured that none of these nutrients were limiting to microbial activity. However, N was still retained in the litter relative to weight loss on plot 2F (Fig. 5) suggesting that this nutrient might also be limiting microbial activity, even with the addition of urea.

It is unlikely that NH4-N toxicity as a result of ureolysis[2] or NH4-N suppression of the ligninolytic enzyme of the microflora[4] would have caused the observed decrease in decomposition rate in that $\bar{k}$ was very constant over the 2 year study period.

The addition of NPK may have caused available carbon to become limiting. The importance of available C in *Pinus banksiana* Lamb. litter layers has been demonstrated[15], as the addition of C as ethanol increased soil respiration five-fold, but C plus urea increased respiration 15-fold. A model of energy and nutrient regulation of decomposition[8], indicates that many of the observed results in nutrient addition/microbial activity/ decomposition studies may be explained in terms of limiting levels of available C as an energy source, as well as limiting levels of nutrients. The possibility that available C is a greater limitation to decomposition at Kielder than NPK levels in the litter, is consistent with the observation that $\bar{k}$ decreased with fertiliser addition. It is also consistent with the observed reduction in CO_2 evolution from Sitka spruce litter after addition of NPK in both laboratory and field incubations[41]. Although not quantified, it was noted that litterbags from plot 2F contained almost no fungal hyphae compared with those from plot 2. It is probable that increased NPK made available C limiting to heterotrophs, leading to a reduction in their activity and number, and hence a decrease in litter decomposition rate. The lowered level of heterotrophic activity, plus addition of excess nutrients, may then have led ultimately to increased activity of nitrifiers, no longer competing with heterotrophs for available nutrients, and not requiring an available carbon source for energy. The delay in nitrate release from urea fertilised forest soils[37] may be a function of other factors limiting nitrifier population growth at the time of fertilisation (temperature, moisture), especially if nitrifiers were constrained by heterotrophic activity to specific microsites within the litter layer.

With an available energy source (NH_4-N) and reduced heterotrophic competitions for nutrients, the nitrifier populations may have expanded at a time of the year (May) of natural population growth into niches from which they were previously excluded by heterotrophs, allowing nitrification to proceed at times of the year when temperature or moisture may have limited pre-fertilisation populations.

Litter nutrients

The dynamics of elements in decomposing litter is dependent on a combination of biotic and abiotic factors. Those not incorporated into organic molecules (*e.g.* Na, K) are readily leached out of coniferous litter[6]. Combined nutrients (*e.g.* N, P) must be released by microbial activity before being leached from decomposing litter. However, nutrients which are limiting for microbial activity are retained, and in the case of nitrogen, three phases (leaching, accumulation, final release) have been described[3]. The application of NPK fertiliser introduces a source of nutrients for microbial populations which might be naturally limiting.

Nitrogen concentrations in brash swathe litter at Kielder rose constantly over the two year study period, as has been observed elsewhere[12,14].

However, although the differences in concentration between plot 2 and plot 2F were not large (Fig. 5), the decreased rate of weight loss on plot 2F led to a fairly constant nitrogen content, while that on plot 2 decreased at a rate slower than weight loss. This suggests that nitrogen was limiting for microbial activity.

Phosphorus concentrations rose sharply with the addition of rock phosphate, but then decreased for 1.5 years before appearing to increase again. This contrasts with the results for the unfertilised treatment, where P concentrations rose steadily over the 2 year study period. Even though total P content on plot 2 decreased slowly, the increasing concentration suggests that P was limiting to microbial activity. This was not the case with plot 2F, where the excess added P was not retained relative to litter weight until 1.5 years after fertilisation. For the final 6 months, P may have become limiting again on the fertilised treatment.

The mobility of K is clearly demonstrated in Figure 6, with both the concentration and content of this nutrient decreasing rapidly in the first year after fertilisation. Thereafter, the K content of brash swathe litter in both treatments remained constant, while its concentration began to increase, suggesting that it may have become limiting. From leachate data for plot 2F (Fig. 2), the start of the K litter concentration increase coincided with a greatly reduced flux of this nutrient in leachate, suggesting that little fertiliser K then remained.

Calcium behaved quite differently from the other nutrients studied. Although the concentration rose immediately with the addition of fertiliser to plot 2F, the concentration then remained constant, as it also did on plot 2. The Ca content of the litter on both plots decreased at approximately the same rate as litter weight, suggesting that this nutrient was neither soluble, nor limiting. The calcium input apparently was bound to needles so that it required litter breakdown by decomposers for release.

Although Mg is mobile within the litter[6] it is clear (Fig. 5 iv) that some microbial activity is required for its release. Magnesium levels in leachate did not change appreciably with the addition of fertiliser (Table 3). If Mg was lost from litter by leaching action alone, then the Mg content of litter from plots 2 and 2F should have been the same. However, the reduced rate of weight loss on plot 2F was matched by a decrease in the rate of loss of magnesium content. As, however, the magnesium concentration decreased in both treatments, leaching clearly played a major role in its release from litter but the concentration differences between treatments indicate that microbial populations were also involved in the release of this element.

The overall results from this study indicate that the addition of NPK fertiliser led to a reduction in the rate of weight loss from needles in brash swathes, perhaps by inducing a reduction in available carbon. However, nutrient results also suggest that N was or became limiting on both plots, while the addition of rock phosphate alleviated any shortages of this nutrient for 1.5 years.

If these results are applicable generally the addition of fertilizer to the forest floor of Sitka spruce stands, either before or after clear felling, might result in a delay in its decomposition or loss. To the extent that regeneration, natural or planted, is dependent on the nutrient pool in decomposing litter its induced retention could be beneficial. On the site studied here, because of adverse physical conditions, newly planted trees rely on adventitious roots operating in a thin band of soil (mainly the LFH of the previous stand). The depth of litter is then important to survival and early growth. Nevertheless, as both the lysimetry and litter bag studies showed, the greater part of the added nutrients was leached from the rootable layer and probably lost to drainage water within two years with little effect on the performance of the planted trees.

Acknowledgements. The authors are grateful for financial support from the University of Edinburgh, the Overseas Research Students Fees Support Scheme, the George Drexler Ofrex Foundation and the Forestry Commission who also provided the experimental site.

References

1   Allen S E (Ed) 1974 Chemical Analysis of Ecological Materials. Blackwell Scientific Publications, Oxford
2   Bååth E, Lundgren B and Söderström 1981 Effects of nitrogen fertilization on the activity and biomass of fungi and bacteria in a podzolic soil. Zbl. Bakt. Hyg., I. Abt. Orig. C2, 90–98
3   Berg B and Staaf H 1981 Leaching, accumulation and release of nitrogen in decomposing forest litter. In Terrestrial Nitrogen Cycles. Eds. F E Clark and T Rosswall. Ecol. Bull. (Stockholm). 33, 163–178
4   Berg B, Wessen B and Ekbohm G 1982 Nitrogen level and decomposition in Scots pine needle litter. Oikos 38, 291–296
5   Berg B, Hannus K, Popoff T and Theander O 1982 Changes in organic chemical components of needle litter during decomposition: long-term decomposition in a Scots pine forest. I Can. J. Bot. 60, 1310–1319
6   Bogatyrev L, Berg B and Staaf H 1983 Leaching of plant nutrients and total phenolic substances from some foliage litters—a laboratory study. Swedish Coniferous Forest Project, Tech. Rep. 33
7   Boggie R and Knight A H 1980 Tracing water movement, using tritium, in a peaty gley soil under Sitka spruce. Forestry 53, 179–186
8   Bosatta E and Berendse F 1984 Energy of nutrient regulation of decomposition: Implications for the mineralization-immobilization response to perturbations. Soil Biol. Biochem. 16, 63–67
9   Cole D W and Gessel S P 1965 Movement of elements through a forest floor as influenced by tree removal and fertilizer additions. In Forest Soil Relationships in North America, Ed. C T Youngberg, pp 95–104. Proc. 2nd N. Am. For. Soil Conf. Oregon State Univ. Press, Corvallis.
10   Crooke W M and Simpson W E 1971 Determination of ammonium in Kjeldahl digests of crops by an automated procedure. J. Sci. Fd. Agric. 22, 9–10
11   Cuttle S P and Malcolm D C 1979 A corer for taking undisturbed peat samples. Plant and Soil 51, 297–300
12   De Catanzaro J B and Kimmins J P 1985 Changes in the weight and nutrient composition of litter fall in three forest ecosystem types on coastal British Columbia. Can. J. Bot. 63, 1046–1056
13   Derome J R M 1980 Urea hydrolysis and ammonia volatilization from the humus layer. Commun. Inst. For. Fenn. 98, 1–23
14   Edmonds R L 1984 Long-term decomposition and nutrient dynamics in Pacific silver fir needles in western Washington. Can. J. For. Res. 14, 395–400
15   Foster N W, Beauchamp E G and Corkey C T 1980 Microbial activity in a *Pinus banksiana* Lamb. forest floor amended with nitrogen and carbon. Can. J. Soil Sci. 60, 199–209
16   Gill R S and Lavender D P 1983 Litter decomposition in coastal hemlock stands: Impact of nitrogen fertilizers on decay rates. Can. J. For. Res. 13, 116–121

17   Haines B L, Waide J B and Todd R L 1982 Soil solution nutrient concentrations sampled with tension and zero-tension lysimeters: report of discrepancies. Soil Sci. Soc. Am. J. 46, 658–661

18   Haveraaen O and Steenberg K 1967 Nedvasking av naeringsstoffer i myrjord. Noen resultater ved bruk av radioaktive isotoper (Leaching of nutrients from peat soil. Some results by use of radioactive isotopes). Meld. Norg. LandbrHogsk. 46, 1–25

19   Hendrickson O, Robinson J B and Chatarpaul L 1982 The microbiology of forest soils—a literature review. DoE, C.F.S. Information Report PI-X-19

20   Henriksen A and Selmer-Olsen A R 1970 Automated methods for determining nitrite and nitrate in water and soil extracts. Analyst 95, 514–518

21   Hill S B, Metz L J and Farrier M H 1975 Soil mesofauna and silvicultural practices. In Forest Soils and Forest Land Management. B Bernier and C H Winget, pp 119–135, Proc. 4th N. Am. Forest Soils Conf. Les Presses de l'Universite Laval, Que.

22   Lehmann P-F 1975 Changes in the fungal succession after application of urea to litter of Scots pine (Pinus sylvestris L.) In Biodegradation et Humification. Eds. G Kilbertus, O Reisinger, A Mourey and J A Cancela de Fonseca, pp 470–476. Rapport du ler colloque International, Nancy.

23   Low A J (ed) 1985) Guide to upland restocking practice. Forestry Commission Leaflet 84, H.M.S.O., 30 pp

24   Mahendrappa M K 1978 Changes in the organic layers under a black spruce stand fertilized with urea and triple super phosphate. Can. J. For. Res. 8, 237–242

25   Malcolm D C 1979 Some effects of the first rotation on site properties. Irish Forestry 36, 76–88

26   Malcolm D C and Cuttle S P 1983 The application of fertilizers to drained peat 1. Nutrient losses in drainage. Forestry 56, 115–174

27   Murphy J and Riley J P 1962 A modified single solution method for the determination of phosphate in natural waters. Anal. Chim. Acta. 27, 31–36

28   McClaugherty C A, Pastor J, Aber J D 19895 Forest litter decomposition in relation to soil nitrogen dynamics and litter quality. Ecology 66, 226–275

29   Ogner G 1972 The composition of forest raw humus after fertilization with urea. Soil Sci. 113, 440–447

30   Olson J S 1963 Energy storage and the balance of producers and decomposers in biological systems. Ecology 44, 322–331

31   Overrein L N 1972 Sulfur pollution patterns observed: leaching of calcium in forest soil determined. Ambio 1, 145–147

32   Popovic B 1977 Effect of ammonium nitrate and urea fertilizers on nitrogen mineralisation, especially nitrification, in a forest soil. Depts. of Forest Ecology and Forest Soils, Stockholm, Research Note 30

33   Pyatt D G 1970 Soil groups of upland forests. Forestry Commission Forest Record 71, H.M.S.O., London

34   Rai B and Srivastava A K 1982 Microbial decomposition of leaf litter as influenced by fertilizers. Plant and Soil 66, 195–204

35   Roberge M R 1976 Respiration rates for determining the effects of urea on the soil-surface organic horizon of a black spruce stand. Can. J. Microbiol. 22, 1328–1335

36   Roberge M R and Knowles R 1967 The ureolytic microflora in a black spruce (Picea mariana Mill.) humus. Soil Sci. Soc. Am. Proc. 31, 76–79

37   Söderström B, Bååth E and Lundgren B 1983 Decrease in soil microbial activity and biomasses owing to nitrogen amendments. Can. J. Microbiol. 29, 1500–1506

38   Staaf H and Berg B 1982 Accumulation and release of plant nutrients in decomposing Scots pine needle litter. Long-term decomposition in a Scots pine forest II. Can. J. Bot. 60, 1561–1568

39   Tamm C O, Holmen H, Popovic B and Wiklander G 1974 Leaching of plant nutrients from soils as a consequence of forestry operations. Ambio 3, 211–221

40   Titus B D 1985 Nutrient dynamics of Sitka spruce stands after clearfelling on peaty gley soils. Unpubl. Ph.D. thesis, University of Edinburgh

41   Williams B L 1983 Nitrogen transformations and decomposition in litter and humus from beneath closed-canopy Sitka spruce. Forestry 56, 17–32

Plant and Soil 100, 323–331 (1987).   Ms. 100-23

The effect of spray volume on spray partitioning between plant and soil

G.B. SHAW, R.B. McKERCHER* and R. ASHFORD
*Department of Crop Science and Plant Ecology, and * Department of Soil Science, University of Saskatchewan, Saskatoon, Canada, S7N 0W0*

Key Words Chlorsulfuron Herbicide Metsulfuron methyl Postemergence applications Spectrofluorometry Spray partitioning Spray volume

Summary Applications of chlorsulfuron ($11.25 \, g \, ai \, ha^{-1}$) were made to wheat, flax, canola and lentils at spray volumes of 48, 108 and $217 \, l \, ha^{-1}$, and with metsulfuron methyl ($6.00 \, g \, ai \, ha^{-1}$) at spray volumes of 48 and $217 \, l \, ha^{-1}$. Applications were made to the shoot only, the soil only and to the plant plus soil. Spectrofluorometric analysis was used to determine spray partitioning within the plant-soil system and foliar retention was related to efficacy. Fresh and dry weights of shoot material were determined 3 weeks after treatment. Flax and wheat were more tolerant of restricted-foliar applications than those made to the soil, the converse being true to canola and lentils. Applications made to the plant and soil were always the most deleterious. Foliar retention and efficacy did not correlate directly. Applications in $217 \, l \, ha^{-1}$ were generally more efficacious than those at $48 \, l \, ha^{-1}$.

Introduction

Chlorsulfuron[11] (2-chloro-N-[(4-methoxy-6-methyl-1,3,5-triazin-2-yl) aminocarbonyl]benzenesulfonamide) and metsulfuron methyl[7] (methyl 2-[[[[(4-methoxy-6-methyl-1,3,5-triazin-2-yl)amino]carbonyl]amino]sulfonyl]benzoate) are the active ingredients in Du Pont's "Glean" and "Ally" herbicides, respectively. Each of these chemicals can be used both pre- and postemergence for the control of a wide spectrum of broad-leaved weeds in cereals[1]. Some reports show that foliar applications of chlorsulfuron are more effective than those made to the soil in the year of application[15]. Similarly, the contribution of soil deposited spray was less important than the foliarly applied spray in the supression of Canada thistle (*Circium arvense* L.) regrowth[8]. However, other species are more susceptible to the root-contacted spray. Flax (*Linum usitatissimum* L.), for example, is tolerant of foliar applications[10] and yet very sensitive to trace amounts of chlorsulfuron in the soil[17].

It is widely accepted that hydraulic spraying is inefficient as only a small proportion of the applied spray strikes the biological target. The trapping efficiency of a plant canopy depends on many features including the microclimate, the physical characteristics of the spray, and the orientation, shape and size of the target[3,20]. Changing spray volume, an application variable which can be manipulated to improve spraying efficiency, requires either an increase in pressure, or a change in nozzle tip or atomizer type. This causes a change in another application variable, the drop size spectra[6]. It seems that herbicides which translocate readily are as effective at low spray volumes as high, whilst contact herbicides, which require good coverage to be effective, are less adaptable[2,19]. This does not always hold true as some herbicides require specific placement on the plant in order to achieve maximum effect. When spray volume is manipulated so also is drop makeup, and this can have implications for the efficacy of a foliar applied spray[1,12,13].

Once a herbicidal spray with foliar activity reaches its target it has to be retained to be effective. Cuticular factors, including hairiness, surface topography and the physical and chemical form of the waxes, govern wettability[9], whilst the physical and chemical properties of the spray further determine its fate[3]. The spray volume chosen influences canopy penetration and the way in which the spray partitions itself between plant and soil[4,21]. The objective of this study was to investigate this effect in different plant canopies using a range of spray volumes.

Materials and methods

A greenhouse experiment was conducted using wheat (*Triticum aestivum* L.) cv. Neepawa, flax cv. McGregor, canola (*Brassica campestris* L.) cv. Tobin and lentils (*Lens culinaris* L.) cv. Laird cultured in 9-cm plastic pots containing a Chernozemic silty clay soil, of pH 7.6 and 4% organic matter content. Seeds were sown at a density of 5 per pot in the case of wheat and lentils, and 10 per pot for flax and canola. Plants were subsequently thinned to 3 (wheat, lentils and canola) or 5 (flax).

Plants were treated when: wheat was at the 2- to 3-leaf stage and at the 4- to 5-leaf stage; flax at a stretch height of 6 to 9-cm (10 to 14 true leaves); lentils at a stretch height of 10 to 12-cm (18 to 22 true leaves): and canola with the first pair of true leaves fully expanded.

There were five spraying events. Chlorsulfuron (75% DF) at a rate of $11.25\,g\,ai\,ha^{-1}$ and metsulfuron methyl (60% DF) at $6.00\,g\,ai\,ha^{-1}$ were each applied at spray volumes of 48 and $2171\,ha^{-1}$ using a bicycle-wheel type sprayer, operated at 290 Kpa and a forward speed of $6\,km\,h^{-1}$, with 800067 and 8003 Teejet nozzles for each of the spray volumes, respectively. Chlorsulfuron was also applied in $1081\,ha^{-1}$ using 80015 Teejet nozzles under the same operating conditions. In all cases the diluent was a solution of 0.1% w/v sodium fluorescein plus o.1% v/v nonylphenoxy polyethoxy ethanol 90% surfactant (Agral® 90) in tap water.

Twelve pots of plants from each of the five plant sets were sprayed in each event. Of these 12, 6 had the soil covered with a 1.5 cm deep layer of polystyrene beads. A further 3 pots had the plants carefully wrapped in a thin plastic film, such that the plants themselves were not damaged nor the canopy shape unduly changed. The remaining 3 pots were sprayed without shielding either the plant or the soil. After spraying the beads were removed from 3 pots, the plants in these represented "shoot only" treatments. The plastic wrap was also removed, and the plants in these pots represented "soil only" treatments. These, plus the "plant and soil" treatments were returned to the growth chamber along with a corresponding set of 3 pots of untreated control plants. The temperature was maintained at $21 \pm 1°C$ day and $18 \pm 1°C$ night, with a 16-h photoperiod of $1100\,\mu E\,m^{-2}\,s^{-1}$. Relative humidity was approximately 40%. Pots were arranged in a completely randomized design and rerandomized at weekly intervals. Soil was maintained at 80 to 100% field capacity moisture by weight by additions of water to the soil surface. The soil was supplemented with 20 ml of a 1% 20-20-20 fertilizer solution at weekly intervals. Plants were maintained for a period of 3 weeks after which time full descriptions were made of the injury symptoms and the shoots were harvested. Shoot fresh weights were determined immediately and dry weights after 48 h at 80°C.

The remaining 3 pots of plants from each plant set were used for spectrofluorometric analyses.

The method used was developed by Merritt[†]. The beads and the plant shoots from each pot were placed in separate polyethylene bags. 50-ml of 0.005 N NaOH aqueous solution was added to each bag and the samples shaken thoroughly to remove the dye. This was carried out within 20 min of each spraying event. The resultant washings were decanted and aliquots placed in the Turner Model 430 spectrofluorometer for fluorescence determination. The peak optima were set at 492 nm excitation and 510 nm fluorescence. Results obtained were as μl of fluorescein and allowed the shoot captured spray deposit to be expressed as μl g dwt^{-1} of plant material after drying the shoots at 80°C for 24 h. Fluorescence readings obtained for the polystyrene beads were expressed as μl cm^{-2} of soil area. Artificial targets, consisting of 9-cm filter paper discs randomly distributed throughout the target area, were also sprayed during each spraying event. These were similarly analyzed for fluorescence to allow the determination of the actual spray volume applied in l ha^{-1}. The spectrofluorometer was precalibrated using the relevant spray solution prior to each set of fluorescence determinations.

Results

Previous tests had established that the diluent was non-phytotoxic. The use of artificial targets showed that each application was uniform both across and along the length of its traverse. As spray volume increased so did retention by the plant shoot except for wheat at the 2- to 3-leaf stage (Table 1); in this case retention dropped when volume rate increased from 108 to 217 l ha^{-1}. Canola, with its relatively large, horizontally projected leaves retained the most spray liquid. Flax retained more than wheat, and retention by lentil shoots was low.

The data for shoot retention and for that portion of the spray which reached the soil under each plant canopy, when expressed as a percentage of that applied (Table 2), showed that retention by the shoot was highest at 108 l ha^{-1} for wheat at the 2- to 3-leaf stage, for flax and for lentils, with little difference in retention occurring between the 48 and 217 l ha^{-1} spray volumes. Canola and wheat at the 4- to 5-leaf stage retained least at 108 l ha^{-1} and canola retained most at 217 l ha^{-1}. The proportion reaching the soil was generally least at the low spray volumes as most of the spray intercepted by the plant was retained. Off-target

Table 1. The effect of spray volume on spray retention (μl g dwt^{-1}) by wheat, flax, canola and lentil shoots

Spray volume l ha^{-1}	Wheat		Flax	Canola	Lentils
	2- to 3-leaf	4- to 5-leaf			
48	39 a	51 a	51 a	169 a	29 a
108	133 b	55 a	152 b	266 b	88 b
217	113 b	119 b	231 c	569 c	117 c

a–c Numbers within columns followed by the same letter are not significantly different at 1% level (Duncan's New Multiple Range).

[†] Merritt C R AFRS Weed Research Organization, Oxford, England (Personal communication).

Table 2. The effect of spray volume on the partitioning of spray liquid within five plant "canopies"

Proportion	Spray volume $l\,ha^{-1}$	Proportion of spray recovered (% of applied)				
		Wheat		Flax	Canola	Lentils
		2- to 3-leaf	4- to 5-leaf			
Soil	48	73.4	81.4	79.2	87.6	78.7
	108	95.2	87.9	92.3	90.1	81.0
	217	83.8	83.6	90.3	83.5	82.0
Shoot	48	11.8	20.0	6.6	11.8	10.6
	108	17.9	15.3	9.4	10.4	17.1
	217	10.1	18.2	8.0	15.7	10.6
"Lost"	48	14.9	−1.4	14.3	0.7	10.7
	108	−13.0	−3.2	−0.3	−0.5	1.9
	217	6.2	−1.8	1.8	0.8	5.9

losses were least at the $108\,l\,ha^{-1}$ spray volume as the spray was coarse enough to ensure placement within the target area, and the drops were not so large and fast that they could rebound from the leaf surface.

Both herbicides were very efficacious, and despite the fact that the chemical rate used for metusulfuron methyl was lower than that for chlorsulfuron, the effects were not separable in terms of shoot dry weight (Table 3); however, there was a difference in phytotoxicity when shoot fresh weight was considered. With the exception of wheat at the 4- to 5-leaf stage, plants treated with chlorsulfuron had a lower fresh weight yield than those treated with metsulfuron methyl, suggesting that water utilization by the plant was impeded most by chlorsulfuron. The interaction between chemical and species (Table 3) showed that lentils and wheat at the 2- to 3-leaf stage were most sensitive to chlorsulfuron, whilst wheat at the 4- to 5-leaf stage was most sensitive to metsulfuron methyl. Flax and canola were generally equally sensitive to both chemicals.

The efficacy of chlorsulfuron did not vary significantly on changing spray volumes although phytotoxicity was lowest at $48\,l\,ha^{-1}$ (Table 3). Metsulfuron methyl was generally more efficacious at $217\,l\,ha^{-1}$ than $48\,l\,ha^{-1}$. Spray volume and species interacted such that the $217\,l\,ha^{-1}$ rate was most efficacious, resulting in the lowest shoot yields, except for chlorsulfuron when wheat was the test plant (Table 4).

Chemical placement caused each of the four species to respond differently (Tables 3 and 5). Applications made to both plant and soil generally proved most deleterious. Wheat and flax were more tolerant of shoot applications than soil applications, the converse being true of lentils and canola. Although wheat and flax were showing greatest sensitivity to the chemicals taken up through the root, the continued vigour of the plants probably would have allowed them to reach full

Table 3. The effect of spray volume, placement, and species on the efficacy of chlorsulfuron and metsulfuron methyl based on shoot[a] fresh and dry weights (g)

	Chlorsulfuron		Metsulfuron methyl		Combined data	
	Fresh wt	Dry wt	Fresh wt	Dry wt	Fresh wt	Dry wt
Spray volume	NS	NS	*	*	*	**
48 l ha^{-1}	3.03	0.72	3.00	0.72	3.00	0.72
108 l ha^{-1}	2.78	0.65	–	–	–	–
217 l ha^{-1}	2.84	0.67	2.70	0.65	2.80	0.66
Placement	**	**	**	**	**	**
Shoot	3.16	0.72	3.10	0.74	3.20	0.74
Soil	3.14	0.72	3.22	0.73	3.18	0.73
Both	2.36	0.60	2.25	0.66	2.33	0.60
Species X chemical interaction					**	*
Wheat 2- to 3-leaf	5.46	1.27	5.61	1.37	5.74	1.37
Wheat 4- to 5-leaf	5.94	1.44	5.00	1.24	5.42	1.32
Flax	1.05	0.22	1.14	0.23	1.08	0.22
Canola	0.72	0.11	0.82	0.12	0.77	0.12
Lentils	1.27	0.36	1.70	0.46	1.47	0.41

NS Not significant.
* Significant at 5% level.
** Significant at 1% level.
[a] Mean weights on a per pot basis where wheat, lentils and canola are at a density of 3 per pot, and flax at 5 per pot.

maturity with just a short maturity delay. In the case of wheat the injury symptoms were chlorosis of the younger leaves, particularly those directly contacted by the spray, and an overall stunting of plant height. Flax exhibited meristematic chlorosis and the very characteristic symptom of leaf reflexing such that the leaves were held upright, close to the stem.

Table 4. The effect of spray volume on the efficacy of chlorsulfuron and metsulfuron methyl to wheat, flax, canola and lentils[a]

Chemical	Spray volume l ha^{-1}	Wheat		Flax	Canola	Lentils
		2- to 3-	4- to 5-			
Chlorsulfuron	48	1.46 a	1.37 a	0.23 a	0.14 a	0.36 a
	108	1.07 b	1.50 b	0.23 a	0.10 a	0.36 a
	217	1.43 a	1.39 a	0.21 a	0.09 a	0.35 a
Metsulfuron methyl	48	1.53 a	1.27 a	0.24 a	0.15 a	0.47 b
	217	1.21 b	1.21 b	0.22 a	0.09 b	0.44 b

[a] Data expressed on a grams dry weight per pot basis where wheat, lentils and canola are at a density of 3 per pot, and flax at 5 per pot.
a–b Numbers within columns for each herbicide followed by the same letter are not significantly different at the 5% level.

Table 5. The effect of spray partitioning on the efficacy of chlorsulfuron and metsulfuron methyl to four plant species (dry weight[a] as % of control)

Placement	Species	Chlorsulfuron			Metsulfuron methyl		Mean	Mean[b] visual score
		$1\,ha^{-1}$						
		48	108	217	48	217		
Shoot	Wheat							
	2- to 3-leaf	95	66	95	91	80	85	7.6
	4- to 5-leaf	110	104	120	104	92	106	8.1
	Flax	71	59	52	71	65	64	7.7
	Canola	7	9	5	11	10	9	0.0
	Lentils	57	52	50	82	57	60	1.9
Soil	Wheat							
	2- to 3-leaf	76	66	82	93	57	75	5.7
	4- to 5-leaf	92	104	99	99	88	97	5.9
	Flax	48	59	52	50	50	52	5.9
	Canola	47	28	25	23	43	33	3.6
	Lentils	66	77	72	89	92	79	4.5
Plant + Soil	Wheat							
	2- to 3-leaf	79	51	66	78	72	69	4.6
	4- to 5-leaf	97	109	75	66	76	85	4.8
	Flax	46	52	49	51	43	48	5.1
	Canola	8	8	11	6	13	9	0.0
	Lentils	46	43	45	52	60	49	0.2
Significant Levels								
Placement			**			**		
Spray volume × placement			NS			NS		
Species × placement			**			*		

Ns Not significant.
  * Significant at 1% level.
** Siginficant at 5% level.
[a] On a per pot basis where wheat, lentils and canola are at a density of 3 per pot, flax at 5 per pot.
[b] Based on combined data for both herbicides and all spray volume, and using a scale of 0 to 9 where 9 is complete tolerance and 0 complete kill.

Lentils and canola, although least sensitive of root uptake, still showed severe injury symptoms. In lentils these consisted of reduction in plant height, leaves became flaccid and took on a blistered appearance and leaf chlorosis and abscission followed soon after. Canola injury symptoms were visibly dramatic. Leaf anthocyanin content rapidly increased, manifesting itself first in the cotyledons and leaf mid-veins and then spreading to the whole plant, giving it a red appearance. The plants also became brittle. Neither lentils nor canola were tolerant enough to withstand contact with these sulfonylurea herbicides.

Discussion

The orientation, shape and size of a target will influence the amount of spray liquid retained by it. Therefore, it was not surprising that canola, given its morphological characteristics, retained the most spray, regardless of the spray volume. It appeared that wheat retained different amounts depending on both its growth stage and the spray volume used. Young leaves are generally more wettable than older ones[5] and this could account for the greater wettability of the 2- to 3-leaf stage wheat at the $108 \, l \, ha^{-1}$ spray volume. Spray retention by any kind of surface will increase up to a point just before runoff then retention drops. The larger the target area the greater the amount that can be retained before this happens. This phenomenom may account for the decrease in retention by wheat, again at the 2- to 3-leaf stage, when spray volume increased from 108 to $217 \, l \, ha^{-1}$. Runoff was not apparent for any of the other plants.

Retention was not directly correlated with efficacy because of the importance of chemical placement within the plant-soil system. However, a greater phytotoxic effect was generally observed at $217 \, l \, ha^{-1}$. In terms of the efficacy of the restricted-foliar applications, lentils retained relatively small amounts but injury symptoms were very pronounced, suggesting a strong innate susceptibility to these herbicides. Although canola retained the most spray and was also highly sensitive to shoot uptake, the response observed was considered far too pronounced to be accounted for by this factor alone. the tolerance of different plant species to the sulfonylurea herbicides is based on their differential ability to metabolize them to non-phytotoxic compounds[10,18]. Earlier work[14] has shown that after treatment of the root system with chlorsulfuron, uptake was good in a range of both tolerant and susceptible species, but the uptake was higher in the susceptible species. Foliar treatments, however, showed a slower export rate to other plant parts. These differences in transport may account for the differences in tolerance levels according to chemical placement.

Wheat and flax both have the ability to metabolize these compounds in the foliage. Uptake via the shoot was, therefore, not causing a seriously debilitating effect. Chemical taken up through the root and lack of metabolism in the root caused greater symptoms of injury to be observed. In lentils and canola the converse was seen, these species being more sensitive to shoot then root uptake. The severe injury caused by these herbicides, regardless of placement, shows that lentils or canola are not well adapted for detoxification of these compounds. Injury to lentils and canola was more severe after shoot than root uptake as the chemicals likely remained in the leaf tissue for a relatively long period, whereas chemical taken up through the root was likely to be rapidly transported throughout the plant, this dilution effect causing a reduction in the severity of the injury.

References

1   Ambach R M and Ashford R 1982 Effects of variations in drop makeup on the phytotoxicity of glyphosate. Weed Sci. 30, 221–224.

2   Ayres P and Cussans G W 1980 The influence of volume rate, nozzle size and forward speed on the activity of three herbicides for the control of weeds in winter cereals. *In* Spraying Systems for 1980's. Ed. J O Walker. pp 57–64. British Crop Protection Council Monograph, 24.

3   Blackman G R, Bruce R S and Holly K 1958 Studies in the principles of phytotoxicity. V. Interrelationships between specific differences in spray retention and selective toxicity. J. Expt. Bot. 9, 175–205.

4   Bryant J E and Courshee R J 1984 The effect of volume of application from hydraulic nozzles on the partitioning of a pesticide spray in a cereal canopy. *In* Application and Biology. Ed. E S E Southcombe, pp 201–210. British Crop Protection Council Monograph, 28.

5   Bukovac M J, Flore J A and Baker E A 1979 Peach leaf surfaces: Changes in wettability, retention, cuticular permeability, and epicuticular wax chemistry during expansion with special reference to spray application. J. Am. Soc. Hort. Sci. 104, 611–617.

6   Combellack J H and Matthews G A 1981 The influence of atomizer, pressure and formulation on the droplet spectra produced by high volume sprayers. Weed Res. 21, 77–86.

7   Doig R I, Carraro G A and McKinley N D 1983 DPX T6376 — A new broad spectrum cereal herbicide. 10th Intern. congr. Plant Prot., Brighton.

8   Hall J C, Bestman H D, Devine M D and Vanden Born W H 1985 Contribution of soil spray deposit from postemergence herbicide applications to control of Canada Thistle (*Circium arvense*). Weed Sci. 33, 836–839.

9   Holloway P J 1970 Surface factors affecting the wetting of leaves. Pesticide Sci. 1, 156–163.

10 Hutchinson J M, Shapiro R and Sweetser P B 1984 Metabolism of chlorsulfuron by tolerant broadleaves. Pesticide Biochem. Physiol. 22, 243–247.

11 Levitt G, Ploeg H L, Weigel Jr and Fitzgerald D J 1981 2 chloro-N[(4-methoxy-6-methyl-1,3,5-triazin-2-yl)aminocarbonyl] benzenesulfonamide, a new herbicide. J. Agric. Food Chem. 29, 418.

12 Merritt C R 1982 The influence of form of deposit on the phytotoxicity of difenzoquat applied as individual drops to *Avena fatua*. Ann. Appl. Biol. 101, 517–525.

13 Merritt C R 1982 The influence of form of deposit on the phytotoxicity of MCPA, paraquat and glyphosate applied as individual drops. Ann. Appl. Biol. 101, 527–532.

14 Muller F, Kang B H and Maruska F T 1984 Fate of chlorsulfuron in cultivated plants and weeds and reasons for selectivity. Med. Fac. Landbouww. Rijksuniv, Gent, 49(36), 1091–1108.

15 O'Sullivan P A 1982 Response of various broadleaved weeds, and tolerance of cereals, to soil and foliar applications of DPX 4189. Can. J. Plant Sci. 62, 715–724.

16 Palm H L, Riggleman J D and Allison D A 1980 Worldwide review of the new cereal herbicide — DPX 4189. pp 1–6. Proc. British Crop Protection Conf. — Weeds.

17 Peterson M A and Arnold W E 1986 Response of rotational crops to soil residues of chlorsulfuron. Weed Sci. 34, 131–136.

18 Sweetser P B, Schow G S and Hutchinson J M 1982 Metabolism of chlorsulfuron by plants: Biological basis for selectivity of a new herbicide for cereals. Pesticide Biochem. Physiol. 17, 18–23.

19 Taylor W A and Merritt C R 1974 Preliminary field trials with 2,4-D ester, barban and triallate applied in spray volumes of 5–20 L/ha. Weed Res. 14, 245–250.

20   Taylor W A and Shaw G B 1983 The effect of drop speed, size and surfactant on the deposition of spray on barley and radish or mustard. Pesticide Sci. 14, 659–665.
21   Western N M, Hislop E C, Herrington P J and Woodley S A 1984 Relationships of hydraulic nozzle and spinning disc spray characteristics to retention and distribution in cereals. *In* Application and Biology Ed. E S E Southcombe. pp 191–199. British Crop Protection Council Monograph, 28.

Plant and Soil 100, 333–343 (1987).
Ms. 100-24

Comparative study of soil bacterial flora as influenced by the application of a pesticide, pentachlorophenol (PCP)

KYO SATO, H. KATO* and C. FURUSAKA**
Institute for Agricultural Research, Tohoku University, Sendai, 980 Japan

Key words PCP-tolerance Pentachlorophenol Soil bacterial flora Soil suspension Water-logged soil

Summary The effects of pentachlorophenol (PCP) applications on the taxonomic composition of bacterial microflora were studied in water-logged soil (WS) and in shake cultures of suspended soil (SS). PCP applications resulted in a predominancy of Gram-negative bacteria over Gram-positive species. Members of the *Acinetobacter* group were the most common in PCP-treated soil although a small portion of the flora were in the *Pseudomonas-Alcaligenes* group or belonged to the *Enterobacteriaceae*. Coryneform bacteria and species of the *Bacillus* were the dominant forms in untreated WS; however, WS cultures treated with PCP at recommended rates ($2.67 \, gm/m^2$) evidenced species of *Pseudomonas, Alcaligenes, Acinetobacter*, and members of the *Enterobacteriaceae* as the predominant bacterial species. The dominance of Gram-negative bacteria in PCP-treated soil was evidenced for 3 months after application of the compound but was not evident after 17 months when PCP had dissipated. Gram-negative bacteria found in PCP-treated soil were highly tolerant of the phenol. In WS cultures coryneform bacteria were the most common although PCP tolerance was heterogenous in nature.

Introduction

Many reports on the relation between pesticides and soil microorganisms are focused on the degradation of pesticides in unsterilized soils[1,19], or by microorganisms isolated from solids[19,20]. In many cases, studies on the effects of pesticides on soil microorganisms have been performed by measuring microbial activities such as CO_2 evolution, ammonification, nitrification, *etc.*[2,15,17] Studies on the effects of pesticides on quantitative changes in microbial composition of soil are limited and typically report changes observed at a single sampling time after addition of pesticides to the soil[3,18]; studies on the succession of quantitative changes in microbial species are even more limited[7,11]. Also, although qualitative changes in broad microbial groups (bacteria, actinomycetes, fungi) in response to applications of pesticides have been conducted[3,18], there are very few studies reporting on changes in bacterial taxonomic composition at the genus level.

Present addresses: * Aichi complex experimental farm of agriculture, Nagakute, Aichi-prefecture, and ** Emeritus professor; Tsurugaya, Sendai, Japan, respectively.

In a previous paper[14], Sato reported that bacterial flora was changed in percolated soils by the addition of pentachlorophenol (PCP), and that the pattern of the bacterial flora was also modified by soil conditions such as the addition of nutrient.

The present report presents results of a study on the effects of PCP on bacterial microflora as affected by different soil conditions.

Materials and methods

Soils

Soil for suspension cultures (SS) was an alluvial garden soil, pH 6.4 (water) with an organic matter content of 6.6% (w/w; by ignition loss). The soil was air-dried, sieved (pass 2-3 mm), and stored at room temperature until used.

Soil for water-logged cultures (WS) was collected from the plow layer of a paddy field and had a pH 5.11, total C 3.62% (w/w), and a clay content of 35% (w/w). The soil was stored in a polyethylene bag at room temperature without any treatment until used.

Soil suspension

One gram of the garden soil was added to each of 100 ml volumes of aqueous solutions containing 0, 40, or 100 μg PCP/ml. The mixtures were shaken (40 rpm) and kept in the dark at 30°C.

Water-logged soil

The soil collected from paddy field was submerged in a concrete pot (60 × 60 cm × 30 cm deep) and depth of surface water was kept at 5 cm. Aqueous solution containing 40 ppm PCP was sprayed on the surface water at rates of 0, 2.76 (RR), and 267 (100RR) g/m^2. The pots were placed in a greenhouse.

Determination of bacterial populations

Ten ml of the suspended soil was sampled at appropriate intervals, and divided into halves. One-half was used for counting and the other half, for analysis of PCP. A plastic tube (10 cm diam × 20 cm long) was inserted carefully into the pot to collect a sample of submerged soil. The surface water inside the tube was pipetted out and after scraping off the brown red surface soil layer (0-3 mm) with a spatula, a soil sample was collected to a depth of 5 cm. Portions of the sample were used to determine PCP content.

Numbers of aerobic bacteria in soil were determined with the dilution agar plate method, and those of anaerobic bacteria with the roll tube technique[4] in which air in the culture tube was displaced with nitrogen gas made oxygen-free by passage through a heated copper column. Total viable bacteria and Gram-negative bacteria were counted on an albumin medium and the albumin medium containing 0.0005% crystal violet, respectively. The albumin medium consisted of (g/l): egg albumin (Wako Pure Chemical Co. Ltd.), 0.25; glucose, 1.0; K_2HPO_4, 0.5; $MgSO_4 \cdot 7H_2O$, 0.2; $Fe_2(SO_4)_3$, trace; yeast extract (Difco), 0.05; and agar, 1. The pH of the medium was adjusted at 7.2. The appropriate dilutions of soil samples were also heat-treated at 80°C for 10 min and inoculated on the albumin medium to count bacterial spores. For anaerobic cultures 0.3 g cystein hydrochloride was added as a reductant to the albumin medium. The anaerobicity of the medium was checked by the reduction of resazurin which had been added to the medium. The inoculated agar plates and roll tubes were incubated at 30°C for 14 days when colonies were counted.

Isolation of bacteria

Isolations were performed randomly from colonies on agar media used for determination of total viable bacteria. The albumin medium was used for isolation and the isolates were grown and maintained on slants of the albumin medium.

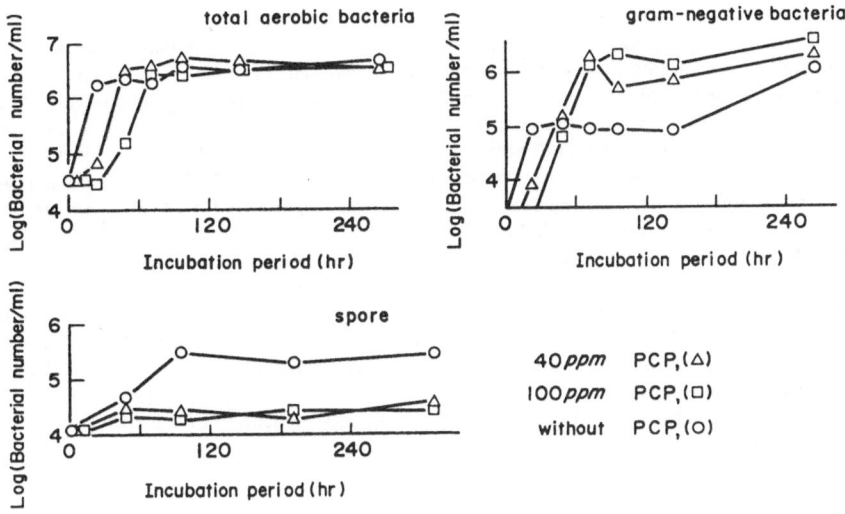

Fig. 1. Changes in numbers of bacteria in soil suspensions. Soil suspensions were cultured with 100 ppm, 40 ppm PCP and without PCP, respectively.

Phenotypic tests of the isolates

Gram-staining, pigment-production, catalase, cytochrome-oxidase, and oxidative or fermentative character of the isolates were tested by the standard method.[5]

PCP-tolerance of the isolated strains

Cells precultured on the albumin medium at 30°C for 72 hours were inoculated into a 10 ml of liquid albumin medium (no agar) containing different concentrations of PCP. After incubating the inoculated medium at 30°C for 7 days, growth was checked as evidenced by turbidity of the cultures.

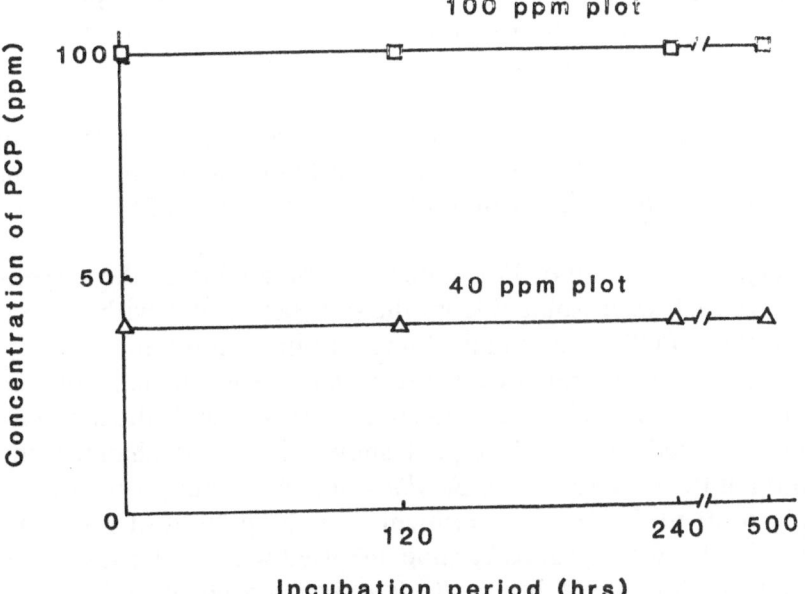

Fig. 2. Persistence of PCP in soil suspensions with 40 or 100 ppm PCP.

Analyses of PCP

PCP in the soil suspensions and the water-logged soils was extracted with benzene by Kuwat-suka's method[9]. Large amounts of PCP (> 20 ppm) were analyzed colorimetrically by absorbance at 575 nm (Hitachi Spectronic 20). A five-ml sample was mixed with 50 ml methyl alcohol and after shaking (40 rpm) for 30 min the mixture was filtered through a filter paper (Toyo Roshi No 1). Five ml of the filtrate was diluted in 95 ml of aqueous 0.0025% (w/v) Na_2CO_3. Ten ml of this dilution was shaken together with 2 ml of 0.2% 4-aminoantipyrin for 1 min and 1 ml of 10% potassium ferricyanide was added to the mixture. The mixture was stirred vigorously for 10 sec. After stirring (within 1 min), the mixture was again shaken with the addition of 10 ml xylene, and was allowed to stand for 15 min. The optical density of xylene layer was then determined colorimetrically.

For small amounts of PCP (< 10 ppm), gas-liquid chromatography, glc, was performed with a Shimadzu GC-4BM equipped with (^{63}Ni) EC detector. Separation was achieved on a 2 m $\times$ 3 mm i.d. glass column containing DEGS + H_3PO_4 (2 + 0.5%) coated on Chromosorb W AW DMCS 80-100 mesh. The operating condition for glc were: N_2 carrier gas 180 ml/min, injection port 190°C, column 190°C and detector 230°C.

PCP powder (phenol free, Merck) was used for both soil-application and for test for PCP-tolerance of bacterial isolates.

Results

Suspended soil

Bacterial populations. PCP retarded the initial increase in the counts of total viable bacteria and Gram-negative ones. The larger the amount of PCP applied, the longer the retardation; however, total number of bacteria reached the same level after 72 hours irrespective of the presence (40 ppm or 100 ppm) or the absence of PCP. The maximum number of Gram-negative bacteria depended on the amount of PCP applied: the number was proportionate to concentration of PCP: numbers of spore-forming bacteria were depressed by PCP (Fig. 1).

Amount of PCP. PCP was not degraded to any appreciable extent throughout the experimental period for 500 hours in soil suspensions with either 40 ppm or with 100 ppm of the chemical (Fig. 2).

Bacterial flora. A total of 175 strains were isolated from the suspended soils; 35 s each were isolated from the soil suspensions with 40 ppm PCP and without PCP after 24 and 72 hours of incubation, respectively. At the starting time 35 strains were also isolated from the control.

Bacterial isolates were classified into 10 groups following the diagnostic scheme depicted in Fig. 3. Fig. 4 shows that PCP changed the composition of the bacterial flora; *Bacillus* spp. and Gram-positive cocci predominated at initiation of experiment, but the pattern of predominancy by a few bacterial groups became distorted with the lapse of time in the control plot; at 72 hours *Bacillus* spp, coryneform bacteria, *Pseudomonas* spp, *Xanthomonas* spp and members of the *Enterobac-*

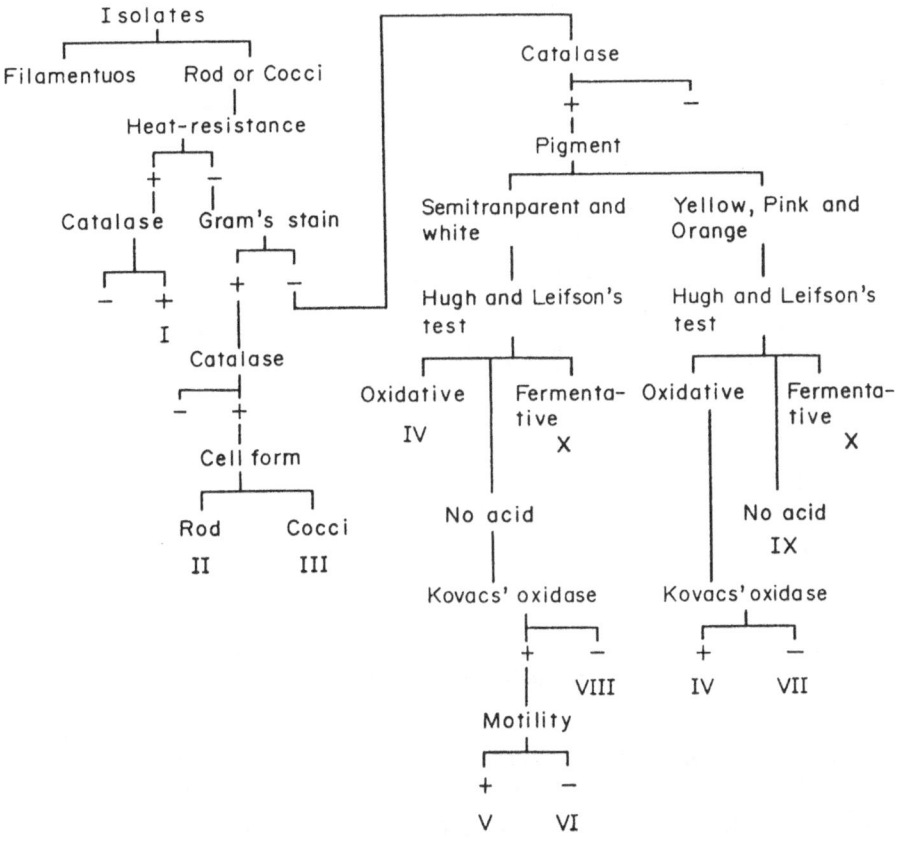

Fig. 3. Diagnostic scheme for grouping bacterial isolates. Group numbers and the corresponding genera: I, *Bacillus*; II, Coryneform bacteria; III, Gram-positive cocci (*Micrococcus, Streptococcus* and *Staphylococcus*); IV, *Pseudomonas*; V, *Pseudomonas-Alcaligenes*; VI, *Moraxella*; VII, *Xanthomonas*; VIII, *Acinetobacter*; IX, *Flavobacterium*; X, *Enterobacteriaceae*.

teriacea were evenly distributed in numbers. In contrast, Gram-negative bacteria such as species of *Acinetobacter* and *Pseudomonas-Alcaligenes* predominated in soil with 40 ppm PCP; these species tolerated high concentration of PCP.

Water-logged soil

Bacterial populations. Although the initial increase in counts of aerobic bacteria retarded in the RR Plot, the count of the RR Plot exceeded slightly that in the control plot for the first 3 months of experiment; however, there was no difference in the counts between the control plot and the RR Plot after 3 months. On the other hand, PCP-application at 100 RR depressed bacterial numbers throughout the experimental period (Fig. 5).

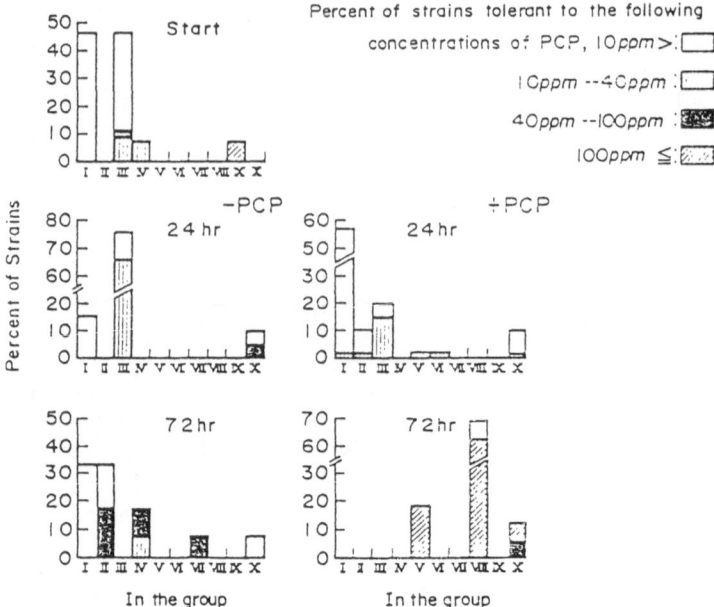

Fig. 4. Bacterial flora in soil suspensions and PCP-tolerance of the bacterial strains. Numbers of groups correspond to those of the diagnostic scheme in Fig. 3.

Changes in numbers of anaerobic bacteria differed from the pattern described for aerobic species. Although population levels of anaerobic bacteria did not differ between plots early in the experiment, the populations were depressed later proportionately to the concentration of PCP (Fig. 5).

Amount of PCP. Concentration of PCP reached the maximum level in both plots one month after application to surface water. PCP decreased rapidly after one month and dissipated after 3 months in the RR Plot; PCP remained in comparatively high concentrations (about 50 ppm) for a long period after an initial decrease in the 100 RR Plot (Fig. 6).

Bacterial flora. A total of 278 strains were isolated from WS: Fifty two isolates were obtained from the control plot at initiation of the experiment. After 1 month, 32 and 71 isolates were recovered from the control plot and the RR Plot, respectively. After 3 months, 41, 56 and 42 isolates, and after 17 months, 25, 24 and 35 isolates were gained from the control plot, the RR Plot and the 100 RR Plot, respectively.

The total number of isolates were classified into 10 groups of bacteria following the scheme depicted in Fig. 3. The composition of the bacterial

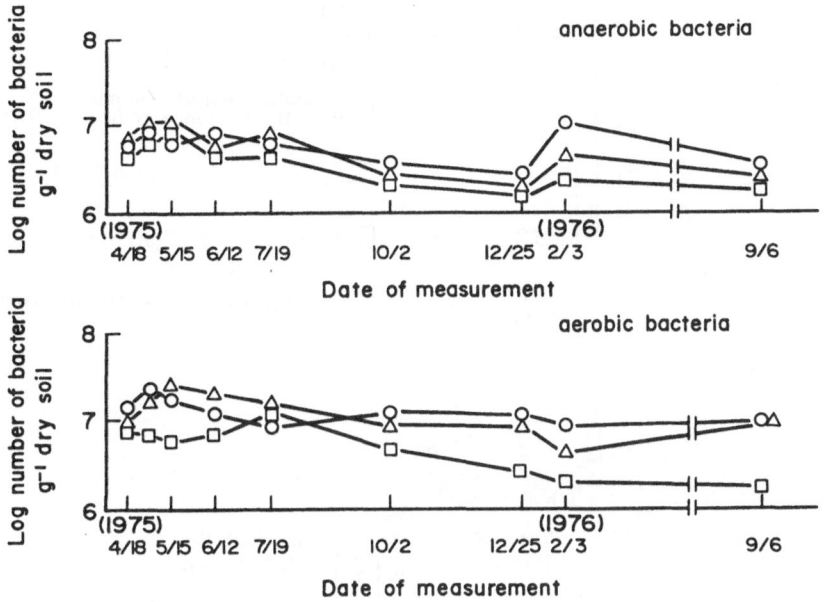

Fig. 5. Changes in numbers of bacteria in water-logged soils. Control (O), Plot RR (△) and Plot 100RR (□).

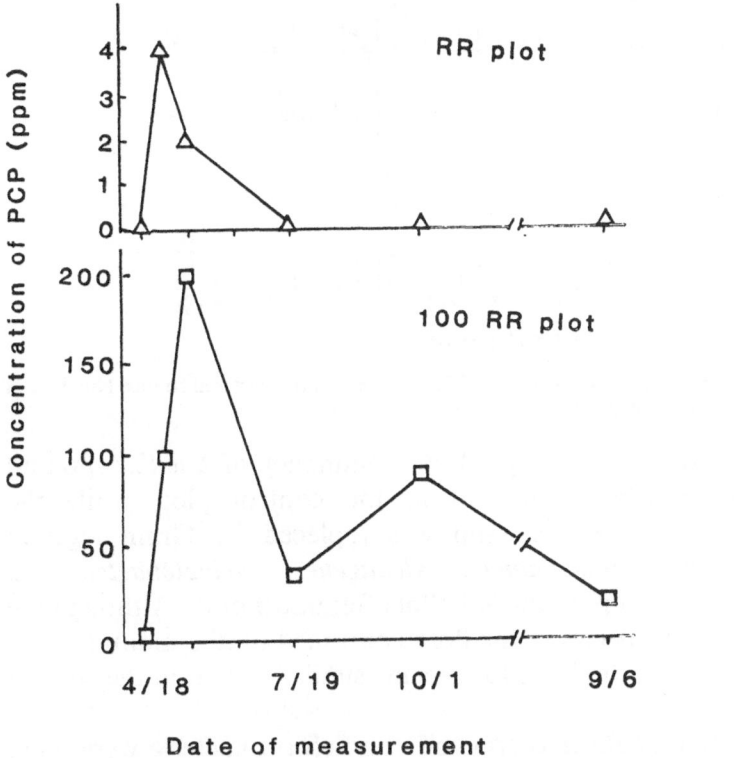

Fig. 6. Changes in concentration of PCP in upper part of reductive soil layer.

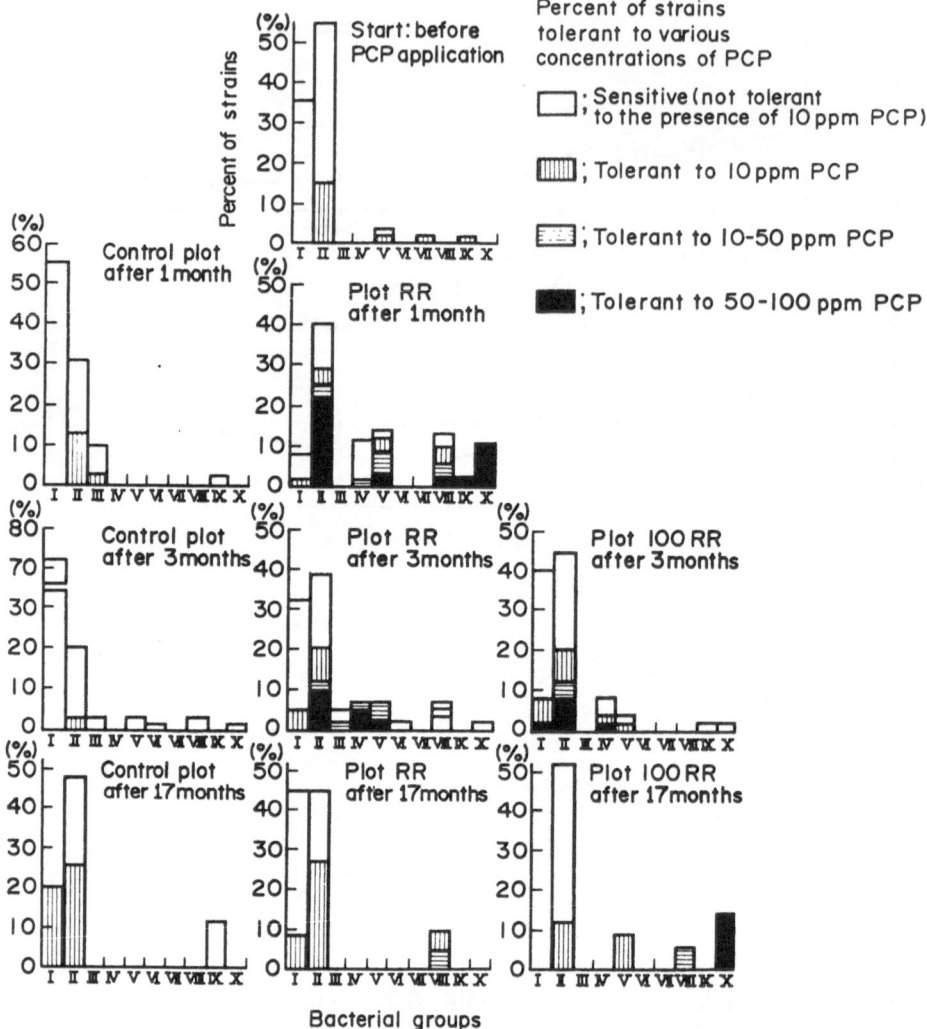

Fig. 7. Bacterial flora in water-logged soils and PCP-tolerance of the bacterial strains. Numbers of groups correspond to those of the diagnostic scheme in Fig. 3.

flora differed between plots (Fig. 7). Predominancy of *Bacillus* spp and of coryneform bacteria continued in the control plot, while the predominancy of the *Bacillus* spp was replaced by Gram-negative bacteria such as *Pseudomonas, Alcaligenes, Acinetobacter* and *Enterobacteriaceae* groups in the RR Plot after one month. Although the diversity in bacterial species in the flora continued until 3 months, after 17 months the flora in the RR Plot became similar to that in the control plot.

In the 100 RR Plot *Bacillus* spp and coryneform bacteria were most common after 3 months, and the predominancy of *Bacillus* spp was replaced by an even distribution of some species of Gram-negative bacteria after 17 months; however, numbers of the bacterial species were fewer than those in the RR Plot.

Many strains in the RR Plot tolerated high concentrations of PCP after one and 3 months, but the high tolerance disappeared after 17 months in the RR Plot. The strains in the 100RR Plot were not endowed with high PCP-tolerance throughout the experimental period compared with those in the RR Plot (Fig. 7).

Discussion

Changes in counts of bacterial colonies in the soil suspension were similar to those observed in the percolated soil[13]. A marked increase in Gram-negative bacteria and depression of spore counts by PCP were phenomena observed in the percolated soils. The percolated soil was incubated aerobically by passing compressed air through it. Thus, the pattern of change in bacterial types observed in the present work may be descriptive of changes in microbial populations following the addition of PCP for aerobic soil conditions.

Microbial populations increase sharply as a result of soil fumigation[12]. This phenomenon is known as "the partial sterilization effect": the fumigant kills portions of the microbial community in soil, and the remaining microorganisms proliferate by utilizing the killed cells. An analogous effect was observed for the Gram-negative bacteria in PCP-treated soil suspensions. However, in water-logged soils the change in microbial population due to PCP application did not follow "the partial sterilization effect". In water-logged soils other explanation must be operative.

Watanabe reported that the MPN of PCP-decomposers in soil was affected by the concentration of PCP in the testing medium[21]: at a concentration of 40 ppm, no PCP was degraded after 4 weeks at 30°C, even in tubes inoculated with a 10^{-1} soil dilution. Although population changes in PCP-decomposing species were not followed in the present study, it is assumed that the decomposers could not operate or proliferate under high concentrations (40 ppm and 100 ppm) of the compound. On the other hand, PCP has been shown to be much less persistent in an upland soil[21]; in such soil more than half of the initial amount applied at 20 kg ha^{-1} disappeared after 2 weeks. PCP is hardly very little ionized in water. Therefore, PCP may be present in insoluble form and distribute heterogeneously in soil. The decomposition of PCP observed in soil applied at 20 kg ha^{-1} was presumed to be due to localization of small amounts brought about by heterogenous distributon of PCP in field soil. If some portion of the decrease in the amount of PCP is attributed to degradation in the 100RR Plot, the degradation may be explained from the same line as mentioned above.

Identification of the isolated bacterial species confirmed the observed stimulation of Gram-negative bacteria by PCP in the suspended soil. Replacement of the *Bacillus* spp by species of Gram-negative bacteria correlates with the observed depression in count of spore-forming bacterial species. Sato[13,14] and Ishizawa et al.[8] reported that PCP stimulated counts of Gram-negative bacteria in soil, and isolated species of the genus *Pseudomonas* as the predominant bacteria. Sato also found that a few species of the genus *Pseudomonas* isolated from percolated soil treated with PCP had high tolerance to PCP (unpublished result). The *Acinetobacter* and *Enterobacteriaceae* species as well as the *Pseudomonas-Alcaligenes* species predominated in the suspended soil following PCP-application, and tolerated high concentration of the compound. *Flavobacterium* spp isolated at the beginning of the experiment also tolerated high PCP concentrations. Coryneform bacteria continued to predominate in water-logged soils irrespective of the presence or the absence of PCP; however, the PCP-tolerance of this group of bacteria varied. The species grouped as coryneform bacteria might include some genera of Gram-positive bacteria. Thus, it is possible that many genera of the Gram-negative bacteria as well as few genera of the Gram-positive bacteria are endowed with PCP-tolerance.

Species of *Pseudomonas* are ubiquitous in aerobic soils such as upland soil, especially in the rhizosphere where nutrients may be supplied through excretion from plant roots[6]. Coryneform bacteria were isolated as a predominant aerobic bacteria from a paddy rice soil[16]. Therefore, the genus *Pseudomonas* and coryneform bacteria are presumed to be predominant aerobic bacteria in aerobic soil suspension and water-logged soil, respectively. It was also reported that PCP-application stimulated PCP-tolerant bacteria[21]. Thus, although PCP modifies soil bacteria flora so that it may tolerate PCP, the pattern of the modification differs depending on soil condition.

References

1   Audus L J 1952 The decomposition of 2,4-dichlorophenoxyacetic acid and 2-methyl-4-chlorophenoxyacetic acid in the soil. J. Sci. Fd Agric. 3, 268–274.

2   Balicka N and Sobieszczanski J 1969 The effect of herbicides on soil microflora. 3. The effect of herbicides on ammonification and nitrification in soil. Acta Microbiol. pol. 1, 7–10.

3   Gunner H B, Zuckerman B M, Walker R W, Miller C W, Deubert K H and Lonley R E 1966 The distribution and persistence of diazinon applied to plant and soil and its influence on rhizosphere and soil microflora. Plant and Soil 25, 249–264.

4   Harrigan W F and MacCance E Margaret 1966 Laboratory Methods in Microbiology. Academic Press, London and New York pp 23–24.

5 Harrigan W F and MacCance E Margaret 1966 Laboratory Methods in Microbiology. Academic Press, London and New York pp 51–68.
6 Holding A J 1960 The properties and classification of the predominant Gram-negative bacteria occurring in soil. J. Appl. Bacteriol. 23, 515–525.
7 Houseworth L D and Tweedy G B 1973 Effect of atrazine in combination with captan or thiram upon fungal and bacterial populations in the soil. Plant and Soil. 38, 493–500.
8 Ishizawa S and Matsuguchi T 1966 Effects of pesticides and herbicides upon microorganisms in soil and water under water-logged condition. Bull. Nat. Inst. Agric. Sci. Series B (Soils and Fertilizers). No. 16, 1–10.
9 Kuwatsuka S and Igarashi M 1975 Degradation of PCP in soils. II. The relationship between the degradation of PCP and the properties of soils, and the identification of degradation products of PCP. Soil Sci. Plant Nutr. 21, 405–414.
10 Loos M A, Roberts R N and Alexander M 1967 Phenols as intermediates in the decomposition of phenoxyacetate by an Arthrobacter species. Can. J. Microbiol. 13, 679–690.
11 Matsuguchi T and Ishizawa S 1968 Effect of agrochemicals on microorganisms and their activity in soil (Part 1) Effect of pentachlorophenol (PCP). J. Sci. Soil and Manure, Japan. 38, 241–246. (in Japanese with English summary).
12 Russell E J and Hutchinson H B 1909 J. Agr. Sci. 3, 111. In Microbial Life in the Soil. Ed. T. Hattori 1973 p. 98, Marcel Decker, Inc, New York.
13 Sato K 1983 Effect of a pesticide, pentachlorophenol (PCP) on soil microflora. I. Effect of PCP on microbiological processes in soil percolated with glycine. Plant and Soil 75, 417–426.
14 Sato K 1985 Effect of a pesticide, pentachlorophenol (PCP) on soil microflora. II. Effect of PCP on bacterial flora in soil percolated with glycine or water. J. Gen. Appl. Microbiol. 31, 197–210.
15 Tu C M 1973 Effects of Mocap, N-serve, Telone and Vorlex at two temperatures on populations and activities of microorganisms in soil. Can. J. Plant Sci. 53, 401–405.
16 Ushigoshi A 1974 Aerobic bacterial flora in a paddy field. Soil Micro. (Bull. Soil Microbiol. Soc. Japan). No. 15, 30–38. (in Japanese).
17 van Schreven D A, Lindenberg D J and Koridon A 1970 Effect of several herbicides on bacterial populations and activity and the persistence of these herbicides in soil. Plant and Soil 33, 513–532.
18 Voets J P, Meerschman P and Verstraete W 1974 Soil microbiological and biochemical effects of long-term atrazine applications. Soil Boil. Biochem. 6, 149–152.
19 Watanabe I and Hayashi S 1972 Degradation of PCP in soils. Microbial depletion of PCP under the dark and submerged conditions. J. Sci. Soil and Manure, Japan. 43, 119–122. (in Japanese with English summary).
20 Watanabe I 1973 Isolation of pentachlorophenol decomposing bacteria from soils. Soil Sci. Plant Nutr. 19, 109–116.
21 Watanabe I 1977 Pentachlorophenol-decomposing and PCP-tolerant bacteria in field soil treated with PCP. Soil Biol. Biochem. 9, 99–103.

Plant and Soil 100, 345–360 (1987).

Response of cassava to water stress

MABROUK A. EL-SHARKAWY and JAMES H. COCK
Centro Internacional de Agricultura Tropical Apartado Aéreo 6713, Cali, Colombia

Key words Leaf area index *Manihot esculenta* Root characteristics Water stress Yield

Summary Cassava (*Manihot esculenta* Crantz) is a staple food for a large sector of human population in the tropics. It is widely produced for its starchy roots by small farmers over a range of environments on poor infertile soils with virtually no inputs. It is highly productive under favorable conditions and produces reasonably well under adverse conditions where other crops fail. The crop, once established, can survive for several months without rain. There is a wide variation within the cassava germplasm for tolerance to prolonged drought and the possibility to breed and select for stable and relative high yields under favorable and adverse conditions does indeed exist. Research with several cassava clones at CIAT has shown that high root yield under mid — term stress is not incompatible with high yield under nonstress conditions. Plant types with high yield potential under both conditions (*e.g.* the hybrid CM 507-37) are characterized by having slightly higher than optimum leaf area index under nonstress conditions, higher leaf area ratio and more intensive and extensive fine root system.

I. Introduction

Cassava (*Manihot esculenta* Crantz) is a major staple crop ;and one of the primary sources of food energy in the diet of several hundred million people in the tropics[4,5]. Its starchy roots are the harvestable product, consumed either fresh or after processing. The leaves are edible, once cooked, and are commonly eaten in several tropical countries[11]. In addition to its use as a human food (approximately 60–70% of total production) in Africa, South America and Asia, cassava is also used for animal feed, starch and alcohol production.

The cassava plant is a perennial shrub, of the Euphorbiaceae, known only in the cultivated form and originally domesticated in South and Central America. After the conquest of the Americas, cassava spread to Africa and Asia, where its current production exceeds that in its center of origin. It is grown over a wide range of environments (between 30°N and S latitude) and at elevations that range from sea level up to 2000 m near the equator. Nevertheless, most cassava is grown in areas where the annual mean temperature is above 20°C and the rainfall is more than 700 mm per year[5,10].

Cassava is often grown in monoculture; however, mixed cropping is the most common production system with tree crops, annual legume and cereal crops[9,12,14]. Cassava is mainly cultivated by small farmers on small plots of poor, infertile acid soils, usually without application of fertilizers and pesticides[8]. In most areas of cassava production, the crop has to

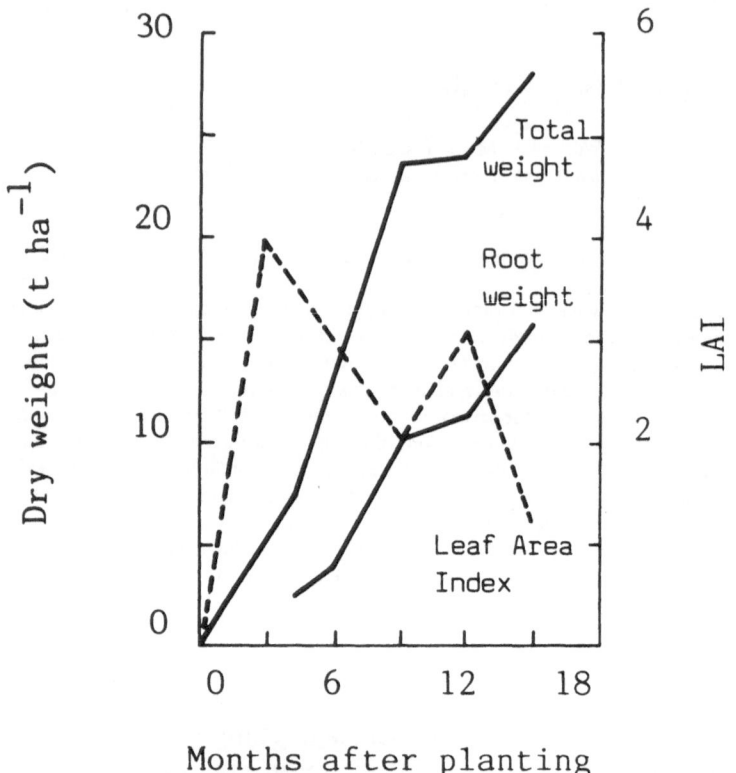

Fig. 1. Growth and development of a typical cassava crop. (Source: Cock, 1985).

endure a prolonged dry period of several months without rain. Under these stressful conditions, cassava can produce from 3 to 5 t ha^{-1} dry roots annually, whereas most other annual food crops would fail. This ability to produce under adverse edaphic and atmospheric conditions has earned cassava the reputation of being a "famine reserve crop"[5].

II. Growth and development

Commercial planting of cassava is from woody stem cuttings (15 to 20 cm long), planted directly after the harvest of mature plants or after several months' storage of long stems under shade. The cuttings may be planted horizontally, vertically or inclined in flat land, ridges or in mounds of soil at a population density of 7000 to 20000 cuttings ha^{-1}. Most cassava planting takes place at the onset of the rainy season; but in some cases, cassava farmers plant towards the end of the rainy season. After planting, one or more of the auxillary buds on the top of the cutting sprout into shoots. Roots initiate mainly from recently formed callus tissues from the base (basal roots) and the lower nodes (nodal

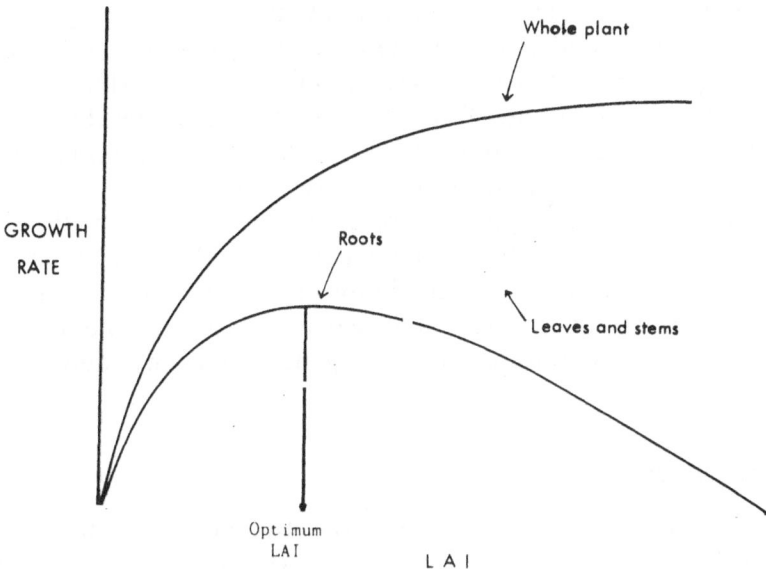

Fig. 2. Schematic representation showing relation between whole plant growth, partitioning of growth between stem (and leaves) or roots, and the leaf area index (LAI). The *vertical arrow* indicates optimum LAI for root growth. (Source: Cock, 1984).

roots) of the cutting[16]. During the first few weeks of crop establishment, the plants form a fibrous root system, mainly in the upper layer (1 m) of soil[7]. About 2 to 3 months after planting, some of the fibrous roots (from 5–15 roots per plant) start to expand rapidly, forming storage roots for starch. These storage roots continue to accumulate starch and to increase in weight during the growing season until the crop is harvested, 8 to 15 months after planting (See Figure 1). However, when the crop is kept for a longer period before harvest, as practiced often by cassava farmers, the quality of fresh roots for human consumption deteriorates and the crop is mainly processed for food products, animal feed and starch extraction.

The formation of leaves in cassava has preference for available assimilates over storage roots in the first 3 months of growth. However, after this period, cassava continues to form new leaves concurrently with storage root filling. The leaf area index (LAI) — leaf area per ground unit area — increases in the first 3 to 6 months and then declines gradually as the older leaves in the lower strata of the canopy fall. The maximum LAI in cassava ranges from 4 to 8, depending upon the cultivar and the atmospheric and edaphic conditions prevailing during crop growth[4].

The continuous development of the source (leaves) and the sink (roots) for carbohydrates in cassava is in contrast with that of cereal crops. In cereal crops such as maize, sorghum, millet, wheat and rice, there are two stages of crop development; namely, vegetative and reproductive. During the vegetative stage (about 70 to 75% of the growth cycle), the leaves, stems and inflorescences develop, followed by a shorter reproductive stage (25 to 30% of the growth cycle), during which grain filling with carbohydrate occurs. With this pattern of crop development and growth in cereals, no competition exists for partitioning the photosynthetic assimilate between source and sink development. In cassava, however, the current photosynthetic assimilates are partitioned between the leaves and the roots, leading to competition between the two organs. This pattern of growth results in an optimum LAI for storage root production (See Figure 2). The balance between formation of leaves and filling of roots is controlled by both genetic and environmental factors.

III. Effect of soil water stress

The response of cassava to water stress was studied using a drainage field lysimeter 30 m × 15 m × 2.3 m deep that was constructed in 1981 at the Santander de Quilichao experiment station of CIAT, Colombia[2]. The bottom of the lysimeter was covered with 0.3 cm layer of black asphalt and drainage tubes were installed 3 m apart and covered with a 10 cm layer of sand and gravel. The tubes drained excess water, due to rainfall or irrigation, to a nearby 5 m × 3 m × 1 m reservoir. The whole area of the lysimeter is protected from lateral water flow by a brick wall extending to within 1.3 m of the soil surface and divided into two equal parts of 15 m × 15 m × 2.3 m. In addition, a 12 m wide strip of land surrounding the lysimeter was reserved as border to be planted with the same crop and treated similarly. The whole area was surrounded by a 2.5 m deep trench.

In the growing season of 1983/1984, the lysimeter and the border area was planted (26 October 1983) with the two cultivars, M Col 1684 and CM 507-37 (a hybrid of M Col 1684 × M Col 1438) at a population density of 12500 plants ha[-1]. The soil was fertilized, prior to planting, with 100, 200 and 100 kg ha[-1] of N, P and K, respectively. Hand weeding and pesticide application were practiced whenever needed. The plots received no water other than the natural rainfall. The water stress treatment was initiated 90 days after planting by covering half of the experimental area with white plastic sheet to exclude rainfall for a period of 3 months. The stress treatment was terminated by removing the plastic sheet at 6 months after planting and the plants were allowed to re-cuperate during the rest of the growth cycle. Six harvests were carried out at 51, 90, 140, 182, 274 and 345 days after planting. Data of LAI, fallen leaves, total biomass and fresh root yield are presented in Figures 3, 4, 5 and 6, respectively.

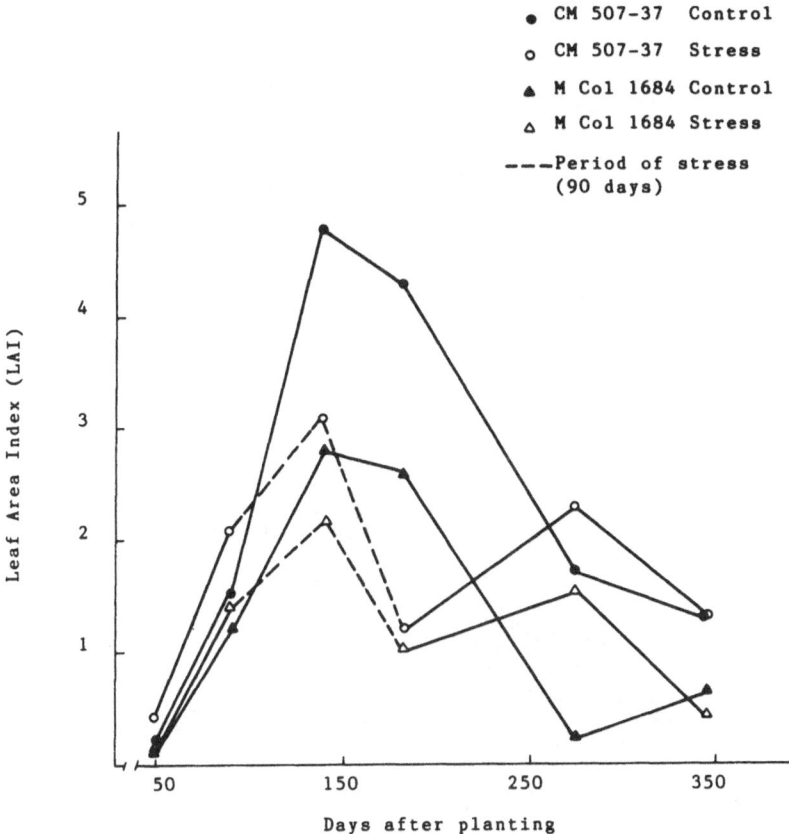

Fig. 3. Leaf area index (LAI) as a function of time after planting under nonstress and mid-term water stress conditions for the hybrid CM 507-37 and the parent M Col 1684. Rainfall was excluded by covering the soil with white plastic sheets from day 90 to day 181 after planting. (Source: CIAT, 1985).

Although cassava is highly productive under high rainfall ($> 1000\,\text{mm year}^{-1}$), it is also highly tolerant of drought and is commonly grown in areas receiving less than 800 mm rainfall per year with a dry season of 4 to 6 months. At the onset of a dry period, the cassava crop reduces its leaf area by producing fewer and smaller leaves and shedding older leaves (See Figures 3 and 4). The reduced leaf area under dry weather could be considered a means by which cassava reduces water loss by transpiration; however, reduction in leaf area during a long period of soil water stress also reduces crop growth rate (See Figure 5). The reduction in crop growth is more pronounced in the shoots than in the roots, particularly in varieties with vigorous vegetative growth (See Table 1).

Upon recovery from water stress, cassava rapidly regenerates new leaves and the LAI of previously stressed plants becomes higher than in

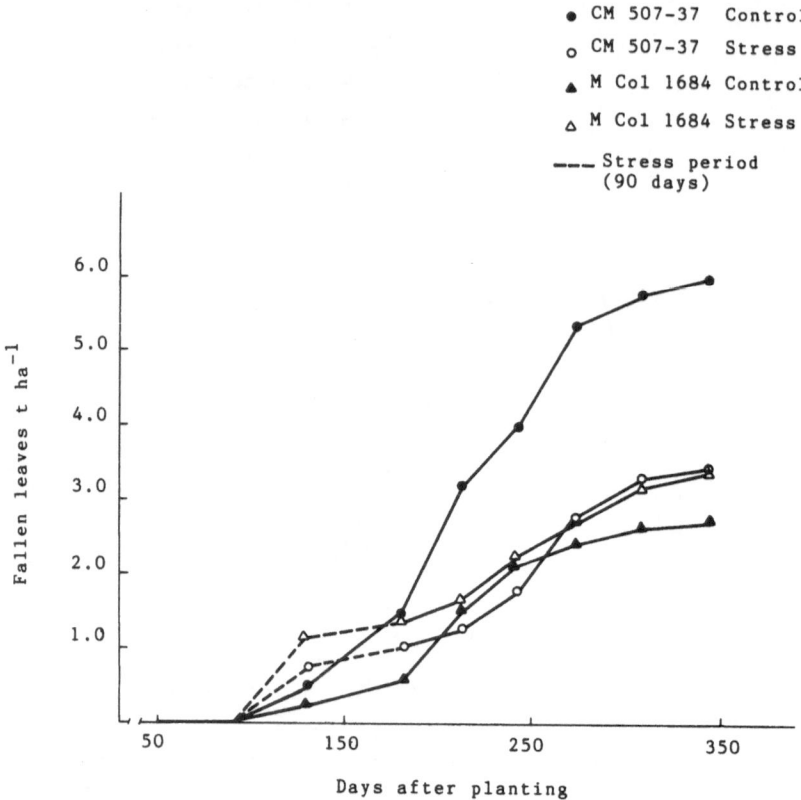

Fig. 4. Amount of fallen leaves as a function of time after planting under nonstress and mid-term water stress conditions for the hybrid MC 507-37 and the parent M Col 1684. Treatments as in Figure 3. (Source: El-Sharkawy *et al.*, unpublished).

nonstressed cassava (See Figure 3). Although the rapid production of new leaves in the stressed cassava occurs temporarily at the expense of stored carbohydrates in stems and storage roots, the increase in LAI leads to a greater accumulation of assimilates in the roots during the few months of recovery (See Figure 6). By the end of the growth cycle, the dry root yield of previously stressed cassava approaches that of the nonstressed crop (See Table 2).

IV. Discussion

Previously[1,6], it was suggested that when water stress occurs in the middle of the growth period, a vigorous variety of cassava may be preferable to a less vigorous type that may yield better under nonstress conditions. In the trials that led to this conclusion, it was not possible to ascertain whether a single plant type could yield well under both stress and nonstress conditions. The data presented in this report clearly

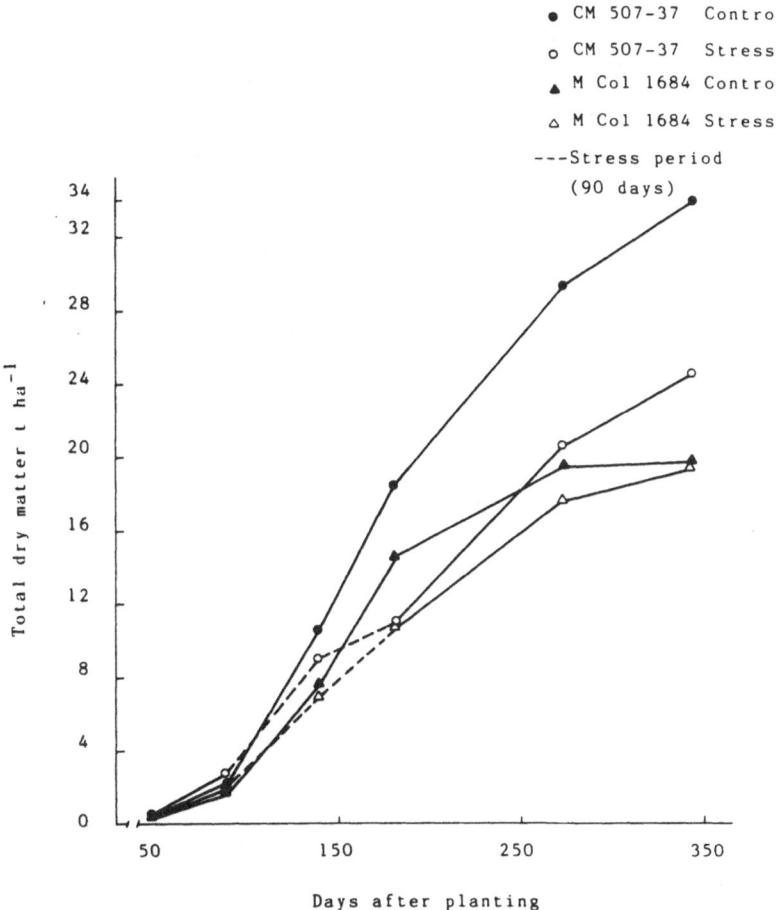

Fig. 5. Total biomass as a function of time after planting under nonstress and mid-term water stress conditions for the hybrid CM 507-37 and the parent M Col 1684. Treatments as in Figure 3. (Source: El-Sharkawy *et al.*, unpublished).

demonstrate that such a plant type does indeed exist in cassava germplasm (See Figures 3, 5, 6 and Table 2).

The hypothesis and the mechanism underlying the response of cassava to water stress are shown in Figure 7. A variety with slightly more than the optimal LAI will reduce its LAI under stress with minimal effect on yield while an optimal plant type for nonstress conditions will yield only very slightly more under these conditions, but will be very sensitive to stress. On the other hand, a variety with very large LAI will produce less under optimal conditions than under stress. It would appear, therefore, that for stable high yield in both stress and nonstress conditions, a variety with above optimal leafiness under good conditions is required. Results from field trials with several cassava varieties grown under

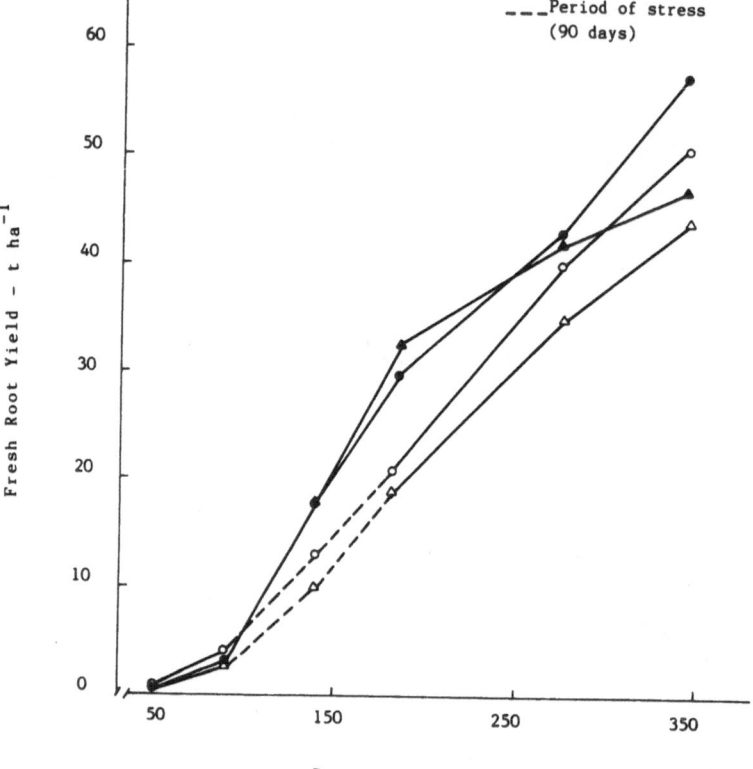

Fig. 6. Fresh root yield as a function of time after planting under nonstress and mid-term water stress conditions for the hybrid CM 507-37 and the parent M Col 1684. Treatments as in Figure 3. (Source: CIAT, 1985).

Table 1. Increase in dry roots relative to total biomass increase of four varieties, with and without water stress during the first 6 months of growth

Variety	Vegetative vigor	Period of water stress initiated at 3 months after planting (days)	Control	Stressed
			(Dry root increase × 100)	
			Total biomass increase	
M Mex 59[a]	Strong	70	32	53
M Col 22[a]	Weak	70	70	87
M Col 1684	Moderate	90	77	79
CM 507-37	Moderate	90	60	80

[a] Source: Connor et al., 1981.

Table 2. Total biomass and dry root yield (t/ha) of cassava at the end of the growth cycle as affected by a period of water stress during the first 6 months of growth

Variety	Period of water stress initiated at 3 months after planting (days)	Age at harvest (days)	Control		Stressed	
			Total biomass[a]	Dry root	Total biomass[a]	Dry root
M Col 1684	90	345	19.72	14.0	19.34	13.0
CM 507-37	90	345	34.0	19.0	24.4	16.0

[a] Fallen leaves are included.

natural rainfall, with and without a period of rain at the early stage of growth confirm this hypothesis[1,3].

The cultivar M Col 1684, a high-yielding, moderately vigorous type, was compared with the leafier CM 507-37 (a hybrid of M Col 1684 × M Col 1438). In the nonstressed plots, CM 507-37 reached LAIs close to five, which are above the optimum; whereas M Col 1684 scarcely reached the optimum levels of 2.5 to 3.5 (See Figure 3). In the later growth stages, the LAIs of both varieties were less than optimal, with very low levels in M Col 1684. At final harvest, CM 507-37, however, maintained significantly higher ($P < 0.01$) LAI than in M Col 1684 despite the much

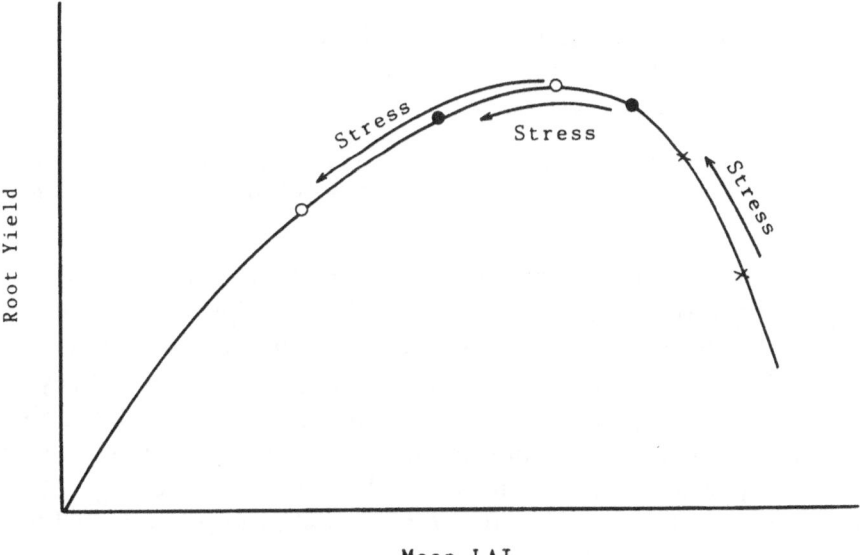

Mean LAI

Fig. 7. Schematic diagram of how a variety can be obtained with good yield under water stress and nonstress conditions. (○ ideal plant type under ideal conditions; ● ideal plant for mid-term stress as well as ideal conditions; × excessively vigorous type, yields better under stress). (Source: CIAT, 1985).

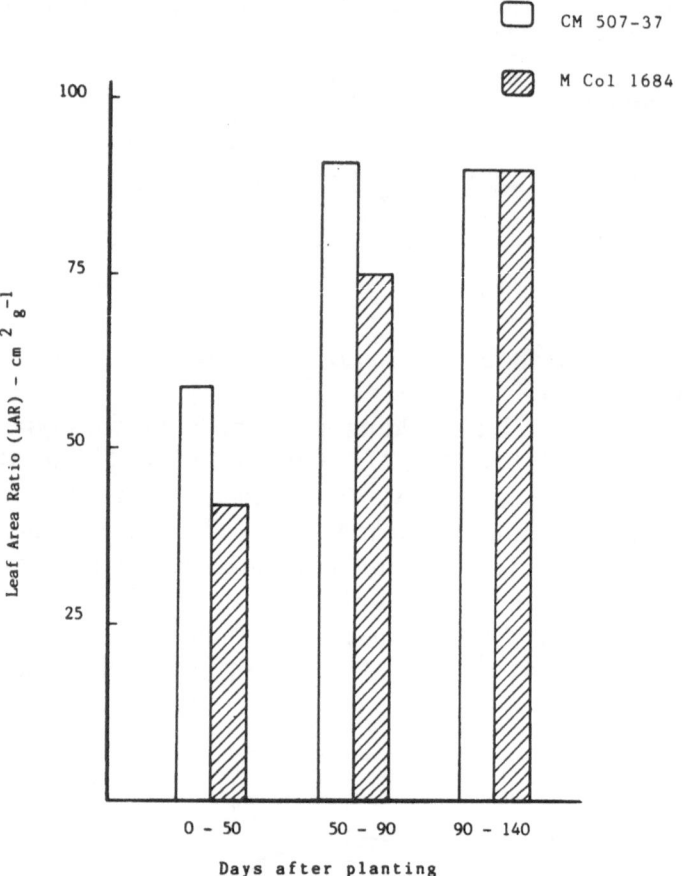

Fig. 8. Leaf area ratio (LAR) in the early growth stage of the hybrid CM 507-37 and the parent M Col 1684. Nonstressed plants. (Source: CIAT, 1985). LAR = leaf area in $cm^2 g^{-1}$ dry weight of stems + leaves.

greater fallen leaves in the former. Both fresh and dry (data not shown for the latter) root yields reflect these trends (See Figure 6). Maximum root growth rates occurred when LAIs were between 2 to 5. When the LAI of M Col 1684 dropped during the last 100 days of the growth cycle, root filling essentially ceased. CM 507-37 maintained an LAI of about 1.5 during this period and substantially outyielded M Col 1684 as a result. The question still remains as to why the early yield (*e.g.*, at six months after planting) of CM 507-37 with an above-optimum LAI was equal to that of M Col 1684. The leaf area ratio (LAR), calculated only for the aboveground portions of the plant, was greater in CM 507-37 than that of M Col 1684 (See Figure 8). This indicates that CM 507-37 used dry matter in the aerial portion of the plant more effectively for producing leaves than M Col 1684, resulting in a higher optimal LAI.

This is presumably related to smaller nodes and internodes per unit leaf area in CM 507-37.

In the stressed plots, the LAIs, total biomass and storage root yields were all significantly reduced ($P < 0.01$) in both varieties after 90 days without rain (see Figures 3, 5 and 6). Compared with the nonstressed plots at 6 months after planting, the percentage decreases in LAI, total biomass, fresh and dry root yields were 61.5, 26.7, 42.4 and 36.4 in M Col 1684, and 72.1, 41.2, 30.0 and 31.6 in CM 507-37. It is apparent, therefore, that in CM 507-37, reductions were greater in LAI and total biomass as compared with M Col 1684. On the other hand, reductions in fresh and dry root yields were greater in M Col 1684. The total biomass and root yields at 6 months after planting were similar for both varieties. CM 507-37, however, maintained significantly higher ($P < 0.01$) LAI during the stress period. This can be attributed, partially, to the ability of the hybrid CM 507-37 to retain its leaves much longer since its fallen leaves were less than in M Col 1684 (See Figure 4). Furthermore, CM 507-37 produced a flush of new leaves during the recovery period and thus maintained a higher LAI, up to the final harvest, than M Col 1684. By the end of the growth cycle, the amount of fallen leaves were the same in both varieties. The significantly higher ($P < 0.05$) LAI of CM 507-37 during the recovery period presumably resulted in significantly higher ($P < 0.05$) root yield at the final harvest (19.3 and 23.1% increases in fresh and dry roots above M Col 1684) (See Figure 6 and Table 2). The same trend existed in final total biomass (26.2% increase above M Col 1684, $P < 0.05$).

Compared with the nonstressed plots, the reduction in dry root yield at the final harvest of M Col 1684 was slight (7.1%) probably because of the large increase in LAI during the recovery period and also because of the lack of root yield increase in the control during the last two months of growth. Similarly, the final total biomass was not different between the two treatments (See Figure 5 and Table 2). On the other hand, the reductions, due to mid — term water stress, in final dry root yield (15.8%) and total biomass (28.2%) in CM 507-37 were significant ($P < 0.05$). Nevertheless, reduction in total biomass in CM 507-37 was substantially greater than reduction in root yield indicating a larger partitioning of assimilates into the roots in this case. This conclusion is further supported by the higher final harvest index (percent dry root to total biomass) in the stressed plots (66%) as compared with the nonstressed plots (56%) (See Table 2). It appears, therefore, that the favorable partitioning of assimilates into the roots that was induced by mid — term water stress (See Table 1) persisted through the recovering period.

The paramount importance of LAI in determining yield even when a long stress period occurs is shown in Figure 9. The yield of four cultivars,

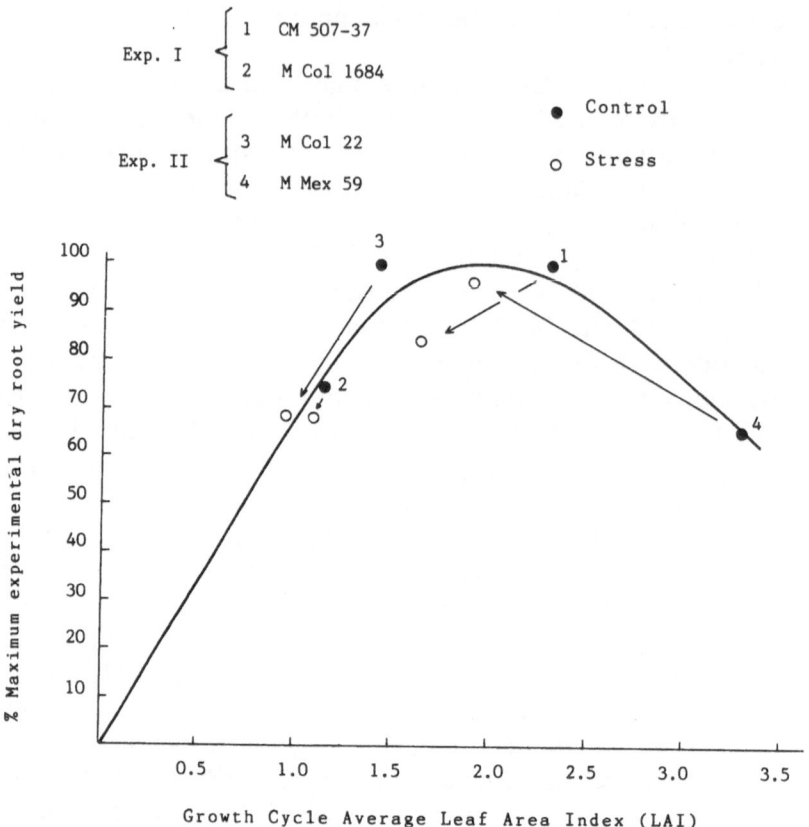

Fig. 9. Dry root yield as a function of growth cycle average leaf area index (LAI) under nonstress and mid−-term water stress conditions for 4 cultivars with different vigors. (Data of experiment I for CM 507-37 and M Col 1684, from El-Sharkawy *et al.*, unpublished; data of experiment II for M Col 22 and M Mex 59, from Connor *et al.*, 1981).

different in their vigor and grown under nonstress and mid−term stress conditions in two separate trials, is closely associated with the average LAI during the growth cycle (LAI is a characteristic of the genotype and is also highly dependent on water conditions). The highest reference (100%) yields in both trials were 19 t dry root ha^{-1} for CM 507-37 under nonstress conditions over 345 days (Experiment I) and 11.2 t ha^{-1} for M Col 22 under nonstress conditions over 306 days (Experiment II). It is apparent that the performance of these cultivars conforms to the suggested hypothesis.

Finally, it is of interest here to comment on another plant characteristic of importance in the soil-water-plant relationship; *i.e.*, the fibrous root system[2,15]. Table 3 summarizes data concerning the characteristics of the fibrous root system and soil water profile of six−month−old, field−grown plants of M Col 1684 and CM 507-37. The two cultivars

Table 3. Characteristics of fibrous root system and soil water profile of six − month − old field − − grown plants of cassava cultivars M Col 1684 and CM 507-37; planted on April 15 and sampled on October 15, 1983, at Santander de Quilichao, Colombia. Soil cores (390 cm³) were taken by hand augor for determination of root characteristics and soil water content. Values are means of 4 profiles

Soil layer (m)	M Col 1684			CM 507-37		
	Root length density $(cm\,cm^{-3})$	Root diameter (mm)	Soil water (% vol)	Root length density $(cm\,cm^{-3})$	Root diameter (mm)	Soil water (% vol)
0.2	0.454	0.970	25.6	0.715	0.520	26.9
0.4	0.189	0.923	39.2	0.195	0.576	37.3
0.6	0.128	0.688	50.2	0.129	0.727	48.8
0.8	0.069	0.670	54.7	0.068	0.553	55.5
1.0	0.051	0.715	58.7	0.068	0.481	58.6
1.2	0.051	0.634	61.5	0.057	0.488	60.5
1.4	−	−	65.2	0.026	0.504	62.7
1.6	−	−	69.0	0.051	0.532	63.3
1.8	−	−	69.0	0.054	0.554	63.3
2.0	−	−	−	0.059	0.536	−

were planted in the same field for commercial production on April 15, 1983, at the CIAT experimental station in Santander de Quilichao, Colombia, and received no water other than natural rainfall. The summer that year was exceptionally dry; from the beginning of June, when the plants were only 45 days old, to the first week of October (130 days), the total amount of rainfall was 180 mm. This amount of rainfall fell far below the potential evaporation (4.4 mm day^{-1}). From the time of planting (April 15) to the end of May, the rainfall was 390 mm, for a total of 570 mm in the first six months of growth[2]. At this age, the total biomass of the hybrid CM 507-37 was 19% greater than that of M Col 1684 (8.0 vs. 6.7 t ha^{-1})[2]. Dry root yield was the same in both cultivars (3.6 and 3.7 t ha^{-1} for CM 507-37 and M Col 1684, respectively). CM 507-37, however, maintained a higher LAI (1.1 as compared with 0.4 for M Col 1684).

These differences in growth and biomass production under naturally stressful environmental conditions were apparently associated with differences in characteristics of fibrous root system and in patterns of soil water depletion (See Table 3). The hybrid CM 507-37 had a finer and more concentrated root density in the upper layer of soil and its root system penetrated into deeper soil layers (perhaps below the 2 meter depth) than M Col 1684. The more intensive and extensive root system of CM 507-37 was advantageous in terms of its ability to withdraw more water (and perhaps more nutrients) from larger and deeper volumes of soil. This might explain, at least partially, the higher leaf area and the greater biomass production by CM 507-37 under severe and prolonged soil-water stress. Varietal differences in rooting characteristics at an early

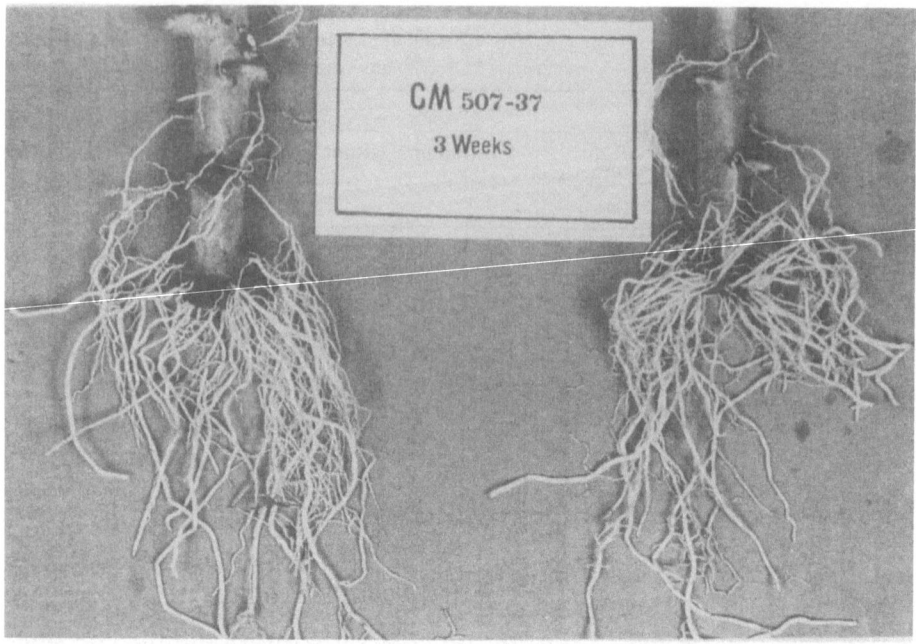

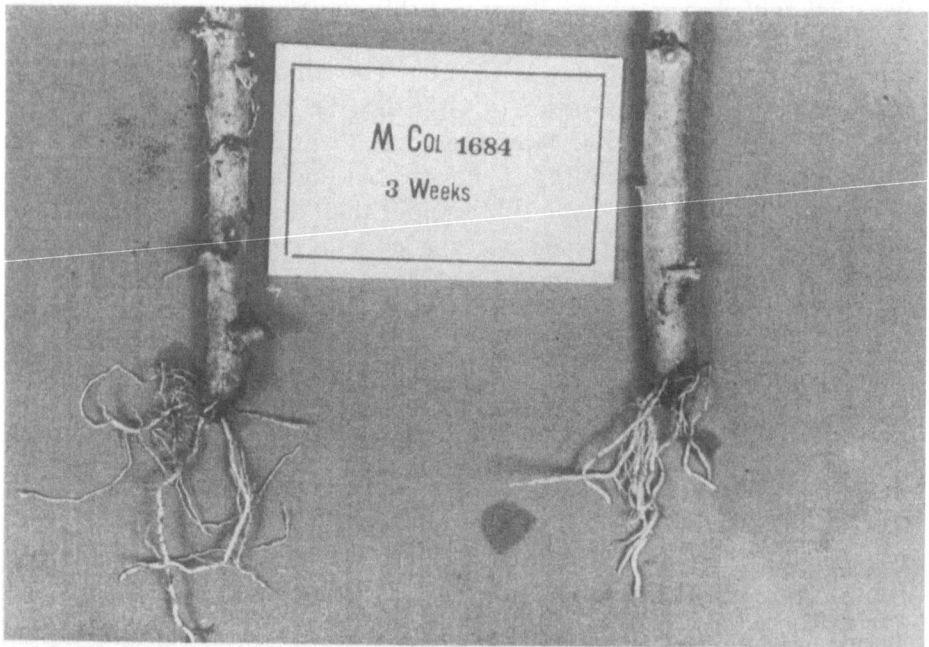

Fig. 10. Rooting characteristics at three weeks after planting of stem cuttings of the hybrid CM 507-37 and the parent M Col 1684 (Source: El-Sharkawy et al., unpublished).

stage were sought, and it was found that CM 507-37 rooted better within the first three weeks of germination (Figure 10). It is possible, therefore, to breed and select for cassava plant types that perform well in both stressful and nonstressful environments, by combining desirable characteristics of leaves and roots.

In conclusion, it can be stated that high yield under mid—term stress conditions is not incompatible with high yield under nonstress conditions. The high yield levels under both conditions (e.g., CM 507-37 with dry root yield of 16 and 19 t ha^{-1} in less than one year, with and without stress) can be obtained by having slightly higher than optimal LAI under nonstress conditions. The extra "energy cost" required for obtaining this type of plant appears to be associated with diverting more dry matter into an extensive fibrous root system. With regard to partitioning dry matter to a fibrous root system, cassava can generally be considered a conservative crop in comparison with other crop species[13,17].

References

1   CIAT 1980 Cassava Program. Annual Report 1979, Centro Internacional de Agricultura Tropical, A.A. 6713, Cali, Colombia.
2   CIAT 1983 Cassava Program. Annual Report 1982–1983, Centro Internacional de Agricultura Tropical, A.A. 6713, Cali, Colombia.
3   CIAT 1985 Cassava Program. Annual Report 1984, Centro Internacional de Agricultura Tropical, A.A. 6713, Cali, Colombia.
4   Cock J H 1984 Cassava. In The Physiology of Tropical Field Crops. Eds. P R Goldsworthy and N M Fisher. pp 529–549. John Wiley & Sons Ltd.
5   Cock J H 1985 Cassava: New Potential for a Neglected Crop. Westview Press. Boulder, Colorado.
6   Connor D J, Cock J H and Parra G E 1981 The response of cassava to water shortage. I. Growth and Yield. Field Crop Res. 4, 181–200.
7   De Sena Z F and Dos Campos R H 1973 Estudio do sistema radicular da mandioca (Manihot esculenta Crantz) submetidas a diferentes frequencias de irrigacao. Univ. Fed. Bahia, Esc. Agron., Ser Pesqui. 1. BRASCAN Nordeste, Cruz das Almas, Brazil, pp 41–52.
8   Howeler R H 1985 Potassium Nutrition of Cassava. In Potassium in Agriculture. pp 819–841. ASA-CSSA-SSSA, South Segoe Road, Madison, WI 53711, USA.
9   IITA 1985 Research Highlights 1984 International Institute of Tropical Agriculture, PMB 5320, Oyo Road, Ibadan, Oyo State, Nigeria.
10 Jones W O 1959 Manioc in Africa. Stanford University Press, Stanford.
11 Lancaster P A and Brooks J E 1983 Cassava leaves as human food. Economic Botany. 37(3), 331–348.
12 Leihner D 1983 Management and Evaluation of Intercropping Systems with Cassava. Centro Internacional de Agricultura Tropical, A.A. 6713, Cali, Colombia.
13 Lesczynski D B and Tanner C B 1976 Seasonal variation of root distribution of irrigated, field grown Russet Burbank Potato. Am. Potato J. 53, 69–78.
14 Nair P K R 1983 Agroforestry with coconuts and other tropical plantation crops. pp 79–102. In Plant Research and Agroforestry. Ed. P A Huxley, International Council for Research in Agroforestry, Nairobi, Kenya.
15 Taylor H M and Klepper B 1978 The role of rooting characteristics in the supply of water to plants. Adv. Agron. 30, 99–128.

16   Wholey D W 1974 Rapid Propagation of Cassava, PhD Thesis, University of the West Indies,
     St. Augustine, Trinidad.
17   Willatt S T and Taylor H M 1978 Water uptake by soya-bean roots as affected by their depth
     and by soil water content. J. Agric. Sci., Camb. 90, 205–213.

Plant and Soil 100, 361–370 (1987).
Ms. 100-29

Activity of root systems of six plant species at different stages of development

L.K. WIERSUM
Institute for Soil Fertility, P.O. Box 30003, 7950 RA Haren, The Netherlands

Key words Active root surface Hydraulic conductivity Influx K^+ NO_3^- Respiration Roots

Summary A number of data on root performance of six different crop species during development were measured. The plants wre cultivated in nutrient solution. Normal plant requirements were in the range of 4 mg O_2 per g dry root per hour, 0.2–4 µg K per cm^2 total root surface per hour,[2] 0.2–2 µg NO_3 per om^2 total root surface per hour.

An attempt was made to establish a ratio betweeen forced water entry and total root surface as a measure of functional root surface. The indication is that the relative surface of permeable root remains dominant during the phase of exponential growth and declines thereafter.

The data collected are considered to be representative of normal requirements. They compare well with results published in the literature.

Introduction

Uptake of available water and nutrients in soil or substrate is the function of roots. For optimum growth of the plant the root medium must be able to supply the substance needed at an adequate rate. It is implied here that the "adequate rate" is expressed as an amount per unit area, and per unit time. This adequate rate depends on the amounts needed and on the size of the root surface area available for uptake. For unimpeded growth of the plant these normal influx values are characteristic of the demands the plant makes on the soil.

Our knowledge concerning these flux values for normal and unimpeded growth of the plant is rather limited. This is due to the fact that although plant or crop requirements per unit time are often well known the actual flux values at the soil/root interface cannot be calculated.

One of the main problems is that the root surface specifically involved in the transfer of water and nutrients or the consumption of oxygen is not known. Determining total root length an especially root surface area is a very laborious task. But more important may be the fact that the root's capacity to function in uptake and transport changes due to anatomical changes upon ageing. A precise indication of the root surface active in uptake is extremely difficult as its transition to older and impermeable regions may be very gradual.

Plant physiologists, however, have provided data on root performance. Parameters such as the value for affinity (C_{min}), maximum influx rate (I_{max}) and permeability can be mentioned[5]. Also oxygen consumption rates per unit of mass have often been reported[1].

Unfortunately research on the behaviour of roots has mainly been performed on very young and active root systems or pieces. The contribution of older parts of root systems in uptake has received little attention in experiments.

Another fact is that plant physiological research and especially a more ecologically oriented approach have demonstrated that root performance is governed by the internal supply status of the plant. There are numerous feedback systems related to conditions in the aboveground parts of the plant — stress or no stress — that regulate rates of root uptake[2].

Also, it has gradually become evident that root permeability for water is variable and can respond to stress conditions[3,5,6].

Despite these limitations the data on root performance have been quite useful and have been utilized in a number of simulation models of root and plant growth in relation to uptake.

To improve the knowledge of soil conditions necessary for unimpeded crop growth, a number of experiments was performed on developing root systems. Especially performance at increasing age was investigated. Measured phenomena were root respiration, uptake of NO_3^- and K^+, root hydraulic conductivity and root surface. It was assumed that root hydraulic conductivity would be an indication of active root surface in relation to influx of water an nutrients.

Methods

All plants were grown in well-aerated Steiner culture solution, with a K/Ca ratio of 60:30. Usually they were grown in a glasshouse at a temperature regime of 20°C by day and 16°C at night. When they were cultivated in a growth chamber a temperature of about 23°C was maintained, with 14 hours of light (about 15,000 Lux).

Root respiration was measured by means of one of two available methods. In both methods closed systems were used (Fig. 1). The varying volume was sufficient for several hours of unimpeded respiration. In one system the oxygen electrode was directly placed in the circulating nutrient solution. This solution circulated at high speed from bottom to top through the vessel containing the root to achieve a thorough replenishment at the root surface and to prevent the formation of stagnant layers. In another system the excised root system or that of an entire plant was placed in one of two interconnected vessels in a see-saw device in which the nutrient solution alternately wetted and flowed away from the roots, 1–2 times per minute for several hours. In this case the oxygen electrode

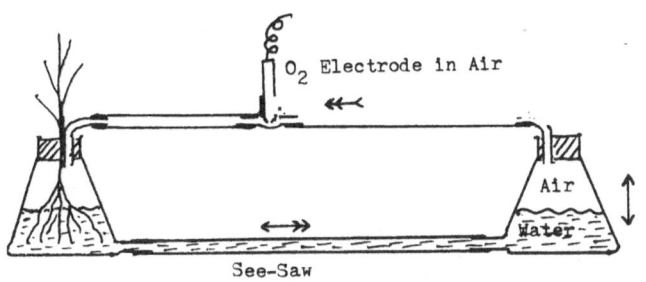

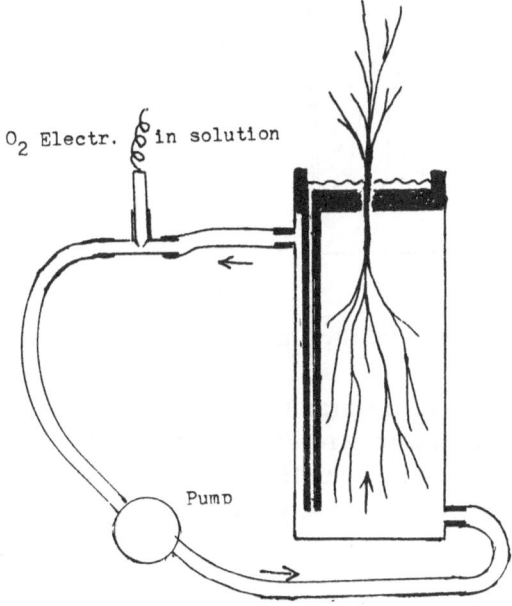

Fig. 1. Schematic drawing of the set up used in oxygen consumption measurement.

was in the enclosed air, which could equilibrate with the solution. The efficiency of replenishment of solution at the root surface was indicated by steady O_2-consumption rate down to about 0.01 bar O_2-pressure in both systems. The respiration of the microbes in the circulating solution was always measured separately and subtracted from total respiration before calculating the "root" (root + rhizosphere) respiration.

Uptake of nitrate and potassium was measured with ion-selective electrodes in a circulation setup, in which a constant volume was maintained (Fig 2).

Root surface was measured by briefly dipping (60 seconds) the root system in a 0.1% safranine solution, washing the roots and eluting the absorbed dye in alcohol. For each species the relationship between

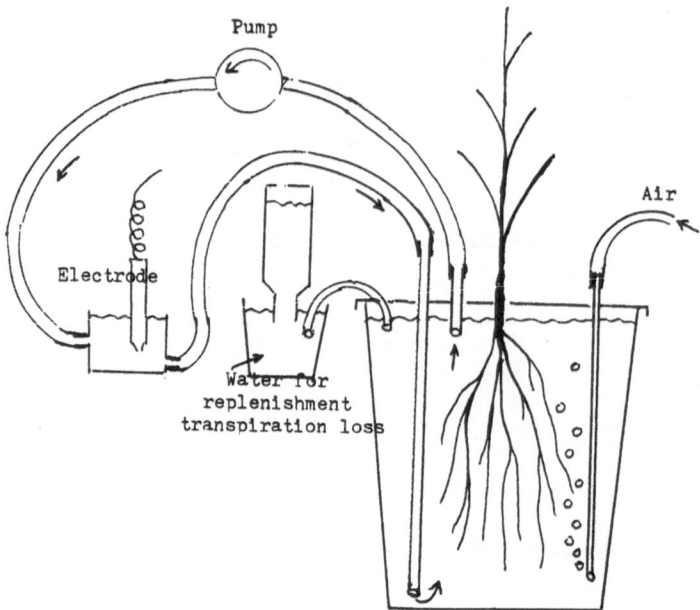

Fig. 2. Measurement of ion uptake in constant volume.

measured surface of a few root samples and the amount of dye eluted was separately established.

Root hydraulic conductivity, *i.e.* forced water influx, was measured by immersing the excised root system in a water-filled pressure bomb and usually applying 1 bar external pressure. The outflow of water at the cut stump was considered to be related to the surface area of water-permeable root tissue.

A decline in the conductivity/root surface ratio was considered to indicate a decrease in percentage physiologically active root surface as regards uptake.

Measurements were nearly always performed in the growth chamber at constant temperatures.

Results

Root respiration

The results are summarized in Table 1. From these figures it can be concluded that root respiration of the investigated species is of the same order of magnitude. Also, the values are in good agreement with those published by other authors as reported in a review by Brouwer and Wiersum[1].

The values, however, would be lower under field conditions, as temperatures during the measurements were usually above 25°C.

Table 1. Root respiration rates measured at temperatures of about 28°C (301 K)

Plant species	Age (days)	Respiration (mg O₂/g dry root/h)	Number of observations
French bean	15–40	3.84 ± 1.04*	18
(*Phaseolus vulgaris*)			
Tomato	15–25	46.0** (appr.)	
(*Lycopersicum*	30–50	4.44 ± 1.58	12
esculentum)	30–61	5.66 ± 3.08	21
Corn	10–17	3.34 ± 1.32	7
(*Zea mays*)			
Lucerne	0–150	1.3 (appr.)	
(*Medicago sativa*)	150–300	0.5 (appr.)	
Brussels sprout	20–70	3.5 (appr.)	
(*Brassica oleracea*)			
Potato	0–15	50 (appr.)	
(*Solanum tuberosum*)	30–90	4.8 (appr.)	
	100–150	1.3 (appr.)	

*   means ± S.E.
** approximate value based on very few and variable observations

In Fig. 3 a typical time course is given for tomato. In very young roots of less than 10 mg dry weight respiration is very high. Here young parenchymatous tissues completely dominate. The same was observed for potato rootlets.

It should be noted that these measurements were performed on roots

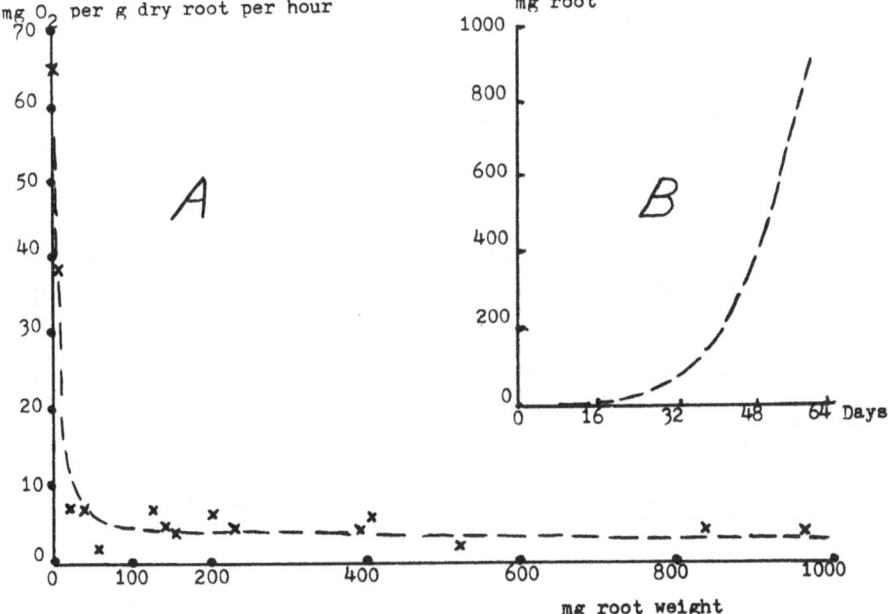

Fig. 3. **A.** Time course of tomato root respiration. **B.** Average root growth in course of time (days).

derived from plants on nutrient solutions, which certainly are not aseptic. The "root" respiration values are thus comprised of root respiration
proper plus respiration of the microbes adhering to the roots.

As roots in soil are always enveloped by their rhizosphere population
the values observed are useful in ecological studies and models and
represent the actual requirement.

Rates of uptake

The results were obtained with six different species of plants. The
laborious investigations were conducted over a number of years. On
account of numerous difficulties in the techniques employed and differences among individual plants, a large variation in the separate values
was often encountered. But although the data should be considered with
caution, the results obtained do show some clear trends and give an idea
of rates normally occurring in transport processes at the root surface.

Peas. During a growth period of about 2 months there was a more or
less exponential increase in root weight, root surface area and amount of
water forced through the root system.

The water flux values declined slowly as the plant became older (Table
2).

Tomato. Again the measured values show a more or less exponential
increase for about 2 months of development with regard to root weight,
root surface area and amount of water forced through the roots (Fig.
3B). It is interesting to note that the induced water flux remained more
or less constant during a single experiment over several hours (Fig. 4).
The flux values for water are of the same magnitude as for peas (Table
2).

French bean. The induced waterflux per unit root surface started to
decrease at the flowering stage and declined still further after pod ripen-

Table 2. Induced water flux at about 1 bar in μl per cm^2 total root surface per hour in approximate
values

Species	Age in days (intervals)															
	0–10	10–20	20–30	30–40	40–50	50–60	60–70	70–80	80–90	90–100	100–125	125–150	150–175	175–200	200–250	250–300
Peas			6.3	4.4	2.8	2.2										
Tomato	5.7		6.1													
French bean			0.44					0.36				0.15				
Potato		4.2				2.5		1.1				0.6				
Lucerne			0.55					0.40				0.39	0.51	0.52		0.07

ing (Table 2). Oxygen consumption shows the same pattern and declines from 1.48 μg O_2 per cm^2 per hour to 0.56 μg per cm^2 per hour.

The influx rates for potassium and nitrate are given in Table 3. The few available data showed a high rate of entry in the period of pod setting (days 40–100). At that time the first evidence of a relative decrease — although small — of permeable root tissue (Table 2) was also noticeable. Later, when the leaves started yellowing (older than 100 days), there was a stronger decline in water influx.

Brussels sprouts. Available data demonstrated that the induced water flux — as measured with the pressure bomb — could remain reasonably constant over a period of 6 hours.

Exponential growth lasted for about 75 days. During this period the rate of potassium remained reasonably constant, as was expected. Nitrate influx behaved differently and declined steadily (Table 3).

Potato. Water flux data (Table 2) declined gradually with age of the plants. In some of these experiments root-pressure exudation was also measured. The exudation values were at about 10% of the pressure-induced flow at 1 bar and showed the same decline. The decrease sets in at an age of about one month, when shoot/root ratio started to increase as a result of storage of starch and the onset of tuber formation.

In table 3 the measured influx rates for potassium and nitrate are given. For both elements the highest values were observed in the period between 60 and 90 days of growth, when root growth lagged behind shoot development. The influx rates for potassium and nitrate were corrected by taking the water influx rates as a measure of active root surface into account. It was then evident that these rates were very high in the third month.

Lucerne. The rates of induced water flux did not decline until 250 days had elapsed (Table 2). However, the uptake rates behaved differently (Table 3): K^+ influx was variable, while NO_3^- uptake rates declined gradually. The high potassium uptake rate in the period of 200–250 days of age might be the result of regenerative growth, after the shoots had been cut at an age of about 150 days. Beyond an age of about 200 days the total root surface remained at the same level, although with rather large individual variations.

In the very first period after cutting away the greater part of the shoots, exudation of potassium instead of uptake from the roots was measured in a few cases.

Table 3. Influx rates for K$^+$ and NO$_3^-$ into roots in μg per cm^2 per hour

Species		0–10	10–20	20–30	30–40	40–50	50–60	60–70	70–80	80–90	90–100	100–125	125–150	150–175	175–200	200–250	250–300	300–350
French bean	K			4.3				15.8										
	NO$_3$			0.38				2.45				0.03						
Brussels sprouts	K				1.5		1.8		1.4									
	NO$_3$				1.8		1.0		0.2									
Potato	K			0.1		1.6			1.9		0.2			0.3				
	NO$_3$								0.5		0.1			0.06				
Potato corrected	K			0.1		2.6			7.8		1.1			2.2				
	NO$_3$								1.9		0.5			0.4				
Lucerne	K								0.1			0.15			0.04	0.3		0.1
	NO$_3$								0.3			0.2			0.1	0.04	0.07	0.05

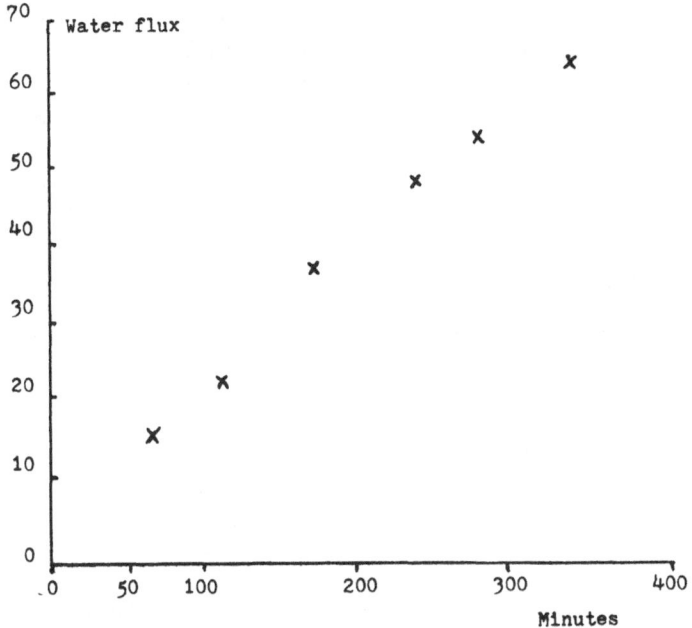

Fig. 4. Induced outflow of water from tomato root system.

Ryegrass. In a few additional experiments on large core samples (appr. 1 l), obtained form a ryegrass culture, soils with and without roots were compared. The "root" respiration values, calculated from the difference, were more or less in the same range.

Discussion

The data on "root" respiration are in good agreement with those published by other authors as summarized in an earlier paper[1]. They also fit in with a more recent investigation on carbon flow through the plant[4].

In considering respiration values of the younger roots as an indication of a requirement that soil conditions have to meet, it should never be forgotten that in arable soils the soil's own respiration can be of the same order.

As regards the time course of respiration intensity, the very young root — consisting entirely of very active tissues — has a very high O_2-requirement. But soon the total consumption rate of the exponentially growing root system levels off. In later stages of development O_2 consumption per unit root mass declines. This decline more or less follows the trends in uptake rates when related to total root surface.

The water flux values, when calculated on total root surface show that root permeability decreases after the exponential growth phase. As the same phenomenon is observed for ion influx, it appears that the ratio of pressure flow to total root surface is in a way a measure of active root surface. For annuals in later stages of development and certainly for perennial plants total root length or surface is not a direct measure of actively performing roots.

The corrected influx rates for potato support the conclusion that indeed the younger root tissues retain their activity at more or less the same level, while older parts of the root decline in activity.

The incidental observation of temporary K^+ exudation fits in with earlier observations on ageing plants, in which "binding" capacity declines.

Acknowledgement   The author is highly indebted to Mr. K Harmanny for the careful execution of the experiments.

References

1   Brouwer R and Wiersum L K 1977 Root aeration and crop growth. *In* Crop Physiology Vol 2. Ed. U S Gupta. pp 157–201. Oxford Univ. Press and IBH Publ. Comp., New Delhi.
2   de Jager A 1985 Response of plants to a localized nutrient supply. Ph.D. Thesis State Univ. Utrecht (The Netherlands).
3   Oosterhuis D M and Wiebe H H 1986 Waterstress preconditioning and cotton root pressure-flux relationships. Plant and Soil 95, 69–76.
4   Paul E A and Kucey R M N 1981 Carbon flow in plant microbial associations. Science 213/4506, 473–474.
5   Wiersum L K 1981 Problems in soil fertility characterization by means of plant requirements. Plant and Soil 61, 259–267.
6   Wiersum L K and Harmanny K 1983 Changes in the water-permeability of roots of some trees during drought stress and recovery, as related to problems of growth in urban environment. Plant and Soil 75, 443–448.